# METHODS IN ENVIRONMENTAL FORENSICS

# METHODS IN ENVIRONMENTAL FORENSICS

EDITED BY

**STEPHEN M. MUDGE**

CRC Press
Taylor & Francis Group
Boca Raton London New York

CRC Press is an imprint of the
Taylor & Francis Group, an **informa** business

CRC Press
Taylor & Francis Group
6000 Broken Sound Parkway NW, Suite 300
Boca Raton, FL 33487-2742


Printed in the United States of America on acid-free paper
10 9 8 7 6 5 4 3 2 1

International Standard Book Number-13: 978-0-8493-5007-8 (Hardcover)



**Library of Congress Cataloging-in-Publication Data**

Methods in environmental forensics / editor, Stephen M. Mudge.
p. cm.
"A CRC title."
Includes bibliographical references and index.
ISBN 978-0-8493-5007-8 (alk. paper)
1. Environmental forensics. I. Mudge, Stephen M. II. Title.

TD193.4.E525 2008
628.5--dc22 2008013349

**Visit the Taylor & Francis Web site at**
**http://www.taylorandfrancis.com**

**and the CRC Press Web site at**
**http://www.crcpress.com**

# Contents

Preface vii
Acknowledgements xi
The Editor xiii
Contributors xv

1 Approaching Environmental Forensics 1
STEPHEN M. MUDGE

2 Radionuclides in the Environment: Tracers and Dating 15
DAVID ASSINDER

3 Chemical Fingerprinting of Petroleum Hydrocarbons 43
ZHENDI WANG AND CARL BROWN

4 Biomarkers and Stable Isotopes in Environmental Forensic Studies 113
R. PAUL PHILP AND TOMASZ KUDER

5 Volatile Organic Compound (VOC) Analysis in Water, Sediments, and Soils and Its Application in Environmental Forensics 171
CLAUDIO BRAVO-LINARES AND STEPHEN M. MUDGE

6 Application of Molecular Microbiology to Environmental Forensics 195
ANDREW S. BALL, JULES N. PRETTY, RAKHI MAHMUD, AND ERIC ADETUTU

7 **Biological Communities as a Forensic Tool in Marine Environments** 219

ANGEL BORJA AND IÑIGO MUXIKA

8 **Normalisation Techniques in the Forensic Assessment of Contaminated Environments** 251

GAVIN F. BIRCH, ANDREW T. RUSSELL, AND STEPHEN M. MUDGE

9 **Multivariate and Geostatistical Methods in Environmental Forensics** 277

STEPHEN M. MUDGE

10 **Identification of Air Pollution Sources via Modelling Techniques** 309

IAN COLBECK

11 **Evidence and Expert Witnesses in Environmental Forensics Cases** 353

ALLAN KANNER

**Index** 379

# Preface

'The best laid schemes o' mice an' men' (Robert Burns, 1786, 'To a Mouse')

This book has been in preparation for much longer than was originally intended. However, I am pleased to say that the chapters presented within are written by experts in their various fields of environmental forensics. This book represents our state of knowledge in these areas and provides a reference for all wishing to practice environmental forensics and, indeed, any environmental investigation.

Environmental forensics (EF) has been around for decades but we have not always called it that. As a scientific community we have been investigating the source and fate of contaminants in the environment and, occasionally, these findings have been used to reduce or mitigate pollution and prosecute offenders. The word *forensics* is derived from the Latin *forum*—a meeting place where judicial issues were presented to the people. Initially, we were concerned about 'crimes against the person', but as we have become more aware of the damage done to our environment by indiscriminate waste disposal, we have strengthened legislation that protects the air we breathe, the water we drink, and the ground we live on. We have also become aware of the toxic nature of the chemicals we had previously taken for granted or thought were benign. It has been suggested that the Roman Empire fell because of the lead in its wine and water; modern food standards agencies would have a field day with that one!

In the past decade, however, there has been a crystallisation of the vague term 'environmental forensics' into a well-disciplined science that integrates sampling design, analytical chemistry, and environmental processes with the legislative framework. As with any science, though, it needs to be rigorously applied and the correct methods used for the study at hand; there is no one ideal method that would solve all problems. There are two journals specifically covering this discipline (*Environmental Forensics,* founded by Bob Morrison and now published by Taylor & Francis, and *Journal of Environmental Monitoring,* published by the Royal Society of Chemistry). If these two august publishing houses are publishing our science, it must have been accepted into the mainstream of scientific advancement.

In some cases of environmental contamination, the EF practitioner is called in rather late and often presented with a very limited budget with which to prove everything. Rigorous science that would stand up in court must also stand up to scientific peer review with replication, errors, significance, and certainty—always difficult to do with a small budget. However, a lot can be achieved without the latest, most expensive piece of scientific equipment; it comes down to the ingenuity of the investigator.

This book outlines the methods that have worked well in past EF cases. The first chapter describes how an environmental case might be approached from inception to court testimony. It is worth noting that in proving that X was responsible, it is almost as important to prove that it could not have been Y or Z. In chapter 2, David Assinder outlines the ways in which natural and artificial radionuclides can be used as tracers of environmental processes and for dating samples from the field, an important aspect when apportioning blame. Zhendi Wang (with Carl Brown) from Environment Canada has provided an excellent review of the methods used for oil spill identification—still a major cause for environmental concern around the world. The ubiquitous nature of oil and its products can make source identification very complex, especially in harbours.

This chemical composition approach is followed by Paul Philp and Tomasz Kuder's chapter on the use of stable isotopes (especially $^{13}C$ and $^{2}H$) to improve source specificity, including with oil spills. This approach has wide application outside oil identification and can be used to track multisource compounds through complex environmental processes. Claudio Bravo-Linares and I have recently developed a significantly more sensitive method for the analysis of volatile organic compounds (VOCs) exploiting the new solid phase microextraction (SPME) technologies. This has been used in tracking the source of VOCs in the atmosphere, waters, sediments, and soils. Chlorinated solvents remain an important contaminant in groundwaters and form the basis of many EF cases in the United States.

In a slight shift away from the chemistry of the environment to the biota that live in it, Andrew Ball (with Jules Pretty, Rakhi Mahmud, and Eric Adetutu) presents a range of methods for the molecular characterisation of soil bacteria that can greatly assist in their identification, treatment regimes, and geographic origin. Angel Borja and Iñigo Muxika show how the macrobiological community or assemblage may be used to classify an area and quantify the degree of stress exerted on the system. These methods are being applied in the implementation of the EU Water Framework Directive. Gavin Birch (with Andy Russell and me) presents a range of methods for the normalisation of data to remove a range of natural effects that may mask environmental processes. Concentration gradients of contaminants can exist purely due to changes in the grain surface area (mud to sand), although these may not represent anthropogenically induced gradients. Similarly, I present a range of statistical methods for the treatment of chemical and biological data to determine the underlying trends within a complex multivariate environment. Geostatistics is often used to present contour maps and to infer a gradient between source and sink, but how frequently are the prerequisite tests conducted to ensure their validity?

Sometimes, measurements on their own are insufficient. Ian Colbeck outlines a range of modelling techniques to identify sources of atmospherically dispersed contaminants. These have had recent usage when determining the source of foot-and-mouth disease outbreaks in the United Kingdom. Finally, Allan Kanner puts the legal perspective to all of these scientific methods: If one's method is unlikely to be accepted by a court, maybe it is not worth pursuing in this particular case. It is noteworthy that the reference structure in this last chapter is different from the others as it principally cites legal cases regarding the admissibility of data and expert testimony.

I thank all of these experts for their input to this book and hope that it will be used many times in the coming years by students and practitioners of environmental forensics. I must also thank the patience of the publishers, Taylor & Francis—especially Jill Jurgensen and Becky Masterman—for their confidence in the book.

**Stephen M. Mudge**

# Acknowledgements

I would like to thank all the contributors to this book; it has taken longer than anticipated but the result is good. I would particularly like to thank Andy Ball from Flinders University, who started the editorial process with me but, due to work pressure, had to drop out. I would also like to thank Bob Morrison, director of the International Society of Environmental Forensics (ISEF), for his encouragement across the years. Finally, I would like to thank Georgina, my wife, and our two children, Xander and Toren, for their understanding when I had to sit at the computer editing the text for the publisher.

# The Editor

**Stephen Mudge** has been conducting environmental forensics investigations for many years; these have principally focused on the identification of the contamination sources, especially in complex, multisource environments. Dr Mudge designed and ran the first undergraduate environmental forensics degree at Bangor University, and students from this course are now active in the commercial sphere. Dr Mudge has acted as an expert witness in several environmental contamination cases and continues to research new methods for the quantification and source apportionment of chemicals around the world.

# Contributors

**Eric Adetutu**
School of Biological Sciences
Flinders University
Adelaide, Australia

**David Assinder**
School of Ocean Sciences
Bangor University
Menai Bridge, Anglesey, United Kingdom

**Andrew S. Ball**
School of Biological Sciences
Flinders University
Adelaide, Australia

**Gavin F. Birch**
Environmental Geology Group
School of Geosciences
The University of Sydney
Sydney, Australia

**Angel Borja**
Marine Research Division
AZTI-Tecnalia Foundation
Pasaia, Spain

**Claudio Bravo-Linares**
Universidad Austral de Chile
Instituto de Química
Valdivia, Chile

**Carl Brown**
Emergencies Science and Technology Division, Environmental Technology Centre
Environment Canada
Ottawa, Ontario, Canada

**Ian Colbeck**
Centre for Environment and Society
Department of Biological Sciences
University of Essex
Colchester, United Kingdom

**Allan Kanner**
Kanner & Whiteley, LLC
New Orleans, Louisiana

**Tomasz Kuder**
School of Geology and Geophysics
University of Oklahoma
Norman, Oklahoma

**Rakhi Mahmud**
Department of Biological Sciences
University of Essex
Colchester, United Kingdom

**Stephen M. Mudge**
School of Ocean Sciences
Bangor University
Menai Bridge, Anglesey, United Kingdom

**Iñigo Muxika**
Marine Research Division
AZTI-Tecnalia Foundation
Pasaia, Spain

**R. Paul Philp**
School of Geology and Geophysics
University of Oklahoma
Norman, Oklahoma

**Jules N. Pretty**
Department of Biological Sciences
University of Essex
Essex, United Kingdom

**Andrew T. Russell**
School of Ocean Sciences
Bangor University
Menai Bridge, Anglesey, United Kingdom

**Zhendi Wang**
Emergencies Science and Technology Division, Environmental Technology Centre
Environment Canada
Ottawa, Ontario, Canada

# Approaching Environmental Forensics

# 1

STEPHEN M. MUDGE

## Contents

Introduction .......... 1
Preparation .......... 2
    Sites .......... 3
    Events and Spills .......... 3
Legal Framework .......... 4
Background versus Baseline .......... 4
Know the Contaminant .......... 6
Sampling .......... 6
    Media .......... 7
    Bias .......... 7
    Number of Samples .......... 7
    Sample Quantity .......... 7
    Security .......... 8
    Cost .......... 8
Analysis .......... 8
Developing the Case .......... 9
    Statistics .......... 9
    Data Presentation .......... 10
Source, Pathway, Sink .......... 10
Proving the Case .......... 11
The Expert Witness Report .......... 11
References .......... 12

## Introduction

Environmental forensics may sound like a glamorous, exciting discipline; it certainly can be, but it can have a lot of routine analyses and report writing as well. The subject must be approached in a scientific manner where hypotheses are rigorously tested. One's duty (in the United Kingdom) is to the court, to help resolve the truth of the situation, rather than to any one party, even if that party might be paying for the work to be done (Civil Procedure Rules, Part 35). The definition of *truth* may also be open to question; as scientists, we generally accept hypotheses to be true until such time as we find either a better hypothesis to describe the observations or we find an exception that disproves it. These truths may be real or just our 'best guess' at the present time. This definition of truth may also be different from what a court defines as true since the former may be based on *belief*.

Environmental forensics is a true multidisciplinary subject where chemical, physical, and biological methods combine within a legal framework to determine the origin and extent of environmental contamination. A logical approach is key to success because the work may need to be defended in a court of law and not just to the scientific community. It should be accepted by practitioners that although they are not experts in all aspects of the environment, they may understand the system's functioning well enough to know what analyses would be most appropriate in each case. It may be that simple chemical analyses would be sufficient, but sampling design and quality assurance must go hand in hand to ensure that results are valid. In other situations, more complex statistical methods, dating techniques, or use of biological community data may be needed; the key factor is to know what to do to answer the question and whom to call.

Society's standards change with time—not only concerning behaviour or morals, but also about what we accept with regard to environmental contamination. This is partly driven by improved understanding of the risks associated with chemicals and also because we demand a cleaner environment in which to live. In response to these societal changes, our laws change to meet our expectations. Higher levels of contamination may have been an acceptable price to pay for rapid industrialization 200 years ago, and several of these chemicals may still be around today in the form of contaminated land, groundwater, or marine systems. In environmental forensics, it is necessary to determine the source of any contamination, place that in context both geographically and legally, demonstrate a pathway to a sink, and then show how much is present above the background. This book provides a series of methods and approaches that can be used to do just that; chapters have been written by experts in each field and logically ordered to provide a guide for all practitioners.

## Preparation

All good scientific studies and legal cases are well planned; 'perfect preparation prevents piss poor performance' and everything that is a necessary part of good environmental forensics. When invited to take up a case, practitioners should plan their approach carefully before leaving their offices. There are two types of cases, however: Sites that have been contaminated over time and are now being investigated may be approached in a slightly more leisurely manner than ongoing acute (spill) events where speed is of the essence. A different approach to each type of case is required.

## Sites

Before approaching a new site, it would be prudent to find out as much as possible about it, if only to direct where samples would be most usefully collected. In this regard, written histories and company records can help a lot. One of the first resources that should be used is maps; this includes standard topographical maps (e.g., U.K. Ordnance Survey) as well as geological maps indicating relief, drainage pattern, and rock and soil type. Not all of these may be available, but efforts should be directed to finding them.

Aerial photographs (Davis et al. 2005) can also have a significant role to play by identifying the assets that were present at the time that the photo was taken. If a series of photographs taken through time is available, key dates can be narrowed down to small ranges (e.g., Davis et al. 2005). This may be of great importance when trying to date particular contamination events or start points for releases. Even Google Earth has been of great help in resolving likely sources of contaminants (Kalin, personal communication).

Physical attributes for sites and past monitoring records can provide an indication of the direction and location of potential sources, contamination plumes, or off-site receptors. Care must be taken when reviewing these data to ensure that no bias is introduced by using the conclusions from previous studies. These studies should be read, but one should draw one's own conclusions from the data.

## Events and Spills

In the case of an ongoing event, a plan should be in place to ensure that statistically meaningful results may be gathered from any samples taken. The message here might be 'be prepared'. This means that appropriate sample collection vessels (e.g., glass for organic contaminants and plastic for metals) have already been cleaned and are ready to go. One should also know something about the chemistry of the contaminant (especially the water solubility) so that the correct phase may be collected.

Some pollutants may be transported via the atmosphere, and access to a Gaussian plume dispersion model may provide a rapid assessment of the likely area of maximum impact under the prevailing weather conditions. Details required for accurate prediction of deposition areas include thermal lift of the contaminant, wind direction and strength, depth of the mixing layer, and effects of buildings. Such simple modelling may not be good enough for other needs, but it should at least point the sampler in the correct

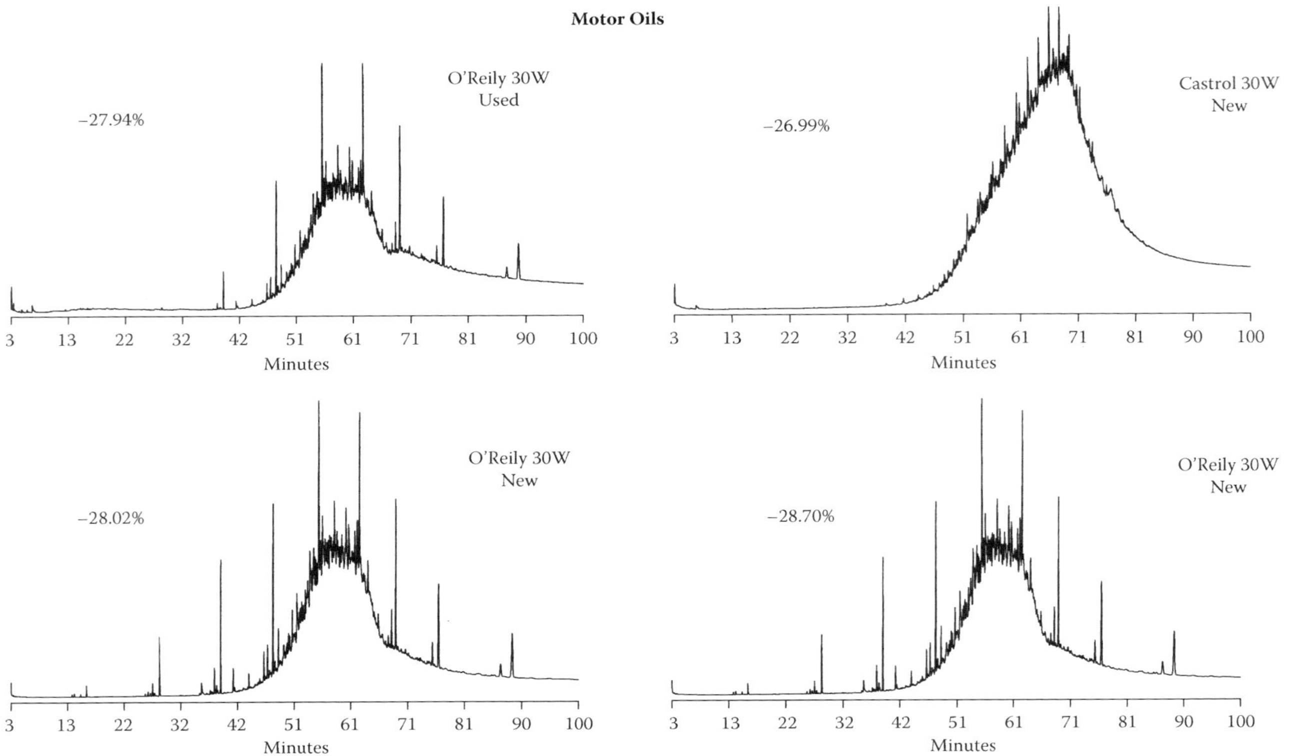

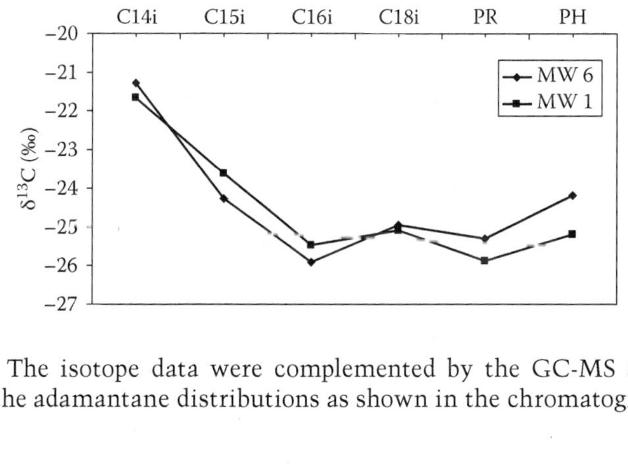

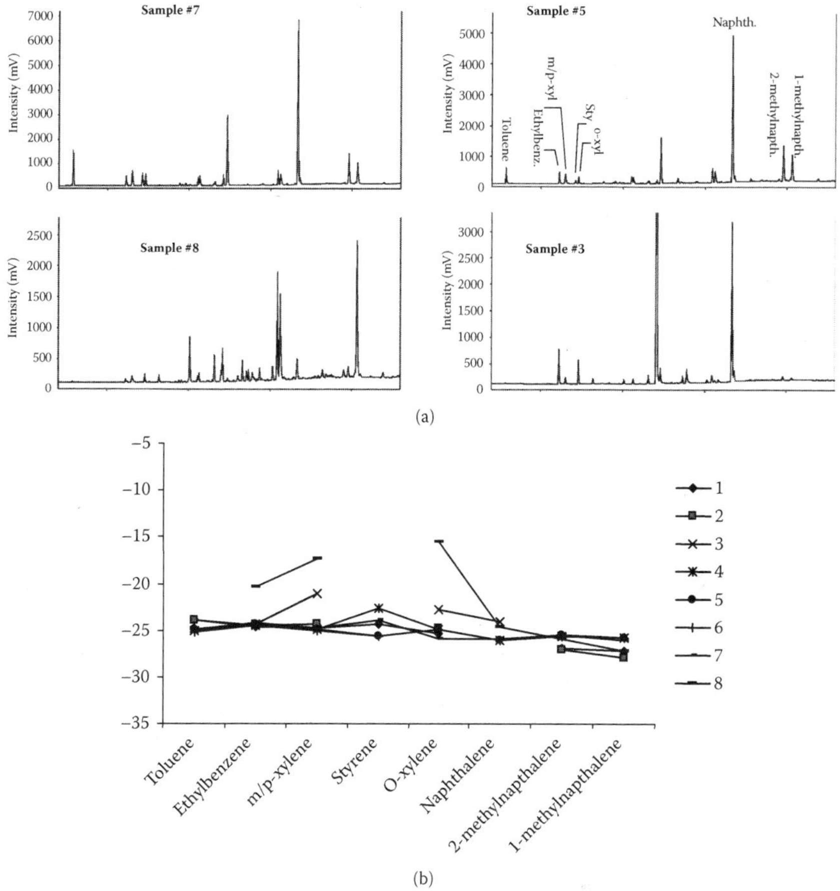

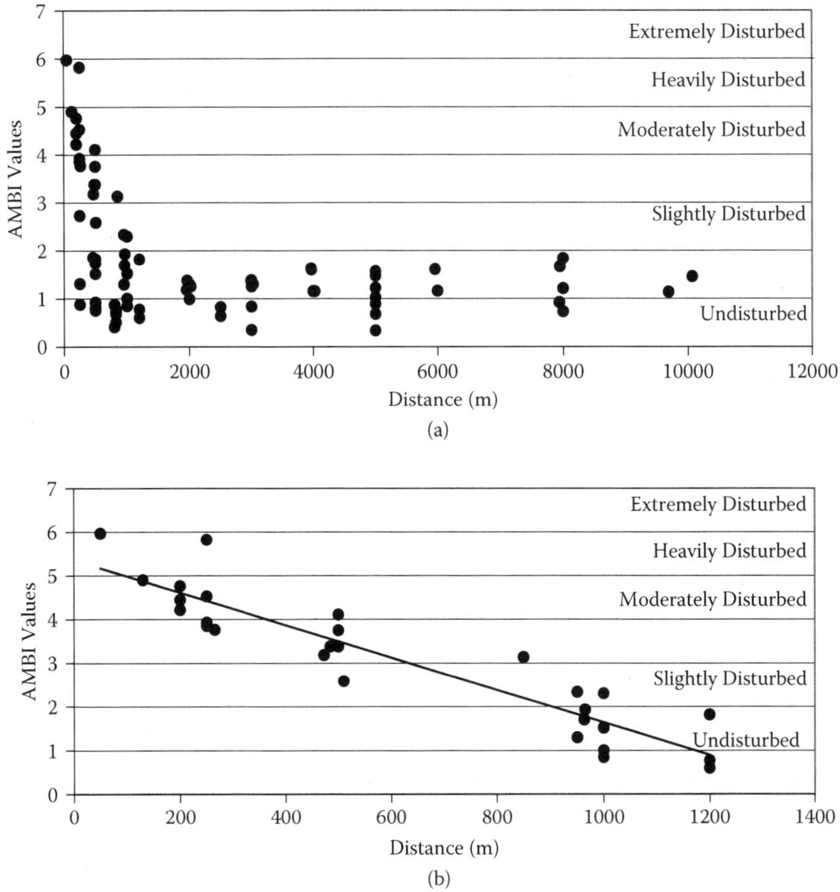

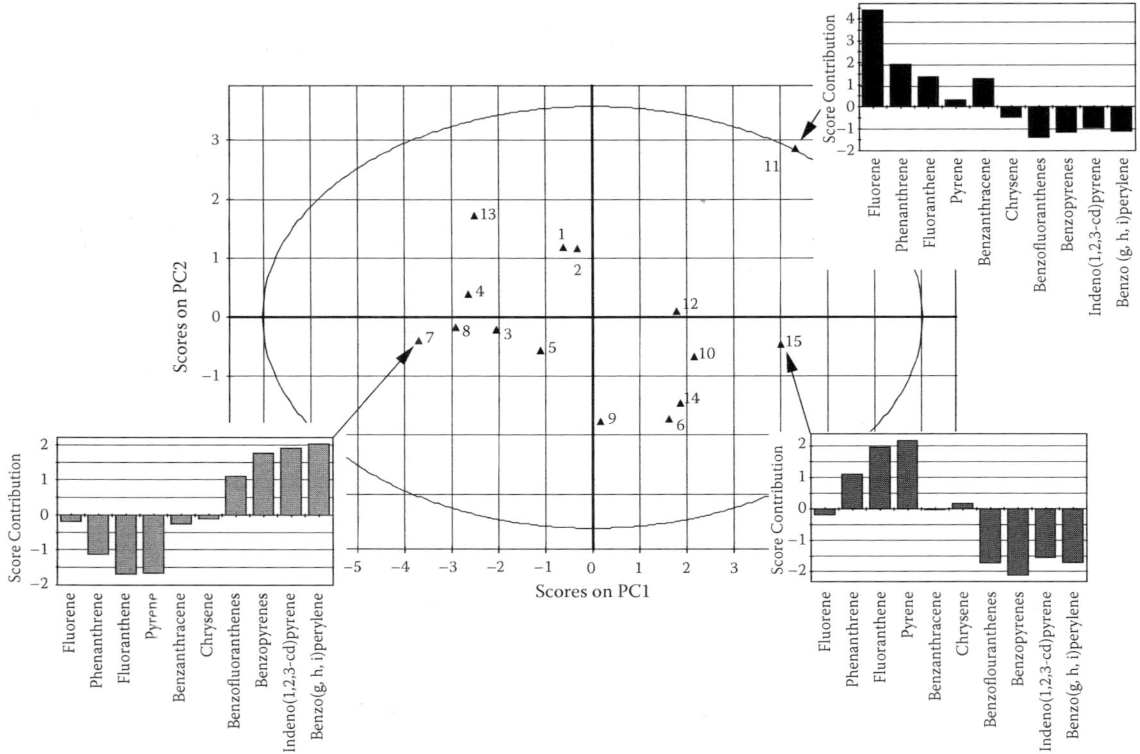

direction and at the right distance from the source to ensure meaningful sample collection.

With spills, it may be prudent to collect more samples than may be needed as it may not be possible to collect them later. Provided they are correctly stored, many materials should be stable long enough for assessment of the analytical needs, although suitable control samples to determine losses should also be included. In some cases, such as oils spills in harbours, there may be several potential sources of hydrocarbons and the responsible party may not be immediately obvious (Hegazi et al. 2004; Staniloac, Petrescu, and Patroeseu 2001). Therefore, as many potential sources as are in the area should be collected and this may require the assistance of the enforcement agencies to facilitate access.

## Legal Framework

For a case to exist in criminal law, some statute must have been contravened and a contamination event must be responsible. Although this may sound easy to assess, many compounds do not have mandatory limits set down in legal texts. Therefore, many of the regulations use catchall statements such as "noxious substance" (e.g., Merchant Shipping and Maritime Security Act 1997) to encompass as many materials as possible. Our laws change with time, especially the secondary instruments underneath the primary legislation (e.g., Statutory Instrument 1998 No. 1153: The Merchant Shipping [Dangerous or Noxious Liquid Substances in Bulk] [Amendment] Regulations 1998), and these should reflect society's acceptance of chemicals in the environment as well as our awareness of the long-term effects of human exposure.

Much of Europe's environmental protection legislation has been derived from EU directives in the last decade. Important pieces of legislation include the Water Framework Directive (2000/60/EC) and the new Environmental Liability Directive (2004/35/CE). EU directives set out the goals, but each member state may implement its own laws to achieve those goals, so there will be differences across the continent.

All this is driving toward a cleaner environment; however, past contamination does not go away just because we have changed the reference values we allow in discharges.

## Background versus Baseline

All elements except some of the radioactive ones existed in the environment long before man was active on the Earth. The concentration of these elements varied widely according to the rock type and physicochemical

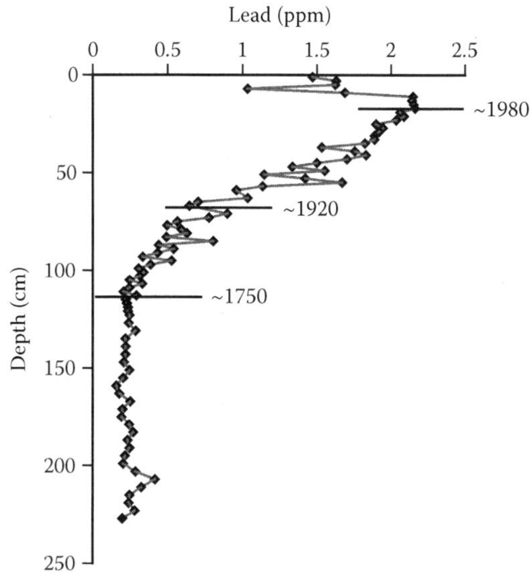


**Figure 1.1** Profile of lead in sediments from a core taken in a Scottish sea loch.

nature of the ecosystem. Therefore, there is a natural range of concentrations we might expect to find, even in pristine environments. Man evolved and, around 1750, began an intense period of industrialization (Clark and Jacks 2007). From this period onward, a much greater change in the concentration of many elements has occurred compared to the millennia before it. Good examples of this include lead (Pb) and copper (Cu). After many years of resource exploitation, society has recognised the potential harm that some of these elements have on human health. Therefore, legislation has altered our usage patterns and we are now discharging less of some of these elements into the environment.

If there is a spillage that includes a naturally occurring element or compound, it is usual to compare the new environmental concentrations to previous ones in order to demonstrate enrichment. The question becomes what values to use to compare the spill concentrations. A supplementary question concerns to what level remediation should be conducted. All of this has to be set in the context of man's previous activities. An example of lead in a core from a relatively remote Scottish sea loch can be seen in Figure 1.1 (Treadwell, unpublished data).

This location does not directly receive wastes from industrial activity; however, like most Northern Hemisphere sites, it does have atmospheric fallout of anthropogenic contaminants. At depths greater than ~110 cm, the Pb concentration is relatively constant with little variation in time. However, this depth coincides with the beginning of the Industrial Revolution and the increased usage of coal and other natural resources. The concentration of

Pb increases in the sediment to ~70 cm, where it increases at an even greater rate; this latter increase is due to the use of Pb in petrol. After 1980, the amount of Pb used in petrol decreased and has been gradually phased out. The sediments have recorded this change, showing significantly lower Pb concentrations in the surface sediments today.

In a remediation programme, should the responsible party be made to clean up to the natural, pre-anthropogenic concentration or to that of the day before the spill? The difference in effort and cost involved between each extreme might be considerable, depending on the nature of the contaminant.

## Know the Contaminant

In any event, it is necessary to understand the environmental chemistry or biology of the materials or organisms involved. Typical questions that will dictate the sampling programme to some extent include what the environmental half-life is and whether the compounds are photodegraded, water soluble, and toxic or ecotoxic.

The design of the sampling programme will have to address these parameters and collect the appropriate media (waters vs. soils or sediments vs. atmosphere) within the correct time frame, especially if the compounds are rapidly degraded or volatilised. Bacteria may die off quickly in certain conditions, so rapid sampling would be required to confirm their presence. The toxicity aspects may be used to consider the health implications to staff collecting the samples and also in determining the effect on the ecosystem. In some cases, the pollutant may have had an effect on the biota, leaving an altered community structure and then moved on. In cases such as these, sampling the community may indicate the presence or magnitude of effect (e.g., Hopkins and Mudge 2004).

When environmental transformations are a key feature of the compound, it may be more appropriate to sample the breakdown products rather than the parent compound. These compounds may also have different water solubilities and this should also be considered. A good example of this is the work on chloroacetamide herbicides that produce relatively uninvestigated degradation products (Hladik, Hsiao, and Roberts 2005).

## Sampling

All sampling programmes need considerable thought to ensure relevant materials are collected (Warren 2005), that they are stored and processed in the correct manner, and that no interferences may have altered the original concentrations. This aspect would form an entire book in itself; a recent publication (Morrison and Murphy 2006) covers a wide range of environmental contaminants and how they may be sampled. A summary of the planning might include the following aspects.

## Media

What sample media are to be collected? Do these represent the phases where the contaminants or their degradation products might accumulate? This aspect can be determined by knowing the contaminant; key resources might include online databases such as Bielstein (Elsevier). The octanol–water partition coefficient ($K_{ow}$) might provide a first 'guesstimate' of the likely partitioning between phases, although care needs to exercised with hydrophobic ionisable organic compounds (HIOCs) (Amiri, Bornick, and Worch 2005; Kolpin et al. 2002).

## Bias

Any scientist's role in an environmental forensics case is to get to the truth. This may involve identifying the polluter or determining the effects a pollutant has had. It is possible in a system that has a heterogeneous nature to select either sample sites or results after analysis that implicates one particular source or another. This may not reflect the truth of the matter, so it is a key condition that any practitioner conducts himself in an unbiased manner to ensure that all possibilities can be explored.

## Number of Samples

How many samples are needed to overcome the natural variability that exists in the environment and to provide an estimate of the concentration with quantifiable errors? Substantial variability can be found in sediments (e.g., Mudge, Assinder, and Russell 2001, 2003) and a suitable protocol needs to be adopted if a value close to the correct one is to be obtained. Frequently, environmental forensics cases seem to be built on a small number of potentially unrepresentative samples, which could be open to challenge in a court. Do not have evidence be declared inadmissible due to inappropriate sampling.

## Sample Quantity

The detection limits for compounds and elements are dependent on the technical capabilities of the analytical instrumentation used (see later discussion). Other factors, such as the extraction efficiency, are also important and should be used together to determine the minimum quantity of sample medium needed to provide a response of at least 10 times the background noise of the instrumentation. It may be appropriate to conduct a range of preliminary extractions before collecting samples if the analyte is not one that has been routinely analysed before. In some cases, if the volume of an aqueous sample needed is large, an in situ concentration method such as solid phase extraction should be considered.

### Security

In most research programmes, security, in its normal interpretation, is not an issue; however, in legal cases where potentially large sums of money are at stake, this aspect should be given greater weight. The definition of security may now be expanded to ensure that the samples collected in the field are indeed the ones analysed; that there has been no potential for tampering, either accidentally or deliberately; that blanks made up in the field remain unaffected by storage or transport; and that results generated and stored on computers can be audited and verified where necessary. This is a completely extra tier of administration and bureaucracy that may be required to ensure the analyses are valid.

### Cost

Conducting sampling and analysis of many chemicals is not cheap. There are many aspects that need to be taken into account when determining the per-analyte or per-sample cost and, to some potentially responsible parties (PRPs), they may seem large. In general, however, they tend to be less than the cost of the legal team or an inappropriate fine. It is necessary to conduct some form of cost/benefit analysis to determine the minimum number of samples needed to accurately reflect the environmental situation. The risk in doing less is that any evidence collated may be rejected because it did not represent the situation; in such events, the legal case may be lost and that money wasted. In many situations, it would be appropriate to oversample to ensure sufficient materials are available when the analyses come in; the exact number used may be determined as the case develops. It would be more cost effective to collect these samples at the outset, store them correctly, and then use what is needed later. There will be an extra cost in doing this, but it would be less than (1) losing the case or (2) analysing everything without regard for the need to do so.

## Analysis

After one has collected samples, stored them in the correct containers, etc., the samples need to be analysed using an instrumental technique. There are many machines available today, each with its own characteristics and price tag. The laboratory conducting the analyses should be aware of the current best practice and methods to be used. In the United States, the Environmental Protection Agency (EPA) methods provide an excellent mechanism to ensure consistency and believability in the results.

When taking measurements, quality assurance (QA) should be paramount. The appropriate surrogate standards should be added to assess yield or recovery, and accredited reference materials should be used routinely to determine the effectiveness of the method and instrumentation. More use should be made of the techniques used in statistical process control, such as Shewhart plots (NIST/SEMATECH 2006), that can assist in determining stability in a series of analyses and rule out long- or short-term changes due to analytical procedure.

It is usual in any batch of samples to include blanks and standards to ensure consistency. Care must be exercised, however, to ensure that materials are not carried over from one analysis to another. Standards are usually present at concentrations significantly greater than those found in the environment; even a small degree of cross-contamination may dramatically affect the results. Standards at a range of concentrations and values close to the expected environmental levels should be used with blanks between them and the samples.

## Developing the Case

Obtaining good data is not the end of the story in an environmental forensics case, but rather the beginning. There are many postanalytical steps that need to be conducted, including determining the data integrity. Errors can creep into data sets in many ways, including mistypes when entering data from a keyboard, poor coding of samples, incomplete analyses, bias or incorrect quantification by analytical software (beware of the black box), or simple computer calculation errors. Many of these can be overcome by good sample security and tracking; others can be readily spotted by visual inspection of the data—do they look right? There are many tools available to assist in this phase of a case, including a range of statistical practices.

### Statistics

Multivariate statistics (MVS) such as principal component analysis (PCA; see chapter 9) can prove very useful in identifying any potential data set inconsistencies because those samples ‘stick out like a sore thumb’. This type of screening can help find data errors and, apart from other functions such as source partitioning, might form the first postanalytical process.

Simple statistics, such as Student's *t*-tests, to prove beyond reasonable doubt that one set of samples has a concentration different from that of another set requires thought and data investigation before use; many of the statistical methods require the data or at least their residuals to be normally distributed. There are many steps that may be taken to either transform data or use alternative methods such as nonparametric tests; however, in many instances, these techniques are less powerful or reliable than their parametric brothers. Care should be exercised in deciding what to use and for what purpose.

Many of the MVS methods can provide some evidence of source partitioning in complex environments (e.g., Prince William Sound; Burns et al. 1997; Mudge 2002) and be used to identify the polluter. All of these methods are tools in the armoury of an environmental forensics expert who uses them as needed.

### Data Presentation

The arbiter (judge, magistrate, jury, etc.) in environmental prosecutions may not have recent scientific training. Therefore, it is the duty of the expert witness to present the data obtained during any investigation in a manner that makes it understandable to the laity. This does not mean 'dumbing down', but rather using the data and providing explanations in an easily understood manner. One of the best ways of accomplishing this is through the use of figures, diagrams, and pictures. In the widest sense, this may also include photographs and video to give the court as close to a first-hand experience of the event or site as possible.

If data can be presented through means of a diagram, maybe they should be—as long as sufficient explanation is given in each case. Good use of graphical representations may enable the arbiters to see and follow the complex scientific case that has been developed.

## Source, Pathway, Sink

For a case to have any chance of success, it is necessary to demonstrate that the *source* exists or existed (which brings up issues of strict liability: In *Environment Agency (formerly National Rivers Authority) v. Empress Car Co. (Abertillery) Ltd.* 1998 2 WLR. 350, examples are given of cases in which strict liability has been imposed for 'causing' events that were the immediate consequence of the deliberate acts of third parties but that the defendant had a duty to prevent or take reasonable care to prevent) and that there was a *pathway* or mechanism by which the source could have travelled to ultimately reach a receptor or *sink*. A case is unlikely to succeed if a source exists but there is no mechanism by which it could have reached the location under investigation. This whole aspect of environmental forensics brings in the need for geomorphological interpretations, hydrological assessment, and modelling. In the case of modelling, it should not be relied on by itself, but rather provide validation for a range of other measures.

Hydrological assessment and demonstrating that a pathway exists between source and sink may use dyes or tracers (Field 2005) added below toxic thresholds or through use of inert proxies. One of the advantages of such testing after the event is the ability to accurately measure flow times and directions, thus dramatically strengthening the link between source and sink.

## Proving the Case

An often overlooked aspect of case development is to think 'outside the box' and look at the system more widely. While a convincing case may be made using the preceding suggestions, it may be made even more convincing if all the possible alternative scenarios are investigated and shown not to be responsible. For example, in a case involving blocking of a sewer with vegetable oil, as well as showing that the most likely source was one particular trader, it was also necessary to demonstrate that the oil could not reasonably have come from all the other potential sources connected to the sewage system, including domestic households (Mudge, unpublished data). Identifying these aspects in advance and being prepared will help when one is cross-questioned.

## The Expert Witness Report

It must be remembered that one's duty in all cases is to the court, as is clearly set out in the guidelines for expert witnesses in the United Kingdom (Practice Directions to the Civil Procedure Rules, Part 35. Experts and Assessors, Department of Constitutional Affairs). Any report produced for the court must follow these rules and the supplementary directions on the structure of a report. According to Practice Direction Supplement CPR Part 35, an expert's report must:

- give details of the expert's qualifications;
- give details of any literature or other material that the expert has relied on in making the report;
- contain a statement setting out the substance of all facts and instructions given to the expert that are material to the opinions expressed in the report or upon which those opinions are based;
- make clear which of the facts stated in the report are within the expert's own knowledge;

say who carried out any examination, measurement, test, or experiment that the expert has used for the report; give the qualifications of that person; and say whether or not the test or experiment has been carried out under the expert's supervision;
summarise the range of opinion and give reasons for the expert's own opinion when there is a range of opinion on the matters dealt with in the report;
contain a summary of the conclusions reached;
state the qualification if the expert is not able to give his opinion without qualification; and
contain a statement that the expert understands his duty to the court and has complied and will continue to comply with that duty.

The overall requirement, notwithstanding the preceding well-structured aspects, is to be clear, logical, truthful, and unbiased. Clarity can be obtained by good scientific writing using as little jargon as necessary, explaining methods and concepts where needed, and having good proofreading from one's counsel. A logical structure should lead the reader through the data, starting with the background to the site or event and leading to the currently observed state. Truth may be subjective in some cases, as different authors will perceive data in different ways and draw different conclusions. Only facts can be stated here and all other interpretations that flow from these facts may be regarded as conjecture. It is appropriate to highlight the range of current thinking on any particular process or concept as well as one's own interpretation. This allows the court to see the extent of different thinking so that any evidence based on that theory may be given the appropriate weight when it considers judgement.

Ultimately, it is the court that decides the outcome of any case, and any report is there to guide the court through the available evidence. One should use that position of trust well.

## References

Amiri, F., H. Bornick, and E. Worch. (2005) Sorption of phenols onto sandy aquifer material: The effect of dissolved organic matter (DOM). *Water Research,* 39(5): 933–941.

Burns, W. A., P. J. Mankiewicz, A. E. Bence, D. S. Page, and K. R. Parker. (1997) A principal-component and least-squares method for allocating polycyclic aromatic hydrocarbons in sediment to multiple sources. *Environmental Toxicology and Chemistry,* 16(6): 1119–1131.

Clark, G., and D. Jacks. (2007) Coal and the Industrial Revolution, 1700–1869. *European Review of Economic History,* 11: 39–72.

Davis, A., B. Howe, A. Nicholson, S. McCaffery, and K. A. Hoenke. (2005) Use of geochemical forensics to determine release eras of petrochemicals to groundwater, Whitehorse, Yukon. *Environmental Forensics,* 6(3): 253–271.

Field, M. S. (2005) Assessing aquatic ecotoxicological risks associated with fluorescent dyes used for water-tracing studies. *Environmental & Engineering Geoscience,* 11(4): 295–308.

Hegazi, A. H., J. T. Andersson, M. A. Abu-Elgheit, and M. S. El-Gayar. (2004) Source diagnostic and weathering indicators of tar balls utilizing acyclic, polycyclic and S-heterocyclic components. *Chemosphere,* 55(7): 1053–1065.

Hladik, M. L., J. J. Hsiao, and A. L. Roberts. (2005) Are neutral chloroacetamide herbicide degradates of potential environmental concern? Analysis and occurrence in the upper Chesapeake Bay. *Environmental Science & Technology,* 39(17): 6561–6574.

Hopkins, F. E., and S. M. Mudge. (2004) Detecting anthropogenic stress in an ecosystem: 2. Macrofauna in a sewage gradient. *Environmental Forensics,* 5(4): 213–223.

Kolpin, D. W., E. T. Furlong, M. T. Meyer, E. M. Thurman, S. D. Zaugg, L. B. Barber, and H. T. Buxton. (2002) Pharmaceuticals, hormones, and other organic wastewater contaminants in U.S. streams, 1999–2000: A national reconnaissance. *Environmental Science & Technology,* 36(6): 1202–1211.

Morrison, R. D., and B. L. Murphy. (2006) *Environmental forensics: A contaminant-specific guide.* Amsterdam: Elsevier.

Mudge, S. M. (2002) Reassessment of the hydrocarbons in Prince William Sound and the Gulf of Alaska: Identifying the source using partial least squares. *Environmental Science & Technology,* 36(11): 2354–2360.

Mudge, S. M., D. J. Assinder, and A. T. Russell. (2001) Microscale variability of contaminants in surface sediments: The implications for sampling. R&D Technical Report P3-057/TR, Environment Agency, UK, 88 pp.

______ (2003). Mesoscale variation of radionuclides in sediments: Normalisation and the implications for sampling. R&D Technical Report P3-093/TR, Environment Agency, UK, 72 pp.

NIST/SEMATECH. (2006) *e-Handbook of statistical methods,* http://www.itl.nist.gov/div898/handbook/.

Staniloac, D., B. Petrescu, and C. Patroeseu. (2001) Pattern recognition based software for oil spills identification by gas-chromatography and IR spectrophotometry. *Environmental Forensics,* 2(4): 363–366.

Warren, J. (2005) Representativeness of environmental samples. *Environmental Forensics,* 6(1): 21 25.

# 2 Radionuclides in the Environment: Tracers and Dating

DAVID ASSINDER

## Contents

Introduction ..... 15
Sources of Radionuclides ..... 16
Inputs to the Terrestrial and Marine Environments ..... 17
Environmental Chemistry of Important Radionuclides ..... 17
Natural Radionuclides ..... 17
Radium Isotopes and $^{222}$Rn ..... 17
$^{210}$Pb ..... 20
Artificial Radionuclides ..... 20
Fission and Activation Products ..... 20
Transuranic Radionuclides ..... 21
Application of Radionuclides for Tracing and Dating ..... 22
Theory of Application and Analysis ..... 22
Particulate Measurements ..... 22
Dissolved Measurements ..... 23
Case Studies ..... 25
Reprocessing Radionuclides as Tracers and for Dating ..... 27
$^{137}$Cs and $^{99}$Tc as Water Mass Tracers ..... 27
$^{237}$Np as a Tracer of Reprocessed Uranium ..... 28
Radium and Radon in Groundwater Studies ..... 28
Radium ..... 29
Radon ..... 29
$^{210}$Pb and Multi-Isotope Sediment Dating ..... 30
Current and Future Roles for Radionuclides in Environmental Forensics ..... 31
References ..... 33

## Introduction

Radionuclides from natural sources are ubiquitous in all materials on Earth. Man-made artificial radionuclides have spread around the globe since their first production in the 1940s. Together they encompass a wide range of elements with differing chemistries and consequently differing environmental behaviour. This, coupled with the fact that each radionuclide decays at a known rate, means they offer powerful tools to trace and date environmental processes and identify when and where releases occurred. These features mean that radionuclides can, in certain situations, provide some key information in the field of environmental forensics. A clear understanding is required of when radionuclides can provide information and what the limitations are on these methods.

## Sources of Radionuclides

Radionuclides in the environment are derived from a variety of natural and artificial sources (Table 2.1). They decay at known rates characteristic of each radionuclide. This is usually quoted as the 'half-life' of the radionuclide—the time taken for half of the original amount to decay. As decay rates can vary between microseconds and billion of years, the possibility of using them as 'clocks' has long been employed in geological and environmental studies.

Sources can be diffuse or localised, depending on the nature of the radionuclide and its mode of generation. Methods of using radionuclides depend on whether the addition is continuous both, temporally and spatially,

**Table 2.1 Sources and Decay of Radionuclides in the Environment**

| Type | Source | Examples |
|---|---|---|
| Natural | Cosmic-ray produced | $^{3}$H, $^{10}$Be, $^{14}$C, $^{26}$Al, $^{36}$Cl |
| | Long-lived radionuclides (decaying directly to stable nuclides) | $^{40}$K, $^{87}$Rb |
| | Decay series (decaying to stable nuclides via a series of radionuclides) | $^{238}$U, $^{235}$U, $^{234}$U, $^{232}$Th, $^{230}$Th, $^{228}$Th, $^{223}$Ra, $^{224}$Ra, $^{226}$Ra, $^{228}$Ra, $^{222}$Rn, $^{210}$Pb, $^{210}$Po |
| Artificial | Fallout and accidents | $^{90}$Sr, $^{131}$I, $^{137}$Cs, $^{238}$Pu, $^{239}$Pu, $^{240}$Pu, $^{241}$Pu |
| | Discharges from nuclear facilities | $^{3}$H, $^{99}$Tc, $^{106}$Ru, $^{134}$Cs, $^{137}$Cs, $^{237}$Np, $^{238}$Pu, $^{239}$Pu, $^{240}$Pu, $^{241}$Pu, $^{241}$Am |

*Notes:* Al = aluminium, Be = beryllium, C = carbon, Cl = chlorine, K = potassium, Rb = rubidium, U = uranium, Th = thorium, Ra = radium, Rn = radon, Pb = lead, Po = polonium, Sr = strontium, I = iodine, Cs = caesium, Pu = plutonium, H = hydrogen, Tc = technetium, Ru = ruthenium, Am = americium.

Decay implies the emission of particles or 'rays' (alpha particles, beta particles, or gamma rays) and conversion to a different radionuclide:

$$\underset{238-\textit{uranium}}{{}^{238}_{92}U} \xrightarrow{\textit{alpha decay}} \underset{234-\textit{thorium}}{{}^{234}_{90}Th} + \underset{\textit{alpha particle}}{{}^{4}_{2}He} \quad \text{Half-life } (t_{1/2}) = 4.5\times10^{9}\ \text{y}$$

$$\underset{241-\textit{plutonium}}{{}^{241}_{94}Pu} \xrightarrow{\textit{beta decay}} \underset{241-\textit{americium}}{{}^{241}_{95}Am} + \underset{\textit{beta particle}}{{}^{0}_{-1}e^{-}} \quad \text{Half-life } (t_{1/2}) = 14.4\ \text{y}$$

or whether specific localised inputs occur. In addition, the physicochemical form of the element will control the occurrence of the radionuclide. For example, because radon is an inert gas, it will be controlled by gaseous diffusion and solubility considerations, whereas caesium is an alkali metal and thus will show solution characteristics appropriate to that group.

### Inputs to the Terrestrial and Marine Environments

Inputs to the environment can be classified as

- natural inputs from diffuse sources (e.g., the natural decay series radionuclides present because the Earth was formed or generated subsequently in a relatively consistent way from parent radionuclides);
- natural inputs from localised sources (e.g., areas of specific concentration of natural radionuclides such as ore deposits);
- artificial inputs from diffuse sources (e.g., nuclear weapons testing or nuclear power plant accidents that produce a diffuse output to large areas); and
- artificial inputs from localised sources (e.g., nuclear installations, testing, or accidents with localised or point source discharges to restricted areas).

Each input type has different potential uses for tracing (using the radionuclide chemical properties) or dating (using the radionuclide decay), or a combination of the two. These may provide a better understanding of environmental processes at different scales that may affect discharged materials to the environment (e.g., dating of sediment deposits in which discharged materials accumulate) or allow direct tracking if the radionuclide itself is the material in question (e.g., unauthorised discharges of radionuclides from a point source). Specific localised artificial inputs may allow detailed tracing of environmental transport routes within the atmosphere, catchment, estuary, sea, or ocean but are often only applicable in that area. Natural radionuclides present globally may allow all areas to be considered but may lack specificity due to their ubiquity.

## Environmental Chemistry of Important Radionuclides

### Natural Radionuclides

#### *Radium Isotopes and $^{222}$Rn*

Although the higher members of the natural decay series such as $^{238}$U, $^{234}$U, $^{228}$Th, $^{230}$Th, and $^{232}$Th have attracted attention for tracing and dating on relatively long time scales (e.g., Milton and Brown 1987), attention has recently

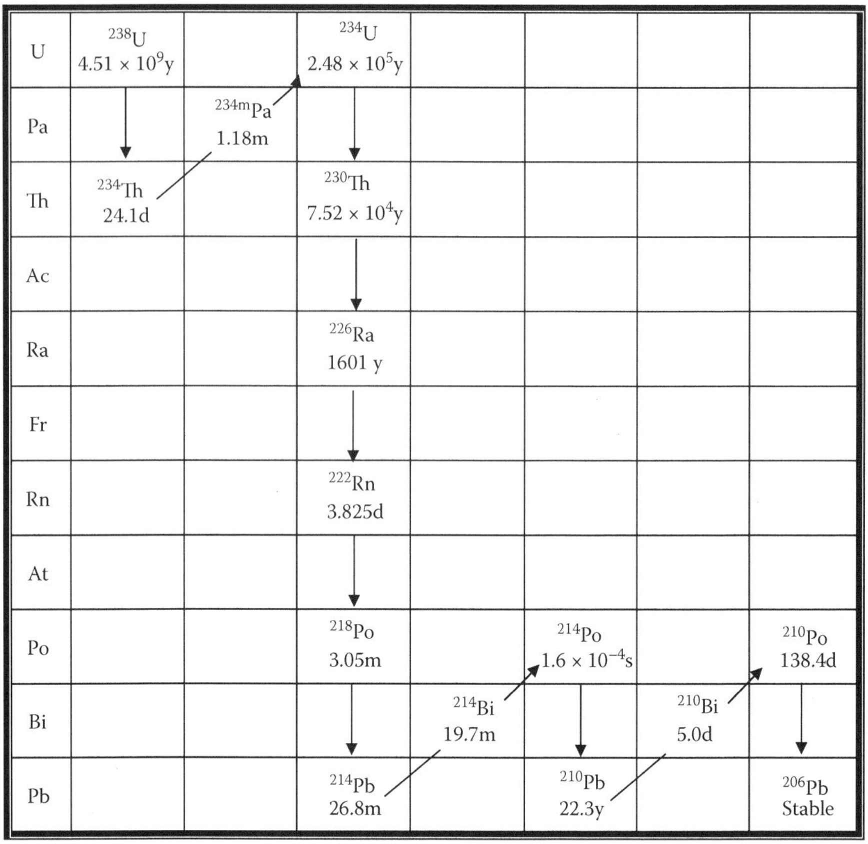


**Figure 2.1** Natural decay series for $^{238}U$ showing radionuclide half-lives and modes of decay. Vertical arrows = alpha decay; oblique arrows = beta decay.

focussed on radium and radon isotopes for shorter time scale events more relevant to environmental forensics, particularly in relation to groundwater flow.

$^{226}Ra$ ($t_{1/2}$ = 1601 y) and $^{222}Rn$ ($t_{1/2}$ = 3.8 days) are part of the natural decay series originating from $^{238}U$ (Figure 2.1). $^{228}Ra$ ($t_{1/2}$ = 5.7 y) and $^{224}Ra$ ($t_{1/2}$ = 3.64 days) are part of the decay series originating from $^{232}Th$ and $^{223}Ra$ ($t_{1/2}$ = 11.1 days) derived from $^{235}U$. All Ra isotopes are generally found out of equilibrium with their thorium parents due to the greater solubility of Ra. $^{222}Rn$ is the decay product of $^{226}Ra$ and is a gas. Due to escape to the atmosphere, it may also not be in equilibrium with its parent, although its short half-life means it is often found 'supported' by its parent and in equilibrium. This state of secular disequilibrium (Box 2.1) is the basis for many dating methods.

The behaviour of radium in the environment has been reviewed by McDowell-Boyer et al. (1980), King et al. (1982), and Kraemer and Genereux (1998). Radium isotopes are generally found in excess of their parents, due to the greater solubility of Ra, and subsequently can diffuse from sediments

**Box 2.1**
**Secular Disequilibrium in the Natural Decay Series**

In a closed system, where no radionuclides enter or leave apart from growth from their parent and decay, an equilibrium is reached, *secular equilibrium*, where the total amount of radioactivity reaches some constant level proportional to the amount of parent uranium or thorium. At that time, each daughter product will have the same activity (i.e., number of decays per second) even though the absolute number of atoms of each radionuclide will differ greatly.

During geological processes such as erosion, sedimentation, melting, or crystallization, different nuclides in the decay series can become fractionated relative to one another, due to variations in their chemistry or the structural site they occupy. This results in a state of secular disequilibrium. Such a situation can be utilized in two different ways as a dating tool, called, respectively, the 'daughter-excess' and 'daughter-deficiency' dating methods.

In the daughter-excess method, a deposit is formed with an excess of the daughter beyond the level that can be sustained by the abundance of its parent nuclide. Over time, the excess or 'unsupported' daughter decays back until secular equilibrium with its parent is restored. If the original fractionation can be estimated, the age of the deposit can be calculated by the progress of decay of the excess (e.g., $^{210}$Pb dating; see '$^{210}$Pb and Multi-isotope Sediment Dating').

In the daughter-deficiency method, chemical fractionation during the formation of a deposit causes it to take up a radioactive parent but effectively none of its daughter. The age of the deposit can then be determined by measuring the growth of the daughter, up to the point when its abundance is within error of secular equilibrium of the parent (e.g., $^{230}$Th accumulation in marine and freshwater carbonates; Gascoyne and Schwarz 1982).

and are detectable in natural waters. However, a fraction of Ra can also be found in soils and sediments due to deposition by ion exchange onto clays and organics (Gascoyne 1982).

The behaviour of radon gas in the soil, water, and atmosphere has been reviewed by King et al. (1982), Gesell (1983), and Wiegand (2001). All radon isotopes are noble or inert gases, occurring as nonpolar, monatomic molecules, and are chemically inert. Radon dissolves in water and is often found in radioactive equilibrium with its parent $^{226}$Ra. Where it escapes to the atmosphere, it is unsupported and acts as a natural break in the $^{238}$U decay series, producing unsupported $^{210}$Pb by its decay, which is suitable for the daughter-excess dating method.

### *$^{210}$Pb*

The behaviour of $^{210}$Pb in the environment has been reviewed by Gascoyne (1982) and Bierman et al. (1998). It is unique among the members of the $^{238}$U decay series because its major pathway to the ocean, rivers, and soils is from the atmosphere. $^{222}$Rn decay, via four short-lived intermediate decay products (Figure 2.1), returns $^{210}$Pb to the land and sea surface by wet and dry fallout.

$^{210}$Pb is highly particle reactive and so tends to occur as excess $^{210}$Pb, out of secular equilibrium with the rest of the decay series, in terrestrial and marine particulates. The return of $^{210}$Pb activity concentrations to secular equilibrium, assuming no additional inputs or losses, provides a time scale of the order of 100–150 years due to the $^{210}$Pb half-life of approximately 22 years. This is a value appropriate for the dating of many recent environmental processes in soils, lakes, and coastal environments and hence the impact of human activities on them (Dearing and Jones 2003). Longer lived natural radionuclide decays (e.g., $^{230}$Th) are more relevant for slowly accumulating deep sea deposits (e.g., Thomson et al. 1984).

## Artificial Radionuclides

### *Fission and Activation Products*

Several radionuclides derived from the fission of uranium atoms or the activation of other nuclides during nuclear weapons tests, accidents, or the power generation cycle have been used for tracing and dating. These include $^{137}$Cs ($t_{1/2}$ = 30 y), $^{134}$Cs ($t_{1/2}$ = 2.2 y), $^{99}$Tc ($t_{1/2} = 2.1 \times 10^5$ y), and $^{129}$I ($t_{1/2} = 1.56 \times 10^7$ y). Case studies involving $^{137}$Cs and $^{99}$Tc will be examined here.

The behaviour of $^{137}$Cs in the environment has been reviewed by Ritchie and McHenry (1990). $^{137}$Cs is a fission product initially introduced into the environment in significant quantities by atmospheric nuclear weapons tests in the 1950s. Major periods of global deposition were seen in 1958 and 1963–1964, and minor periods occurred in 1971 and 1974 due to variations in the extent of testing. Periods of lower fallout can be related to moratoria on testing (1958–1961) and the signing of the Limited Test Ban Treaty of 1963. Global fallout rates have since declined overall. $^{137}$Cs is also derived from nuclear fuel reprocessing inputs to marine environments and was a major component of radioactivity derived from the Chernobyl accident in 1986. This complex input function from a variety of sources on both global and local scales has allowed several tracing and dating applications to be introduced.

$^{137}$Cs is rapidly adsorbed to clay minerals, especially illite, in freshwater systems and is transported with eroded soils into freshwater catchment basins or into the sea in association with suspended particulate matter. In the sea,

$^{137}Cs$ remains largely in solution as competing cations limit its adsorption. The soluble fraction has found uses in oceanic water mass tracing, whilst the particulate-bound fraction has been used for sediment dating in terrestrial and marine settings. The extent to which Cs adsorption is reversible has led to many studies that question the validity of Cs dating due to possible post-depositional migration in sediment core profiles.

The environmental behaviour of $^{99}Tc$ has become of importance in recent years due to the increased releases of this radionuclide from nuclear fuel reprocessing (McCubbin et al. 2002). Tc remains largely in the dissolved fraction in seawater (Aarkrog et al. 1987), making it similar to Cs for tracing water mass movements (Oliver, Perkins, and Mudge 2006).

### *Transuranic Radionuclides*

Several transuranic radionuclides derived from neutron activation of uranium atoms during nuclear weapons tests, accidents, or the power generation cycle have been used for tracing and dating. These include $^{238}Pu$ ($t_{1/2} = 87.7$ y), $^{239}Pu$ ($t_{1/2} = 2.4 \times 10^4$ y), $^{240}Pu$ ($t_{1/2} = 6570$ y), $^{241}Pu$ ($t_{1/2} = 14.4$ y), $^{241}Am$ ($t_{1/2} = 432$ y), and $^{237}Np$ ($t_{1/2} = 2.1 \times 10^6$ y).

Four principal isotopes of plutonium are found in the environment from artificial sources: $^{238}Pu$ to $^{241}Pu$. These radionuclides are derived from weapons test fallout of unfissioned plutonium and Pu produced by neutron irradiation of $^{238}U$ (Eisenbud and Gesell 1997), nuclear fuel reprocessing, and a limited localised input after Chernobyl (Ketterer et al. 2004). In addition, $^{238}Pu$ was introduced from the abortive re-entry of a navigational satellite carrying a SNAP-9A power generator. The behaviour of Pu isotopes in the environment has been reviewed by Morse and Choppin (1991).

Plutonium isotopes are particle reactive and have the same dating applications as the particulate fraction of $^{137}Cs$. However, their analysis is generally more time consuming and expensive; hence, their application has been more limited. Concerns similar to those for Cs have been raised regarding the postdepositional migration of isotopes in sediment core profiles.

$^{241}Am$ and $^{237}Np$ are also neutron activation products of uranium with the potential for tracing and dating in the environment. $^{241}Am$ is also derived from the decay of $^{241}Pu$; like Pu, it is highly particle reactive and has been used for particulate fraction studies. Np is also derived from the decay of $^{241}Am$ and is less particle reactive than Pu and Am but more so than Cs (Assinder 1999). Both radionuclides are derived from weapons testing and nuclear fuel cycle discharges, particularly from fuel reprocessing. During reprocessing, unwanted fission and transuranic activation products are removed from uranium so that it can be reused. A small fraction of Np can remain with the reprocessed uranium until the production of new fuel rods, when it is released. This has proved useful in tracing the use of reprocessed uranium.

## Application of Radionuclides for Tracing and Dating

### Theory of Application and Analysis

Radionuclide concentrations or ratios between selected radionuclides can be used to trace environmental processes in the same way as other, non-radioactive elements or compounds. If the radionuclide activity concentration or ratio is sufficiently distinguished from the local background values, the nature of this distinction can be tracked and monitored to trace the process responsible for any alteration. Multiple end-member modelling is possible with sufficient characterisation. In addition, the added dimension of radioactive decay may allow tracing and time scale of the process to be examined in some instances. In others, only dating of an environmental deposit or process with the radionuclide is possible.

The majority of early measurements of alpha-emitting radionuclides (e.g., Pu isotopes, radium, neptunium) were made by alpha spectrometry, which detects the alpha particle emission to assess activity concentration. Recent advances in mass spectrometric techniques mean that more measurements of long half-life (i.e., large numbers of atoms but less radioactivity) radionuclides are now made by these methods, which measure the concentration (number of atoms) rather than the activity (number of atoms decaying) (e.g., Yamamoto et al. 1995; Kuwabara et al. 1999; Ketterer et al. 2004). Short half-life alpha emitters (i.e., small number of atoms but more radioactivity) continue to be measured by radioactive alpha counting techniques. Gamma-emitting radionuclides or those that produce an associated gamma emission during their decay (e.g., $^{137}Cs$, $^{241}Am$) are conventionally measured by gamma spectroscopy (e.g., Assinder et al. 1997). Beta emitters (e.g., $^{241}Pu$, $^{99}Tc$) are measured by various beta counting techniques, including liquid scintillation, or by mass spectrometry for the long half-life $^{99}Tc$.

Alpha measurements require separation of the alpha emitter from the associated matrix (e.g., Assinder et al. 1997) as alpha particles have a very short range and penetration and are stopped by most environmental matrices. Gamma analysis can often be carried out on the whole sample as gamma rays are highly penetrating, although preconcentration may be required if the activity concentrations are low.

### Particulate Measurements

For many tracing and dating methods there is still a reliance on sample collection and laboratory analysis. Only limited in situ detection methodologies exist for gamma emitters (e.g., Macdonald, Smith, and Assinder 1996; Macdonald et al. 1999; Jones 2001; Tyler 2004). This leads to inherent problems with spatial and temporal heterogeneities in soil and sediment properties, and hence radionuclide activity concentrations, at a variety of scales (Mudge, Assinder, and Russell 2002, 2003). Assinder (2002) has summarised the problems of collecting a core profile of soil or sediment for dating purposes in relation to contaminant variations over time. Problems include:

- collection of a 'representative' vertical profile of sediment at a site, bearing in mind that there will be lateral and vertical variations in both contaminant and dating element concentrations;
- assessment of whether the profile shows a true record of contaminant concentration variations over time due to sediment deposition;
- assessment of whether the sediment deposition has been constant or has varied in a way that can be measured using a natural or artificial radionuclide such that a sedimentation rate can be assigned and the core 'dated';
- assessment of whether there are sediment compositional changes including grain size variations down the profile; and
- assessment of whether there has been any postdepositional remobilisation or redistribution affecting the activity concentration profile due to physical (e.g., erosion), chemical (e.g., redox changes), or biological (e.g., mixing of the sediment by organism activity, known as bioturbation) processes.

When using $^{210}$Pb to date a sediment profile—perhaps to study erosional changes in a catchment or the change in anthropogenic inputs over time—a variety of modelling strategies has been used to examine the core profile and negate many of the problems outlined earlier (Box 2.2). However, questions still remain—particularly about postdepositional changes (Smith 2001)—and multi-isotope profiling or measurement of other time-dependent features to check for internal consistency is a common procedure.

$^{210}$Pb dating makes use of the decay of the radionuclide over time. $^{137}$Cs and dating with Pu isotopes do not make use of the decay itself but, rather, rely on matching the known input functions of these radionuclides with the core profile. For example, $^{137}$Cs dating may make use of its first appearance in the 1950s coupled with global fallout peaks in the 1960s or the Chernobyl Cs input in 1986 to date sections of the profile. In certain areas, point source discharges with known input functions, such as the Sellafield discharges, may be used.

## Dissolved Measurements

As for particulate studies, most measurements rely on sample collection for subsequent analysis; hence, problems exist with regard to spatial and temporal heterogeneity. Most soluble fraction methods involve the measurement of

**Box 2.2**
**Modelling $^{210}Pb$ Inputs and Profiles**

$^{210}Pb$ formed by decay of $^{222}Rn$ in the atmosphere is removed by rain, snow, or dry fallout. It is scavenged by environmental particulates such as soil, river, lake, and marine sediments and deposited with a flux that varies due to local or regional meteorological factors. The unsupported $^{210}Pb$ activity concentration in each sediment layer in a core profile declines with age due to radioactive decay. Ideal conditions for $^{210}Pb$ measurements are high rates of sediment accumulation (>0.1 cm $y^{-1}$) so that subsampling can be made with adequate resolution, low rates of sediment mixing by physical and biological means, and little variation in sediment grain size or composition down the core (Bierman et al. 1998; Assinder 2002). Variations from these ideals can often be interpreted by mathematical modelling.

If the processes controlling the arrival of particulates at the core surface give rise to a constant rate of particulate accumulation, then it is reasonable to suppose that every sediment layer will have the same initial unsupported $^{210}Pb$ activity concentration. In this case, the unsupported $^{210}Pb$ activity concentration will decline exponentially with the cumulative dry mass of sediment (Appleby and Oldfield 1983). A graphical representation of unsupported $^{210}Pb$ on a logarithmic versus cumulative dry mass of sediment will give a linear profile that can be fitted to produce a sedimentation rate and hence dates of deposition of sediment layers as:

$$A_{(z)} = A_{(0)}\, e^{-bz}$$

where $A_{(z)}$ is the excess $^{210}Pb$ activity concentration at cumulative dry mass $z$, $A_{(0)}$ is the activity at the sediment surface (or bottom of the uniformly mixed layer), and $b$ is the slope of the linear plot. The rate of sediment accumulation is equal to $\lambda/b$ in grams per square centimetre per year, where $\lambda$ is the decay constant $((\ln 2)/t_{1/2})$ for $^{210}Pb$. This is referred to as the constant flux:constant sedimentation rate (cf:cs) model. A typical $^{210}Pb$ profile is illustrated in Figure 2.2.

In many cases, however, the sedimentation rate varies due to changes in climate, anthropogenic activity, etc.; this results in nonlinear plots and this model is not appropriate. Other models such as the constant initial concentration (c.i.c.) model, where particles will have the same initial unsupported $^{210}Pb$ activity concentration irrespective of any variations in sediment accumulation rate, have been used; however, often more realistically, a constant rate of supply (c.r.s.) or constant flux (c.f.) method has been used. This model assumes that there is constant fallout of $^{210}Pb$ to the sediments irrespective of any variations that may have occurred in the sediment accumulation rate. Examples of these methods are shown later in this chapter.

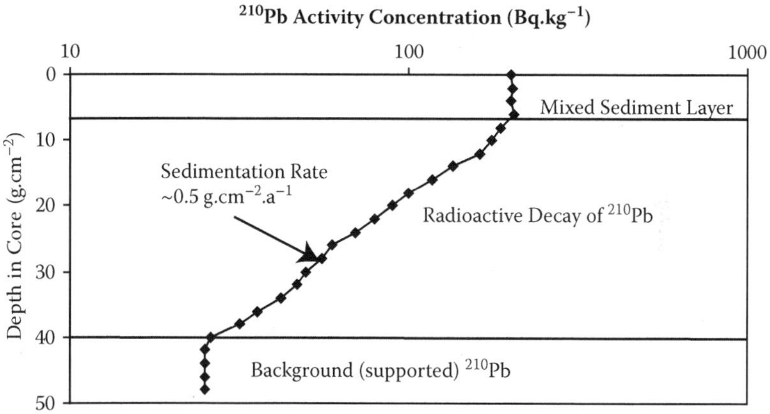


**Figure 2.2** Hypothetical $^{210}Pb$ profile modelled using the cf:cs model.

radionuclide activity concentrations or radionuclide ratios, establishing the unique nature of any radionuclide presence and then tracking this distinction away from a source.

Modelling of the distribution of the radionuclide with allowance for diffusion, reaction with particulates, gaseous escape (if relevant), dilution, and radioactive decay allows identification of water mass movement. Again, the decay of the radionuclide may be used for dating (e.g., $^{222}Rn$) or a known input function can be tracked over time (e.g., $^{99}Tc$ from Sellafield). The radionuclide itself may be the focus of the environmental forensic study (e.g., Jefferies and Steele 1989) or it may act as a tracer of the soluble phase processes involved at that location—for example, dilution effects that allow environmental forensic studies of other contaminants (e.g., Charette and Buesseler 2004).

## Case Studies

Relevant case studies that contain pertinent examples for environmental forensics are numerous and diverse. Recent papers make the link between isotope tracing or dating and environmental perturbation explicit as studies on environmental changes due to human activities increase. Relevant studies include:

- $^{137}Cs$ for measuring soil erosion, sediment accumulation patterns, and related particulate bound contamination (e.g., Pennington et al. 1976; Dearing, Elner, and Happey-Wood 1981; Smith, Ellis, and Nelson 1987; Ritchie and McHenry 1990; Eades et al. 1998; Albrecht et al. 1998; Ketterer et al. 2000; Muller et al. 2000; Owen and Sandhu 2000; Weis, Callaway, and Gersberg 2001; Yan et al. 2002; Yasuhara et al. 2003; Schuller et al. 2004; Wihlborg, Danielsson, and Klingberg 2004; Belyaev et al. 2005; Howarth et al. 2005; Madsen et al. 2005);
- $^{210}Pb$ dating for catchment processes or assessment of anthropogenic environmental changes (e.g., Smith and Walton 1980; Kim 2003; Sonke et al. 2003; Smoak and Swarzenski 2004; Kim 2005; Moore et al. 2005);
- multi-isotope studies, often employing $^{210}Pb$ and fallout $^{137}Cs$ (e.g., Oktay et al. 2000; Steiner, Hanselmann, and Krahenbuhl 2000; Benoit and Rozan 2001; Lee and Cundy 2001; Ketterer et al. 2002; Cundy et al. 2003; Appleby 2004; Pfitzner, Brunskill, and Zagorskis 2004; San Miguel, Bolivar, and Garcia-Tenorio 2004);
- tracing and dating in ice deposits (e.g., Jaworowski et al. 1997; Kudo et al. 1998; Zeeberg, Forman, and Polyak 2003);
- catchment denudation rate studies using cosmogenic radionuclides (e.g., Bierman et al. 1998; Zapata 2003; Jull et al. 2004);
- groundwater studies with radium and radon (e.g., Charette, Buesseler, and Andrews 2001; Kelly and Moran 2002; Charette and Buesseler 2004);
- atmospheric studies using either artificial inputs to trace transport paths (e.g., Kudo et al. 1998; Gallagher et al. 2005) and the assessment of regional fluxes of climatically sensitive gases or general atmospheric transport paths using $^{222}Rn$, $^{7}Be$, and $^{210}Pb$ gas measurements (e.g., Gerasopoulos et al. 2001; Zahorowski, Chambers, and Henderson-Sellers 2004; Zheng et al. 2005); and
- point source discharged radionuclide research (e.g., Jefferies, Steele, and Preston 1982; Prandle 1984; Aarkrog et al. 1987; Assinder et al. 1991; Kudo et al. 1998; Lindahl et al. 2003; Karcher et al. 2004).

This section will focus on three types of study: (1) the use of discharged artificial radionuclides as tracers for marine water mass movement on a large scale and the movement of reprocessed uranium on a localised scale, (2) the use of natural radionuclides in groundwater tracing studies, and (3) $^{210}Pb$ and multi-isotope studies for terrestrial and marine sediment dating for assessment of local or global environmental change.

## Reprocessing Radionuclides as Tracers and for Dating

### *$^{137}Cs$ and $^{99}Tc$ as Water Mass Tracers*

British Nuclear Fuels (BNFL) Sellafield nuclear fuel reprocessing plant has been discharging authorised quantities of various fission and activation products to the Irish Sea since the 1950s. These include $^{137}Cs$, $^{134}Cs$, and $^{99}Tc$, which do not occur naturally in the environment; peak discharges of radioactive Cs occurred in the 1970s and peak discharges of $^{99}Tc$ in the 1990s (BNFL 2002). The appearance of large quantities (compared to previous fallout) of these artificial tracers, which tended to remain largely soluble in the water column, allowed a series of tracing studies concentrating on $^{137}Cs$ and $^{134}Cs$ movement in the Irish Sea and further afield. This was of importance for examining the fate of the Cs itself; in addition, the isotope ratio between the two caesium isotopes was used for dating the transport times of the water.

Early work (Jefferies, Preston, and Steele 1973; Jeffries et al. 1982) examined the initial spread of Cs and established the residence half-times of water in the Irish Sea to be of the order of a year between 1970 and 1976 but considerably less than a year from 1976 to 1978. Prandle (1984) and Prandle and Beechey (1991) extended this modelling to show the northerly movement of Cs out of the Irish Sea, around the Scottish coast, and hence into the North Sea, with an advective travel time from Sellafield of approximately 2 years, and finally exiting approximately 6 years after discharge along the Norwegian coast. Aarkrog et al. (1983) similarly estimated a transit time from Sellafield to the east Greenland current of approximately 7 years. Jefferies and Steele (1989) summarised and modelled a large data set and established a northerly flow rate of between 2.2 and 4.4 $km^3$ $day^{-1}$.

With improvements in radiochemical techniques to measure $^{99}Tc$ and an increase in its discharges from Sellafield in 1994 due to enhanced waste throughput following the opening of the Enhanced Actinide Removal Plant (EARP), a number of studies began to make use of Tc alone (e.g., Nawakowski et al. 2004) or Tc in combination with other isotopes to extend this work. Aarkrog et al. (1987) found Sellafield-derived $^{134}Cs$ and $^{99}Tc$ in Arctic waters, including Baffin Bay, with a transit time of 8 years.

McCubbin et al. (2002) examined $^{99}Tc$ data for the North Sea indicating that the leading edge of the first EARP pulse, entering via the Scottish coastal current, may have migrated to the limit of the current flowing south along the British coastal margin within 9 months. Transit times from Sellafield to the Pentland Firth and Lowestoft of approximately 9 and 24 months, respectively, were significantly shorter than previous estimates based on $^{137}Cs$ data from the 1970s and 1980s. Kershaw et al. (2004) studied the transport of $^{99}Tc$ from the Irish Sea into the North Sea post-EARP and confirmed the more rapid transport times than reported for other radionuclides prior to EARP. Their results showed that the EARP-related $^{99}Tc$ contamination had reached Arctic waters by 2000. It was found that following an initial rapid transport of $^{99}Tc$ from the Irish Sea into the North Sea in 1994 and 1995, the transport rate from the North Sea and northwards with the Norwegian coastal current and West Spitsbergen current slowed markedly, in apparent correspondence with variations in the North Atlantic oscillation (NAO) winter index. Other studies have examined the movement of this material into Arctic areas (Dahlgaard et al. 2004) and shown the typical pathway of dissolved radionuclides from the Irish Sea via the North Sea along the Norwegian coast, with subsequent separation into three branches, of which the two Arctic branches bear the potential for future monitoring of the signal in the next decades (Karcher et al. 2004).

Similar types of studies on the particulate fraction have allowed sediment transport rates in the Irish Sea to be quantified (e.g., Aston, Assinder, and Kelly 1985; Aldridge et al. 2003).

### *$^{237}Np$ as a Tracer of Reprocessed Uranium*

An example of a highly specific use of isotopes as tracers is the use of $^{237}Np$ as a tracer for reprocessed uranium production. Assinder et al. (1991) measured activity concentrations of $^{237}Np$ and other radionuclides in surface sediments from around the Irish Sea, including in the Ribble Estuary, United Kingdom. This estuary is known to be contaminated with radionuclides derived from both nuclear fuel production at BNFL Springfields in the Ribble catchment and by transport of radionuclides through the Irish Sea from BNFL Sellafield.

$^{237}Np$ activity concentrations were detected around the Irish Sea at levels, in general, consistent with dilution of contaminated sediment with other sediments during transport. Anomalously high $^{237}Np$ activity concentrations were detected in the Ribble sediments that could not be explained by the expected level of $^{237}Np$ contamination from Sellafield. Further studies revealed that the excess $^{237}Np$ was derived from BNFL Springfields discharges, despite the fact that this is not a reprocessing facility. However, Springfields is now known to release $^{237}Np$ intermittently due to its use of some reprocessed uranium from Sellafield, which still contains low levels of $^{237}Np$. Approximately 80–90% of $^{237}Np$ in the Ribble can be attributed to Springfields, with the remainder from Sellafield (Assinder 1999).

$^{237}Np$ can, therefore, act as a highly specialised and location-specific tracer of the use of reprocessed uranium in nuclear fuel fabrication facilities. Other similarly specialised radionuclide tracers exist—for example, $^{234}Th$ and $^{234m}Pa$ as tracers of the use of weapons containing depleted uranium (Assinder, unpublished data).

## Radium and Radon in Groundwater Studies

A developing area in tracing and dating is the use of radium and radon isotopes in groundwater studies. This area is likely to expand further (Kraemer and Genereux 1998) and has potential uses in assessing the flow of groundwater contaminated by discharges and the contribution of contaminated groundwater to surface water bodies.

### *Radium*

Radium can be used as a chemical tracer (relying on its activity concentration) in areas where sufficient distinction is present between Ra in separate water masses and where the radium behaves conservatively (i.e., with no adsorption to particles or other processes abstracting or adding radium). Additionally, the use of multi-isotope Ra techniques with the calculation of isotope ratios is possible where no activity concentration difference exists or where Ra is behaving nonconservatively, since the same process will affect each Ra isotope in the same way.

Radium activity concentrations and isotope ratios have been used to trace groundwater input to estuarine systems (e.g., Miller, Kraemer, and McPherson 1990, Charette et al. 2001; Kelly and Moran 2002; Charette and Buesseler 2004) and the ocean (Charette et al. 2003; Moore 2003; Purkl and Eisenhauer 2004), quantify groundwater inputs to a stream (Kraemer and Genereux 1998) and a lake (Kraemer 2005), identify the source of water issuing from springs (Kraemer and Genereux 1998), and as tracers of riverine plumes as they mix into the ocean (Moore and Krest 2004).

As an example, Kraemer (2005) used $^{223}$Ra, $^{224}$Ra, $^{226}$Ra, and $^{228}$Ra to assess groundwater fluxes to lake and tributary water of Cayuga Lake, New York, during the course of a vernal inflow event in the spring of 2001. A large influx of groundwater entered the lake at its extreme southern end early in the vernal inflow event and, due to its low $^{228}$Ra/$^{226}$Ra activity ratio compared with bulk lake water, allowed its identification through time, its spread, and its mixing to be examined. Groundwater inflow to the lake around the delta of a major tributary was also detected on the basis of $^{223}$Ra and $^{224}$Ra activity of lake and tributary water. It was concluded that radium isotopes can be valuable new tools in limnological investigations, allowing detection and monitoring of events and processes such as water inflow and mixing, determining sources of inflowing water, and monitoring introduced water masses as they move within the lake (Kraemer 2005). These aspects of water flow are of particular relevance for forensics studies.

### *Radon*

Radon has been detected at elevated levels in spring water, and Rogers (1958) first used this to locate areas of groundwater discharge to a stream by examining the change in activity concentration of the radon during mixing. Since then, radon has been employed in similar studies (e.g., Genereux, Hemind, and Mulholland 1993; Cook et al. 2003; Schwartz 2003; Wu, Wen, and Zhang 2004) to examine submarine groundwater discharge (SGD) into the near-shore ocean (e.g., Cable, Bugna, et al. 1996; Cable, Burnett, et al. 1996; Lambert and Burnett 2003; Purkl and Eisenhauer 2004; Stieglitz 2005). In addition, the rate of radon decay has been used in a dating sense for estimation of groundwater flow rates (e.g., Cook, Love, and Dighton 1999; Hamada 2000; Kafri 2001).

Groundwaters mixing with surface flows are modelled with assumptions made about thoroughness of mixing, constancy of radon source activity concentrations, and known rates of degassing, which can be corrected for (Kraemer and Genereux 1998). As an example, Lambert and Burnett (2003) estimated the changing flux of groundwater discharge into a coastal area in the northeast Gulf of Mexico (Florida) based on continuous measurements of radon concentrations over a period of several days. Radon inventories were converted to fluxes after accounting for losses due to atmospheric evasion and mixing. Radon fluxes were converted to groundwater inflow rates by estimating the radon concentration of the fluids discharging into the study domain. The results suggested that the flow was highly variable, with flows ranging from approximately 5 to 50 cm $day^{-1}$, and that it was strongly influenced by the tides, with spikes in the flow every 12 hours. As with radium, potential groundwater contaminant contributions to surface waters can be assessed with these methods.

## $^{210}$Pb and Multi-Isotope Sediment Dating

$^{210}$Pb dating, using a variety of modelling tools (Box 2.2), has been employed since the 1960s (Goldberg 1963). In recent years, the importance of checking $^{210}$Pb values with another isotope profile, often $^{137}$Cs or occasionally $^{239,240}$Pu, has been widely discussed (Smith 2001) and introduced. This follows many investigations on relative Cs and Pb mobility in core profiles and potential problems with the use of single isotope profiles (e.g., Benoit and Rozan 2001; Abril 2004). Other authors have validated their dating against other indicators such as pollen or palaeomagnetic information (e.g., Robbins, Edgington, and Kemp 1978; Brush et al. 1982; Appleby, Dearing, and Oldfield 1985).

$^{210}$Pb profiles have been employed in a variety of environments to calculate sedimentation rates and hence allow dates to be attached to different sediment layers. These include rivers, bogs, reservoirs, and lakes (e.g., Aston et al. 1973; Appleby et al. 1979; Appleby and Oldfield 1983; El-Daoushy and Tolonen 1984; Murchie 1985; Jha et al. 1999; Brenner, Schelske, and Keenan 2001; Kim and Rejmankova 2001; Al-Masri et al. 2002; Jha et al. 2002; Harrison, Heijnis, and Caprarelli 2003; Kim 2003; Sonke et al. 2003; Smoak and Swarzenski 2004; Kim 2005; Moore et al. 2005) and estuaries and coastal and marine areas (e.g., Smith and Loring 1981; Brush et al. 1982; San Miguel et al. 2001; Ramesh et al. 2002; Bay et al. 2003; Ram et al. 2003; Varekamp et al. 2003; Ip et al. 2004; Edgar and Sampson 2004; Hung and Hsu 2004).

Multi-isotope studies including $^{210}$Pb have been applied to lakes and rivers (e.g., Robbins et al. 1978; Dominik, Mangini, and Muller 1981; Appleby et al. 1985; Kumar et al. 1999; Steiner et al. 2000; Cisternas et al. 2001; Guevara et al. 2002; Appleby 2004; Davies et al. 2004; Lima et al. 2005) and estuaries and coastal and marine areas (e.g., Smith and Ellis 1982; Ligero et al. 2002; Cundy et al. 2003; San Miguel et al. 2004).

The purposes for which $^{210}Pb$, $^{137}Cs$, and Pu isotope dates have been used vary and, from a forensics viewpoint, include studies of soil erosion due to natural or anthropogenic changes (e.g., Appleby and Oldfield 1983; Murchie 1985; Cundy et al. 2003; Zapata 2003) and impacts of industrialisation and human environmental changes to metal (e.g., Aston et al. 1973; Dominik et al. 1981; Smith and Loring 1981; El-Daoushy and Tolonen 1984; Cundy et al. 2003; Sonke et al. 2003; Ram et al. 2003; Hung and Hsu 2004; Ip et al. 2004), nutrient (e.g., Brenner et al. 2001; Kim 2003; Moore et al. 2005), natural and artificial organic (e.g., Bay et al. 2003; Lima et al. 2005), and radionuclide fluxes (e.g., Smith et al. 1987, 2000).

Davies et al. (2004) have provided an example of how radionuclide dating using multi-isotope techniques can be used in a forensic sense to examine the impacts of industrial activity in the environment, as well as natural changes over time. This study measured a range of parameters to establish core chronology ($^{210}Pb$, $^{137}Cs$, $^{241}Am$, $^{14}C$, tephra layers) using independent age markers in Lago de Zirahuen situated within a montane basin in highland Michoacán, Mexico. The aim was to examine Late Holocene environmental change including recent metal (Fe, Pb, Mn, Cu, and Zn) anthropogenic inputs and sources. Dating of various climatic events in the core was possible and changes due to land use. Copper smelting and agricultural developments could be identified and dated as well as very recent tourist developments and commercial agriculture. Overall, the palaeoenvironmental evidence suggested that the lake was responding rapidly to land use intensification and diversification in the basin.

In coastal areas, the study by Lee and Cundy (2001) illustrates another multi-isotope core study that specifically examined anthropogenic impacts, sources, and time-scales. Cores were collected from the Humber Estuary in the United Kingdom and analysed for $^{210}Pb$, $^{137}Cs$, and a range of trace elements including Pb, Zn, Cu, Al, Mn, and Fe. Metal fluxes were calculated using the measurements of sedimentation rate and related to mid-twentieth century industrialisation for Cu, Pb, and Zn, and Ti-oxide processing facilities for Ti, Al, and Fe. Salt marsh sediments thus provided a (time-integrated) record of historical pollutant inputs.

## Current and Future Roles for Radionuclides in Environmental Forensics

Natural and artificial radionuclides have been employed since the 1960s in studies that potentially have an environmental forensics dimension. These have included tracing and dating applications in the atmosphere and terrestrial and marine environments in particular:

- tracing of water mass and sediment movement in terrestrial and marine environments using variations in radionuclide activity concentrations or activity concentration ratios either as true tracers (i.e., the radionuclide is not the element of concern) or as self-tracers where the radionuclide is the subject of forensic study;
- dating water mass and sediment movement to examine transport rates and fluxes using either radioactive decay or matching known input signals with environmental records, again as either a true or self-tracer;
- tracing and dating overall catchment processes and rates, including their alteration due to anthropogenic activity; and
- dating sediment, ice, and other deposited material to provide a geochronology for examination of anthropogenic inputs.

Improvements in analytical capabilities with lower detection limits have allowed smaller sample masses to be analysed, making these types of studies more feasible and widespread. Studies such as those outlined in this chapter will continue to provide useful transport path and transport rate information in the future, with more information obtained as detection limits lower. Additional improvements and changes in the future would be beneficial in the areas of:

- in situ measurements: Currently, a limited amount of in situ gamma spectrometric work is carried out on land (e.g., Macdonald et al. 1996; Tyler 2004), from the air (e.g., Sanderson et al. 2004), and underwater (e.g., Povinec, Osvath, and Baxter 1996; Osvath et al. 2001, 2005). Advances have been made in understanding how the measurement relates to gamma radionuclide activity concentration in three dimensions (e.g., Macdonald et al. 1999; Tyler 2004), theoretically allowing a rapid assessment to be made in the field of gamma transport paths and rates. These procedures need to develop further before becoming a reliable tracing tool.
- fallout Pu studies: $^{137}Cs$ has a half-life of 30 years, allowing the fallout peak of 1963 and Chernobyl input of 1986 to be readily measured at present, although questions still remain about the environmental mobility of Cs. However, the 1963 fallout peak will become increasingly unreliable; by 2023 only a quarter of the activity concentration will remain. In contrast, Pu from fallout has a much longer half-life and, with improvements and simplifications in analytical techniques, may become more useful in the future. For example, ICP-MS studies of Pu (e.g., Ketterer et al. 2002, 2004) have been used to rapidly establish

the chronology of recent aquatic sediments via measurements of the activity concentrations of $^{239}Pu$, $^{240}Pu$, and the atom ratio $^{240}Pu/^{239}Pu$. The Pu activity profiles, obtained in approximately 6 h of instrumental measurement time, were in agreement with a gamma spectrometric $^{137}Cs$ profile requiring longer measurement times.

- multi-isotope or alternative geochronometer verification: Future studies are likely to use, more consistently, multi-isotope or other aging techniques in sediment core studies. Reliance on one dating technique has been found to cause errors in interpretation of the anthropogenic time line being investigated.
- expansion of the possibilities provided by natural radionuclides in tracing and dating studies: The instrumental ability to discriminate between smaller variations in activity concentration and activity concentration ratios is increasing. The ubiquitous nature of natural radionuclides and the potential for these small, but now traceable, variabilities to be measured should lead to an expansion in their use and an increase in their potential effectiveness in environmental forensics investigations.

## References

Aarkrog, A., S. Boelskifte, H. Dahlgaard, S. Duniec, L. Hallstadius, E. Holm, and J. N. Smith. (1987) Technetium-99 and $^{134}Cs$ as long distance tracers in Arctic waters. *Estuarine, Coastal and Shelf Science,* 24, 637–647.

Aarkrog, A., H. Dahlgaard, L. Hallstadius, H. Hansen, and E. Holm. (1983) Radiocaesium from Sellafield effluents in Greenland waters. *Nature,* 304, 49–51.

Abril, J. M. (2004) Constraints on the use of Cs-137 as a time-marker to support CRS and SIT chronologies. *Environmental Pollution,* 129: 31–37.

Albrecht, A., R. Reiser, A. Luck, J. M. A. Stoll, and W. Giger. (1998) Radiocesium dating of sediments from lakes and reservoirs of different hydrological regimes. *Environmental Science and Technology,* 32: 1882–1887.

Aldridge, J. N., P. Kershaw, J. Brown, D. McCubbin, K. S. Leonard, and E. F. Young. (2003) Transport of plutonium (Pu-239/240) and caesium (Cs-137) in the Irish Sea: Comparison between observations and results from sediment and contaminant transport modelling. *Continental Shelf Research,* 23: 869–899.

Al-Masri, M. S., A. Aba, H. Khalil, and Z. Al-Hares. (2002) Sedimentation rates and pollution history of a dried lake: Al-Oteibeh Lake. *Science of the Total Environment,* 293: 177–189.

Appleby, P. G. (2004) Environmental change and atmospheric contamination on Svalbard: Sediment chronology. *Journal of Paleolimnology,* 31: 433–443.

Appleby, P. G., J. A. Dearing, and F. Oldfield. (1985) Magnetic studies of erosion in a Scottish lake catchment. 1. Core chronology and correlation. *Limnology and Oceanography,* 30: 1144–1153.

Appleby, P. G., and F. Oldfield. (1983) The assessment of $^{210}Pb$ data from sites with varying sediment accumulation rates. *Hydrobiologia,* 103: 29–35.

Appleby, P. G., F. Oldfield, R. Thompson, P. Huttunen, and K. Tolonen. (1979) $^{210}Pb$ dating of annually laminated lake sediments from Finland. *Nature,* 280: 53–55.

Assinder, D. J. (1999) A review of the occurrence and behaviour of Np in the Irish Sea. *Journal of Environmental Radioactivity,* 44: 335–347.

______. (2002) Potential problems with the long-term monitoring of radionuclides in the environment. *The Nuclear Engineer,* 43: 76–79.

Assinder, D. J., S. M. Mudge, and G. S. Bourne. (1997) Radiological assessment of the Ribble Estuary. I. Distribution of radionuclides in surface sediments. *Journal of Environmental Radioactivity,* 36: 1–19.

Assinder, D. J., M. Yamamoto, C. K. Kim, R. Seki, Y. Takaku, Y. Yamauchi, K. Komura, K. Ueno, and G. S. Bourne. (1991) Neptunium in intertidal coastal and estuarine sediments in the Irish Sea. *Journal of Environmental Radioactivity,* 14: 135.

Aston, S. R., D. J. Assinder, and M. Kelly. (1985) Plutonium in intertidal coastal and estuarine sediments in the northern Irish Sea. *Estuarine Coastal and Shelf Science,* 20: 761–771.

Aston, S. R., D. Bruty, R. Chester, and R. C. Padgham. (1973) Mercury in lake sediments, a possible indicator of technological growth. *Nature,* 241: 450–451.

Bay, S. M., E. Y. Zeng, T. D. Lorenson, K. Tran, and C. Alexander. (2003) Temporal and spatial distributions of contaminants in sediments of Santa Monica Bay, California. *Marine Environmental Research,* 56: 255–276.

Belyaev, V. R., P. J. Wallbrink, V. N. Golosov, A. S. Murray, and A. Y. Sidorchuk. (2005) A comparison of methods for evaluating soil redistribution in the severely eroded Stavropol region, southern European Russia. *Geomorphology,* 65: 173–193.

Benoit, G., and T. F. Rozan. (2001) Pb-210 and Cs-137 dating methods in lakes: A retrospective study. *Journal of Paleolimnology,* 25: 455–465.

Bierman, P. R., A. Albrecht, M. H. Bothner, E. T. Brown, T. D. Bullen, L. B. Gray, and L. Turpin. (1998) Erosion, weathering and sedimentation. In *Isotope tracers in catchment hydrology,* ed. C. Kendall and J. J. McDonnell, 647–678. Amsterdam: Elsevier.

BNFL (British Nuclear Fuels). (2002) *Discharges and monitoring of the environment in the U.K.* Annual Report 2002. Warrington, U.K.: BNFL, 148 pp.

Brenner, M., C. L. Schelske, and L. W. Keenan. (2001) Historical rates of sediment and nutrient accumulation in marshes of the Upper St. Johns River Basin, Florida, USA. *Journal of Paleolimnology,* 26: 241–257.

Brush, G. S., E. A. Martin, R. S. DeFries, and C. A. Rice. (1982) Comparisons of $^{210}Pb$ and pollen methods for determining rates of estuarine sediment accumulation. *Quaternary Research,* 18: 196–217.

Cable, J. E., G. C. Bugna, W. C. Burnett, and J. P. Chanton. (1996) Application of Rn-222 and $CH_4$ for assessment of groundwater discharge to the coastal ocean. *Limnology and Oceanography,* 41: 1347–1353.

Cable, J. E., W. C. Burnett, J. P. Chanton, and G. L. Weatherly. (1996) Estimating groundwater discharge into the northeastern Gulf of Mexico using radon-222. *Earth and Planetary Science Letters,* 144: 591–604.

Charette, M. A., and K. O. Buesseler. (2004) Submarine groundwater discharge of nutrients and copper to an urban subestuary of Chesapeake Bay (Elizabeth River). *Limnology and Oceanography,* 49: 376–385.

Charette, M. A., K. O. Buesseler, and J. E. Andrews. (2001) Utility of radium isotopes for evaluating the input and transport of groundwater-derived nitrogen to a Cape Cod estuary. *Limnology and Oceanography,* 46: 465–470.

Charette, M. A., R. Splivallo, C. Herbold, M. S. Bollinger, and W. S. Moore. (2003) Salt marsh submarine groundwater discharge as traced by radium isotopes. *Marine Chemistry,* 84: 113–121.

Cisternas, M., A. Araneda, P. Martinez, and S. Perez. (2001) Effects of historical land use on sediment yield from a lacustrine watershed in central Chile. *Earth Surface Processes and Landforms,* 26: 63–76.

Cook, P. G., G. Favreau, J. C. Dighton, and S. Tickell. (2003) Determining natural groundwater influx to a tropical river using radon, chlorofluorocarbons and ionic environmental tracers. *Journal of Hydrology,* 277: 74–88.

Cook, P. G., A. J. Love, and J. C. Dighton. (1999) Inferring ground water flow in fractured rock from dissolved radon. *Ground Water,* 37: 606–610.

Cundy, A. B., I. W. Croudace, A. Cearreta, and M. J. Irabien. (2003) Reconstructing historical trends in metal input in heavily disturbed, contaminated estuaries: Studies from Bilbao, Southampton Water and Sicily. *Applied Geochemistry,* 18: 311–325.

Dahlgaard, H., M. Eriksson, S. P. Nielsen, and H. P. Joensen. (2004) Levels and trends of radioactive contaminants in the Greenland environment. *Science of the Total Environment,* 331: 53–67.

Davies. S. J., S. E. Metcalfe, A. B. MacKenzie, A. J. Newton, G. H. Endfield, and J. G. Farmer. (2004) Environmental changes in the Zirahuen Basin, Michoacán, Mexico, during the last 1000 years. *Journal of Paleolimnology,* 31: 77–98.

Dearing, J. A., J. K. Elner, and C. M Happey-Wood. (1981) Recent sediment flux and erosional processes in a Welsh upland lake-catchment based on magnetic susceptibility measurements. *Quaternary Research,* 16: 356–372.

Dearing, J. A., and R. T. Jones. (2003) Coupling temporal and spatial dimensions of global sediment flux through lake and marine sediment records. *Global and Planetary Change,* 39: 147–168.

Dominik, J., A. Mangini, and G. Muller. (1981) Determination of recent deposition rates in Lake Constance with radioisotopic methods. *Sedimentology,* 28: 653–677.

Eades, L. J., J. G. Farmer, A. B. MacKenzie, A. Kirika, and A. E. Bailey-Watts. (1998) High-resolution profile of radiocaesium deposition in Loch Lomond sediments. *Journal of Environmental Radioactivity,* 39, 107–115.

Edgar, G. J., and C. R. Samson. (2004) Catastrophic decline in mollusc diversity in eastern Tasmania and its concurrence with shellfish fisheries. *Conservation Biology,* 18: 1579–1588.

Eisenbud, M., and T. Gesell. (1997) *Environmental radioactivity. From natural, industrial and military sources.* San Diego: Academic Press, 656 pp.

El-Daoushy, F., and K. Tolonen. (1984) Lead-210 and heavy metal contents in dated ombrotrophic peat hummocks from Finland. *Nuclear Instruments and Methods in Physics Research,* 223: 392–399.

Gallagher, D., E. J. Mcgee, P. I. Mitchell, V. Alfimov, A. Aldahan, and G. Possnert. (2005) Retrospective search for evidence of the 1957 windscale fire in NE Ireland using I-129 and other long-lived nuclides. *Environmental Science and Technology,* 39: 2927–2935.

Gascoyne, M. (1982) Geochemistry of the actinides and their daughters. In *Uranium series disequilibrium: Applications to environmental problems,* ed. M. Ivanovich and R. S. Harmon. Oxford, England: Clarendon Press.

Gascoyne, M., and H. P. Schwarz. (1982) Carbonate and sulphate precipitates. In *Uranium series disequilibrium: Applications to environmental problems,* ed. M. Ivanovich and R. S. Harmon. Oxford, England: Clarendon Press.

Genereux, D. P., H. F. Hemind, and P. J. Mulholland. (1993) Use of radon-222 and calcium as tracers in a three-end-member mixing model for streamflow generation on the west fork of Walker Branch watershed. *Journal of Hydrology,* 142: 167–211.

Gerasopoulos, E., P. Zanis, A. Stohl, C. S. Zerefos, C. Papastefanou, W. Ringer, L. Tobler, et al. (2001) A climatology of Be-7 at four high-altitude stations at the Alps and the Northern Apennines. *Atmospheric Environment,* 35: 6347–6360.

Gesell, T. F. (1983) Background atmospheric Rn-222 concentrations outdoors and indoors—A review. *Health Physics,* 45: 289–302.

Goldberg, E. D. (1963) Geochronology with $^{210}$Pb. In *Radioactive dating.* Vienna: IAEA.

Guevara, S. R., J. Massaferro, G. Villarosa, M. Arribere, and A. Rizzo. (2002) Heavy metal contamination in sediments of Lake Nahuel Huapi, Nahuel Huapi National Park, Northern Patagonia, Argentina. *Water Air and Soil Pollution,* 137: 21–44.

Hamada, H. (2000) Estimation of groundwater flow rate using the decay of $^{222}$Rn in a well. *Journal of Environmental Radioactivity,* 47: 1–13.

Harrison, J., H. Heijnis, and G. Caprarelli. (2003) Historical pollution variability from abandoned mine sites, Greater Blue Mountains World Heritage Area, New South Wales, Australia. *Environmental Geology,* 43: 680–687.

Howarth, R. J., G. Evans, I. W. Croudace, and A. B. Cundy. (2005) Sources and timing of anthropogenic pollution in the Ensenada de San Simon (inner Ria de Vigo), Galicia, NW Spain: An application of mixture modelling and nonlinear optimization to recent sedimentation. *Science of the Total Environment,* 340: 149–176.

Hung, J. J., and C. L. Hsu. (2004) Present state and historical changes of trace metal pollution in Kaoping coastal sediments, southwestern Taiwan. *Marine Pollution Bulletin,* 49: 986–998.

Ip, C. C. M., X. D. Li, G. Zhang, J. G. Farmer, O. W. H. Wai, and Y. S. Li. (2004) Over one hundred years of trace metal fluxes in the sediments of the Pearl River Estuary, South China. *Environmental Pollution,* 132: 157–172.

Jaworowski, Z., P. Hoff, J. O. Hagen, and W. Maczek. (1997) A highly radioactive Chernobyl deposit in a Scandinavian glacier. *Journal of Environmental Radioactivity,* 35: 91–108.

Jefferies, D. F., A. Preston, and A. K. Steele. (1973) Distribution of $^{137}$Cs in British coastal waters. *Marine Pollution Bulletin,* 4: 118–121.

Jefferies, D. F., A. K. Steele, and A. Preston. (1982) Further studies on the distribution of Cs-137 in British coastal waters. 1. Irish Sea. *Deep Sea Research,* 29: 713–738.

Jefferies, D. F., and A. K. Steele. (1989) Observed and predicted concentrations of caesium-137 in seawater of the Irish Sea 1970–1985. *Journal of Environmental Radioactivity,* 10: 173–189.

Jha, S. K., S. B. Chavan, G. G. Pandit, B. S. Negi, and S. Sadasivan. (2002) Behaviour and fluxes of trace and toxic elements in creek sediment near Mumbai, India. *Environmental Monitoring and Assessment,* 76: 249–262.

Jha, S. K., T. M. Krishnamoorthy, G. G. Pandit, and K. S. V. Nambi. (1999) History of accumulation of mercury and nickel in Thane Creek, Mumbai, using Pb-210 dating technique. *Science of the Total Environment,* 236: 91–99.

Jones, D. G. (2001) Development and application of marine gamma-ray measurements: a review. *Journal of Environmental Radioactivity,* 53: 313–333.

Jull, A. J. T., G. S. Burr, L. R. McHargue, T. E. Lange, N. A. Lifton, J. W. Beck, D. J. Donahue, and D. Lal. (2004) New frontiers in dating of geological, paleoclimatic and anthropological applications using accelerator mass spectrometric measurements of C-14 and Be-10 in diverse samples. *Global and Planetary Change,* 41: 309–323.

Kafri, U. (2001) Radon in groundwater as a tracer to assess flow velocities: two test cases from Israel. *Environmental Geology,* 40: 392–398.

Karcher, M. J., S. Gerland, I. H. Harms, M. Iosjpe, H. E. Heldal, P. J. Kershaw, and M. Sickel. (2004) The dispersion of Tc-99 in the Nordic Seas and the Arctic Ocean: A comparison of model results observations. *Journal of Environmental Radioactivity,* 74: 185–198.

Kelly, R. P., and S. B. Moran. (2002) Seasonal changes in groundwater input to a well-mixed estuary estimated using radium isotopes and implications for coastal nutrient budgets. *Limnology and Oceanography,* 47: 1796–1807.

Kershaw, P. J., H. E. Heldal, K. A. Mork, and A. L. Rudjord. (2004) Variability in the supply, distribution and transport of the transient tracer Tc-99 in the NE Atlantic. *Journal of Marine Systems,* 44: 55–81.

Ketterer, M. E., K. M. Hafer, C. L. Link, D. Kolwaite, J. Wilson, and J. W. Mietelski. (2004) Resolving global versus local/regional Pu sources in the environment using sector ICP-MS. *Journal of Analytical Atomic Spectrometry,* 19: 241–245.

Ketterer, M. E., B. R. Watson, G. Matisoff, and C. G. Wilson. (2002) Rapid dating of recent aquatic sediments using Pu activities and Pu-240/Pu-239 as determined by quadrupole inductively coupled plasma mass spectrometry. *Environmental Science and Technology,* 36: 1307–1311.

Ketterer, M. E., W. C. Wetzel, R. R. Layman, G. Matisoff, and E. C. Bonniwell. (2000) Isotopic studies of sources of uranium in sediments of the Ashtabula River, Ohio, USA. *Environmental Science and Technology,* 34: 966–972.

Kim, J. G. (2003) Response of sediment chemistry and accumulation rates to recent environmental changes in the Clear Lake watershed, California, USA. *Wetlands,* 23: 95–103.

______. (2005) Assessment of recent industrialization in wetlands near Ulsan, Korea. *Journal of Paleolimnology,* 33: 433–444.

Kim, J. G., and E. Rejmankova. (2001) The paleoecological record of human disturbance in wetlands of the Lake Tahoe Basin. *Journal of Paleolimnology,* 25: 437–454.

King, P. T., J. Michel, and W. S. Moore. (1982) Ground water geochemistry of $^{228}Ra$, $^{226}Ra$ and $^{222}Rn$. *Geochimica et Cosmochimica Acta,* 46: 1173–1182.

Kraemer, T. F. (2005) Radium isotopes in Cayuga Lake, New York: Indicators of inflow and mixing processes. *Limnology and Oceanography,* 50: 158–168.

Kraemer, T. F., and D. P. Genereux. (1998) Applications of uranium- and thorium-series radionuclides in catchment hydrology studies. In *Isotope tracers in catchment hydrology,* ed. C. Kendall and J. J. McDonnell, 679–722. Amsterdam: Elsevier.

Kudo, A., J. Zheng, R. M. Koerner, D. A. Fisher, D. C. Santry, Y. Mahara, and M. Sugahara. (1998) Global transport rates of $^{137}Cs$ and $^{239,240}Pu$ originating from the Nagasaki A-bomb in 1945 as determined from analysis of Canadian Arctic ice cores. *Journal of Environmental Radioactivity,* 40: 289–298.

Kumar, U. S., S. V. Navada, S. M. Rao, R. P. Nachiappan, B. Kumar, T. M. Krishnamoorthy, S. K. Jha, and V. K. Shukla. (1999) Determination of recent sedimentation rates and pattern in Lake Naini, India by Pb-210 and Cs-137 dating techniques. *Applied Radiation and Isotopes,* 51: 97–105.

Kuwabara, J., M. Yamamoto, S. Oikawa, K. Komura, and D. J. Assinder. (1999) Measurements of $^{99}Tc$, $^{137}Cs$, $^{237}Np$, Pu isotopes and $^{241}Am$ in sediment cores from intertidal coastal and estuarine regions in the Irish Sea. *Journal of Radioanalytical and Nuclear Chemistry,* 240: 593–601.

Lambert, M. J., and W. C. Burnett. (2003) Submarine groundwater discharge estimates at a Florida coastal site based on continuous radon measurements. *Biogeochemistry,* 66: 55–73.

Lee, S. V., and A. B. Cundy. (2001) Heavy metal contamination and mixing processes in sediments from the Humber Estuary, Eastern England. *Estuarine Coastal and Shelf Science,* 53: 619–636.

Ligero, R. A., M. Barrera, M. Casas-Ruiz, D. Sales, and F. Lopez-Aguayo. (2002) Dating of marine sediments and time evolution of heavy metal concentrations in the Bay of Cadiz, Spain. *Environmental Pollution,* 118: 97–108.

Lima, A. L., J. B. Hubeny, C. M. Reddy, J. W. King, K. A. Hughen, and T. I. Eglinton. (2005) High-resolution historical records from Pettaquamscutt River basin sediments: 1. Pb-210 and varve chronologies validate record of Cs-137 released by the Chernobyl accident. *Geochimica et Cosmochimica Acta,* 69: 1803–1812.

Lindahl, P., C. Ellmark, T. Gafvert, S. Mattsson, P. Roos, E. Holm, and B. Erlandsson. (2003) Long-term study of Tc-99 in the marine environment on the Swedish west coast. *Journal of Environmental Radioactivity,* 67: 145–156.

Macdonald, J., C. J. Gibson, P. J. Fish, and D. J. Assinder. (1999) An experimental comparison of in-situ gamma spectrometric methods for quantifying Cs-137 radioactive contamination in the ground. *IEEE Transactions in Nuclear Science,* 46: 429–432.

Macdonald, J., P. H. Smith, and D. J. Assinder. (1996) The development and use of an in situ gamma-ray spectrometry system in North Wales. *Journal of Radiological Protection,* 16: 115–127.

Madsen, A. T., A. S. Murray, T. J. Andersen, M. Pejrup, and H. Breuning-Madsen. (2005) Optically stimulated luminescence dating of young estuarine sediments: A comparison with Pb-210 and Cs-137 dating. *Marine Geology,* 214: 251–268.

McCubbin, D., K. S. Leonard, J. Brown, P. J. Kershaw, R. A. Bonfield, and T. Peak. (2002) Further studies of the distribution of technetium-99 and caesium-137 in U.K. and European coastal waters. *Continental Shelf Research,* 22: 1417–1445.

McDowell-Boyer, L. M., A. P. Watson, and C. C. Travis. (1980) A review of parameters describing food-chain transport of lead-210 and radium-226. *Nuclear Safety,* 21: 486–495.

Miller, R. L., T. F. Kraemer, and B. F. McPherson. (1990) Radium and radon in Charlotte Harbor Estuary, Florida. *Estuarine, Coastal and Shelf Science,* 31: 439–457.

Milton, G. M., and R. M. Brown. (1987) Uranium series dating of calcite coatings in groundwater flow systems of the Canadian Shield. *Chemical Geology,* 65: 57–65.

Moore, T., C. Blodau, J. Turunen, N. Roulet, and P. J. H. Richard. (2005) Patterns of nitrogen and sulphur accumulation and retention in ombrotrophic bogs, eastern Canada. *Global Change Biology,* 11: 356–367.

Moore, W. S. (2003) Sources and fluxes of submarine groundwater discharge delineated by radium isotopes. *Biogeochemistry,* 66: 75–93.

Moore, W. S., and J. Krest. (2004) Distribution of Ra-223 and Ra-224 in the plumes of the Mississippi and Atchafalaya Rivers and the Gulf of Mexico. *Marine Chemistry,* 86: 105–119.

Morse, J. W., and G. R. Choppin. (1991) The chemistry of transuranic elements in natural waters. *Reviews in Aquatic Sciences,* 4: 1–22.

Mudge, S. M., D. J. Assinder, and A. T. Russell. (2002) *Microscale variability of contaminants in surface sediments: The implications for sampling.* R&D Technical Report P3-057/TR. Report to the Environment Agency, U.K., 88 pp.

______. (2003) *Mesoscale variability of radionuclides in sediments: Normalization and implications for sampling.* R&D Technical Report P3-093/TR. Report to the Environment Agency, U.K., 88 pp.

Muller, J., H. Ruppert, Y. Muramatsu, and J. Schneider. (2000) Reservoir sediments—A witness of mining and industrial development (Malter Reservoir, eastern Erzgebirge, Germany). *Environmental Geology,* 39: 1341–1351.

Murchie, S. L. (1985) $^{210}$Pb dating and the recent geologic history of Crystal Bay, Lake Minnetonka, Minnesota. *Limnology and Oceanography,* 30: 1154–1170.

Nawakowski, C., M. D. Nicholson, P. J. Kershaw, and K. S. Leonard. (2004) Modelling Tc-99 concentrations in *Fucus vesiculosus* from the northeast Irish Sea. *Journal of Environmental Radioactivity,* 77: 159–173.

Oktay, S. D., P. H. Santschi, J. E. Moran, and P. Sharma. (2000) The $^{129}$I bomb pulse recorded in Mississippi River Delta sediments: Results from isotopes of I, Pu, Cs, Pb and C. *Geochimica et Cosmochimica Acta,* 64: 989–996.

Oliver, L. R., W. T. Perkins, and S. M. Mudge. (2006) Detection of technetium-99 in *Ascophyllum nodosum* from around the Welsh coast, *Chemosphere,* 65: 2297–2303.

Osvath, I., P. P. Povinec, M. S. Baxter, and L. Huynh-Ngoc. (2001) Mapping of the distribution of Cs-137 in Irish Sea sediments. *Journal of Radioanalytical and Nuclear Chemistry,* 248: 735–739.

Osvath, I., P. P. Povinec, H. D. Livingston, T. P. Ryan, S. Mulsow, and J. F. Commanducci. (2005) Monitoring of radioactivity in NW Irish Sea water using a stationary underwater gamma-ray spectrometer with satellite data transmission. *Journal of Radioanalytical and Nuclear Chemistry,* 263: 437–440.

Owen, R. B., and N. Sandhu. (2000) Heavy metal accumulation and anthropogenic impacts on Tolo Harbour, Hong Kong. *Marine Pollution Bulletin,* 40: 174–180.

Pennington, W., R. S. Cambray, J. D. Eakins, and D. D. Harkness. (1976) Radionuclide dating of the recent sediments of Blelham tarn. *Freshwater Biology,* 6: 317–331.

Pfitzner, J., G. Brunskill, and I. Zagorskis. (2004) Cs-137 and excess Pb-210 deposition patterns in estuarine and marine sediment in the central region of the Great Barrier Reef Lagoon, north-eastern Australia. *Journal of Environmental Radioactivity,* 76: 81–102.

Povinec, P. P., I. Osvath, and M. S. Baxter. (1996) Underwater gamma-spectrometry with HPGe and NaI(Tl) detectors. *Applied Radiation and Isotopes,* 47: 1127–1133.

Prandle, D. (1984) A modelling study of the mixing of Cs-137 in the seas of the European continental shelf. *Philosophical Transactions of the Royal Society of London Series A—Mathematical Physical and Engineering Sciences,* 310: 407–436.

Prandle, D., and J. Beechey. (1991) Marine dispersion of caesium-137 released from Sellafield and Chernobyl. *Geophysical Research Letters,* 18: 1723–1726.

Purkl, S., and A. Eisenhauer. (2004) Determination of radium isotopes and Rn-222 in a groundwater affected coastal area of the Baltic Sea and the underlying subsea floor aquifer. *Marine Chemistry,* 87: 137–149.

Ram, A., M. A. Rokade, D. V. Borole, and M. D. Zingde. (2003) Mercury in sediments of Ulhas estuary. *Marine Pollution Bulletin,* 46: 846–857.

Ramesh, R., R. Purvaja, S. Ramesh, and R. A. James. (2002) Historical pollution trends in coastal environments of India. *Environmental Monitoring and Assessment,* 79: 151–176.

Ritchie, J. C., and J. R. McHenry. (1990) Application of radioactive fallout $^{137}$Cs for measuring soil erosion and sediment accumulation rates and patters: A review. *Journal of Environmental Quality,* 19: 215–233.

Robbins, J. A., D. N. Edgington, and A. L. W. Kemp. (1978) Comparative $^{210}$Pb, $^{137}$Cs and pollen geochronologies of sediments from lakes Ontario and Erie. *Quaternary Research,* 10: 256–278.

Rogers, A. S. (1958) Physical behaviour and geologic control of radon in mountain streams. *U.S. Geological Survey Bulletin,* 1052-E: 187–212.

Sanderson, D. C. W., A. J. Cresswell, E. M. Scott, and J. J. Lang. (2004) Demonstrating the European capability for airborne gamma spectrometry: Results from the ECCOMAGS exercise. *Radiation Protection Dosimetry,* 109: 119–125.

San Miguel, E. G., J. P. Bolivar, and R. Garcia-Tenorio. (2004) Vertical distribution of Th-isotope ratios, Pb-210, Ra-226 and Cs-137 in sediment cores from an estuary affected by anthropogenic releases. *Science of the Total Environment,* 318: 143–157.

San Miguel, E. G., J. P. Bolivar, R. Garcia-Tenorio, and J. E. Martin. (2001) Pollutant concentrations in a sediment core dated by Th-isotopic ratios and the Pb-210 dating method. *Radiochimica Acta,* 89: 811–814.

Schuller, P., D. E. Walling, A. Sepulveda, R. E. Trumper, J. L. Rouanet, I. Pino, and A. Castillo. (2004) Use of Cs-137 measurements to estimate changes in soil erosion rates associated with changes in soil management practices on cultivated land. *Applied Radiation and Isotopes,* 60: 759–766.

Schwartz, M. C. (2003) Significant groundwater input to a coastal plain estuary: assessment from excess radon. *Estuarine Coastal and Shelf Science,* 56: 31–42.

Smith, J. N. (2001) Why should we believe Pb-210 sediment geochronologies? *Journal of Environmental Radioactivity,* 55: 121–123.

Smith, J. N., and K. M. Ellis. (1982) Transport mechanism for Pb-210, Cs-137 and Pu fallout radionuclides through fluvial-marine systems. *Geochimica et Cosmochimica Acta,* 46: 941–954.

Smith, J. N., K. M. Ellis, and D. M. Nelson. (1987) Time-dependent modelling of fallout radionuclide transport in a drainage basin: Significance of 'slow' erosional and 'fast' hydrological components. *Chemical Geology,* 63: 157–180.

Smith, J. N., K. M. Ellis, L. Polyak, G. Ivanov, S. L. Forman, and S. B. Moran. (2000) Pu-239, -240 transport into the Arctic Ocean from underwater nuclear tests in Chernaya Bay, Novaya Zemlya. *Continental Shelf Research,* 20: 255–279.

Smith, J. N., and D. H. Loring. (1981) Geochronology for mercury pollution in the sediments of the Saguenay Fjord, Quebec. *Environmental Science and Technology,* 15: 944–951.

Smith, J. N., and A. Walton. (1980) Sediment accumulation rates and geochronologies measured in the Saguenay Fjord using the Pb-210 dating method. *Geochimica et Cosmochimica Acta,* 44: 225–240.

Smoak, J. M., and P. W. Swarzenski. (2004) Recent increases in sediment and nutrient accumulation in Bear Lake, Utah/Idaho, USA. *Hydrobiologia,* 525: 175–184.

Sonke, J. E., W. C. Burnett, J. A. Hoogewerff, S. R. van der Laan, J. Vangrosveld, and D. R. Corbett. (2003) Reconstructing 20th century lead pollution and sediment focusing in a peat land pool (Kempen, Belgium), via Pb-210 dating. *Journal of Paleolimnology,* 29: 95–107.

Steiner, B., K. W. Hanselmann, and U. Krahenbuhl. (2000) Dating and heavy metal contents of sediment cores of a high-alpine, remote lake: Jorisee (Switzerland). *International Journal of Environmental Analytical Chemistry,* 78: 131–148.

Stieglitz, T. (2005) Submarine groundwater discharge into the near-shore zone of the Great Barrier Reef, Australia. *Marine Pollution Bulletin,* 51: 51–59.

Thomson, J., M. S. N. Carpenter, S. Colley, T. R. S. Wilson, H. Elderfield, and H. Kennedy. (1984) Metal accumulation rates in northwest Atlantic pelagic sediments. *Geochimica et Cosmochimica Acta,* 48: 1935–1948.

Tyler, A. N. (2004) High accuracy in situ radiometric mapping. *Journal of Environmental Radioactivity,* 72: 195–202.

Varekamp, J. C., B. Kreulen, M. R. B. ten Brink, and E. L. Mecray. (2003) Mercury contamination chronologies from Connecticut wetlands and Long Island Sound sediments. *Environmental Geology,* 43: 268–282.

Weis, D. A., J. C. Callaway, and R. M. Gersberg. (2001) Vertical accretion rates and heavy metal chronologies in wetland sediments of the Tijuana Estuary. *Estuaries,* 24: 840–850.

Wiegand, J. (2001) A guideline for the evaluation of the soil radon potential based on geogenic and anthropogenic parameters. *Environmental Geology,* 40: 949–963.

Wihlborg, P., A. Danielsson, and F. Klingberg. (2004) Mercury in Lake Vanern, Sweden. Distribution in surface sediment and catchment budget. *Water Air and Soil Pollution,* 154: 85–99.

Wu, Y., X. Wen, and Y. Zhang. (2004) Analysis of the exchange of groundwater and river water by using radon-222 in the middle Heihe Basin of northwestern China. *Environmental Geology,* 45: 647–653.

Yamamoto, M., I. Syarbain, K. Kofuji, A. Tsumura, K. Komura, K. Ueno, and D. J. Assinder. (1995) Determination of low-level Tc-99 in environmental samples by high-resolution ICP-MS. *Journal of Radioanalytical and Nuclear Chemistry,* 197: 185–194.

Yan, P., P. J. Shi, S. Y. Gao, L. Chen, X. B. Zhang, and L. X. Bai. (2002) Cs-137 dating of lacustrine sediments and human impacts on Dalian Lake, Qinghai Province, China. *Catena,* 47: 91–99.

Yasuhara, M., H. Yamazaki, T. Irizuki, and S. Yoshikawa. (2003) Temporal changes of ostracode assemblages and anthropogenic pollution during the last 100 years, in sediment cores from Hiroshima Bay, Japan. *Holocene,* 13: 527–536.

Zahorowski, W., S. D. Chambers, and A. Henderson-Sellers. (2004) Ground based radon-222 observations and their application to atmospheric studies. *Journal of Environmental Radioactivity,* 76: 3–33.

Zapata, F. (2003) The use of environmental radionuclides as tracers in soil erosion and sedimentation investigations: Recent advances and future developments. *Soil and Tillage Research,* 69: 3–13.

Zeeberg, J., S. L. Forman, and L. Polyak. (2003) Glacier extent in a Novaya Zemlya fjord during the 'Little Ice Age' inferred from glaciomarine sediment records. *Polar Research,* 22: 385–394.

Zheng, X. D., G. J. Wang, J. Tang, X. C. Zhang, W. Yang, H. N. Lee, and C. S. Wang. (2005) Be-7 and Pb-210 radioactivity and implications on sources of surface ozone at Mt. Waliguan. *Chinese Science Bulletin,* 50: 167–171.

# Chemical Fingerprinting of Petroleum Hydrocarbons

# 3

ZHENDI WANG
CARL BROWN

## Contents

Introduction ........ 44
Oil Hydrocarbon Fingerprinting Methodologies ........ 47
  Source-Specific Target Hydrocarbons ........ 47
  Tiered Analytical Approach ........ 51
  Revised Nordtest Methodology for Oil Spill Identification ........ 58
    Level 1 ........ 58
    Level 2 ........ 59
    Level 3 ........ 59
    Conclusion ........ 59
  Two-Dimensional GC: An Emerging Technique for Fingerprinting Hydrocarbons ........ 60
Chemical Composition of Oil and Petroleum Products and Spill Identification ........ 61
  Chemical Composition Features of Crude Oil ........ 62
    Background Hydrocarbons: Distinguishing Biogenic Hydrocarbons from Petrogenic Hydrocarbons in Oil-Contaminated Samples ........ 64
  General Chemical Composition Features of Refined Products ........ 67
    Light Distillates ........ 67
    Midrange Distillates ........ 68
    Classic Heavy Residual Fuel ........ 70
    Lubricating Oil ........ 70
    Waste Oil ........ 71
  PAH Fingerprints of Oils and Petroleum Products ........ 71
    Distribution of Alkylated PAH Homologues and Other EPA Priority PAHs ........ 71
    Recommended Diagnostic Ratios of PAHs ........ 72
    PAH Isomer and Cluster PAH Analysis ........ 75
    Methyl Phenanthrenes ........ 76
    Methyl Dibenzothiophenes ........ 76
    Other Relative Ratios of PAH Isomers ........ 77

Cluster PAHs at *m/z* 216 ........ 77
Distinguishing Pyrogenic Hydrocarbons from Petrogenic Hydrocarbons ........ 78
Biomarker Fingerprints of Oils and Petroleum Products ........ 79
Distributions and Quantification of Biomarkers ........ 79
Low-Boiling Sesquiterpanes in Oils and Lighter Petroleum Products ........ 86
Diagnostic Ratios (Indices) of Biomarkers ........ 89
Weathering Check Using Hydrocarbon Fingerprints ........ 94
Oil Weathering ........ 94
Weathering Check Using *n*-Alkanes and GC Traces ........ 95
Weathering Check Using PAHs ........ 97
Weathering Check Using Biomarkers ........ 98
A Case Study: Using a Multicriterion Approach for Source Identification of Unknown Spill Samples ........ 99
Product Type Screen and Determination of Hydrocarbon Groups ........ 100
Determination of Oil-Characteristic Alkylated PAH Homologues and Their Diagnostic Ratios ........ 102
Input of Pyrogenic PAHs to the Spill Samples ........ 103
Characterisation of Biomarker Compounds ........ 105
Conclusions ........ 106
Acknowledgements ........ 106
References ........ 107

## Introduction

Petroleum plays an extremely important role in modern society. As the population of the world increases and developing countries become more industrialised, the demand for energy keeps growing worldwide. Just fewer than 2 billion barrels (1 barrel = 159 L) of crude oil was processed by refiners in the United States in 2004. Consumption worldwide was about 30 billion barrels in 2004. Table 3.1 presents the worldwide petroleum demand and supply from 1970 to 2004 (DOE 2004). Currently, oil is the dominant energy source and is expected to remain so over the next several decades (NRC 2002). In addition to natural seeps, which are purely natural phenomena that occur when crude oil seeps from the geologic strata beneath the seafloor to the seawater column, the worldwide extraction, transportation, and consumption of petroleum inevitably result in its release to the environment. Waterborne oil spills of unknown origin from continuous leaks or illegal discharge often occur in rivers, open waters, and coastal waterways. Petroleum and its com-

**Table 3.1 Worldwide Petroleum Demand and Supply (1970–2004)**

| Year | Demand (×1000 barrels/day) | Supply (×1000 barrels/day) |
|---|---|---|
| 1970 | 46,808 | 48,986 |
| 1971 | 49,416 | 51,766 |
| 1972 | 53,094 | 54,574 |
| 1973 | 57,237 | 59,300 |
| 1974 | 56,677 | 59,391 |
| 1975 | 56,198 | 56,511 |
| 1976 | 59,673 | 61,121 |
| 1977 | 61,826 | 63,665 |
| 1978 | 64,158 | 64,225 |
| 1979 | 65,220 | 66,973 |
| 1980 | 63,108 | 64,135 |
| 1981 | 60,944 | 60,728 |
| 1982 | 59,543 | 58,199 |
| 1983 | 58,779 | 58,008 |
| 1984 | 59,822 | 59,607 |
| 1985 | 60,087 | 59,234 |
| 1986 | 61,825 | 61,391 |
| 1987 | 63,104 | 62,084 |
| 1988 | 64,963 | 64,394 |
| 1989 | 66,092 | 65,552 |
| 1990 | 66,443 | 66,472 |
| 1991 | 67,061 | 66,419 |
| 1992 | 67,273 | 66,781 |
| 1993 | 67,372 | 67,290 |
| 1994 | 68,679 | 68,313 |
| 1995 | 69,955 | 70,056 |
| 1996 | 71,522 | 71,680 |
| 1997 | 73,292 | 73,905 |
| 1998 | 73,932 | 75,407 |
| 1999 | 75,826 | 74,583 |
| 2000 | 76,954 | 77,484 |
| 2001 | 78,105 | 77,514 |
| 2002 | 78,439 | 76,858 |
| 2003 | 79,892 | 79,462 |
| 2004 | 82,631 | 82,972 |

bustion-derived hydrocarbons are often some of the most frequently discovered chemicals of concern at contaminated sites on land.

Based on analysis of data from a wide variety of sources, each year on average about 260,000 t of petroleum are released to the waters off North America. Annual worldwide estimates of petroleum input to the sea exceed 1,300,000 t (NRC 2002). Most oils spilled into the sea are fuels (48%) and crude oils (29%). The fuel category consists primarily of bunker oils and intermediate fuel oils (IFOs), which consist of bunker oils mixed with lighter fuels such as diesel. A list of the major oil spills in the last 40 years has been provided by Fingas (2001). The spills are listed according to their volume, beginning with the largest spill to date—the release of oil during the Gulf War in 1991 (800,000 tons). According to the spill volume, the most influential *Exxon Valdez* spill ranks at number 52 (37,000 tons), while the most recent two large-scale marine spills, the 1999 *Erika* spill (occurred about 110 km south of Brest, France) and the 2002 *Prestige* spill (occurred on water about 240 km off the northwest coast of Spain), rank only at number 124 (12,000 tons) and number 90 (24,000 tons), respectively. Although most of the large oil spills are from tankers, these spills make up only about 5% of all oil pollution entering the sea. Most oil pollution in the oceans comes from the runoff of oil and fuel from land-based sources rather than from accidental spills. In Canada, about 12 spills of more than 4000 L are reported each day, of which about 1 spill is into navigable waters. Most spills take place on land, including oil spills from pipelines, underground storage tanks, and aboveground storage containers. In the United States, about 25 such spills occur each day into navigable waters and about 75 occur on land (Fingas 2001).

Oil poses a range of environmental risks and causes wide public concerns when released into the environment, whether as catastrophic spills or chronic discharges. Oil spills have led to legal battles resulting in billions of dollars in damage awards and punitive fines. Therefore, to precisely characterise spilled oil hydrocarbons in complex environmental samples and to defensibly identify the sources of hydrocarbons entering the environment are extremely important for site contamination assessment, for prediction of the potential long-term impact of spilled oils on the ecosystem, and for determining responsibility for the spill. In addition, successful forensic investigation and analysis of hydrocarbons in contaminated sites and receptors yield a wealth of chemical fingerprinting data. These data, in combination with historic, geological, environmental, and any other related information, can in many cases help to settle legal liability and support litigation against the spillers.

This chapter focuses on development of hydrocarbon fingerprinting techniques for environmental forensic applications. The hydrocarbon fingerprinting and data interpretation techniques discussed include spill oil identification protocol, tiered analytical approach, ancillary/emerging techniques for fingerprinting hydrocarbon, generic features and chemical composition of oils, understanding effects of refining processes on hydrocarbon fingerprinting, the effects of weathering on distribution of hydrocarbons once released into the environment, hydrocarbon screening and spill oil source identification by analysis of 'source-specific marker' compounds, and determination of diagnostic ratios. The issues of distinguishing biogenic and pyrogenic hydrocarbons from petrogenic hydrocarbons are also briefly addressed.

## Oil Hydrocarbon Fingerprinting Methodologies

### Source-Specific Target Hydrocarbons

Crude oil and many other petroleum-related hydrocarbons such as combustion-derived mixtures are the compounds of concern most often discovered at many contaminated sites. Oils consist of complex mixtures of hydrocarbons and nonhydrocarbons that range from small, volatile compounds to large, nonvolatile ones. Comprehensive characterisation of total petroleum composition across a wide boiling range remains a challenge. Recently, ultrahigh-resolution Fourier transform ion cyclotron resonance mass spectrometry (Marshall and Rodgers 2004) has revealed that crude oil contains heteroatom-containing (N, O, S) organic components having more than 20,000 distinct elemental compositions ($C_cH_hN_nO_oS_s$).

In the last two decades, a wide variety of instrumental techniques, particularly gas chromatography equipped with a flame ionization detector (GC-FID) or mass spectrometer (GC-MS), has been extensively used for analysis of various organic compounds in environmental samples. Regulatory bodies such as the U.S. Environmental Protection Agency (EPA) and Canadian Council of Ministries of the Environment (CCME) have developed and codified a series of methods based on the GC techniques. The EPA methods, including the EPA 418.1 (total recoverable petroleum hydrocarbons by infrared spectroscopy), 1664 (*n*-hexane extractable material and silica-gel treated *n*-hexane extractable material by extraction and gravimetry), 600 series (method standards for wastewater), and 8000 series methods (SW-846 methods for solid waste analyses), have been widely used as routine procedures for determination of organic volatile and semivolatile compounds.

However, there is fundamental barrier for environmental forensic scientists and investigators: These methods were originally designed for measuring industrial chemicals in wastewater and solid waste and none of them provides information on detailed chemical components that comprise the complex spilled oil or petroleum-derived samples. The data generated from these methods are generally insufficient to answer the fundamental questions (such as type and source, weathering status of spilled oil, potential spillers, and so on) raised in an oil spill liability investigation. Of the more than 160 EPA priority pollutant organic compounds determined by these methods, only 20 are petroleum-related hydrocarbons (including benzene, toluene, ethyl-benzene, xylenes, and 16 EPA priority PAH compounds) that would be useful for oil spill and contamination studies. Further, only half of these 20 compounds are found in significant quantities in oils and petroleum products.

Also, the PAH compounds in oils are dominated almost exclusively by the $C_1$–$C_4$ alkylated homologues of the parent PAHs—in particular, naphthalene, phenanthrene, dibenzothiophene, fluorene, and chrysene—none of which are measured by these standard EPA methods. Other important classes of petroleum hydrocarbons (e.g., biomarkers and *n*-alkanes) are not measured by these methods at all. In recent years, many EPA and ASTM methods have been modified (such as the modified EPA methods 8015, 8260, and 8270 and the modified ASTM methods 3328-90, 5037-90, and 5739-95) to allow flexibility in the deployment of the standard analytical methods and to improve specificity and sensitivity for measuring spilled oil and petroleum products in soils and waters by environmental chemists. For example, EPA method 8270 has been modified to increase analytical sensitivity and to expand the analyte list to include petroleum-specific compounds such as the alkylated PAHs, sulphur- and nitrogen-containing PAHs, and biomarker triterpane and sterane compounds. The principal modification to EPA method 8270 is the use of the high-resolution GC-MS selected ion mode (SIM) analysis that offers increased sensitivity relative to the full-scan mode. Many environmental laboratories have used the modified EPA method 8270, combined with column cleanup and rigorous quality assurance (QA) measures, to identify and quantify low levels of hydrocarbons.

Hydrocarbon-contaminated site investigation and oil spill identification require further elaboration of oil target analytes to include determination of the individual specific target compounds and isomeric groups. The selection of appropriate target oil analytes is dependent mainly on the type of oil spilled, the particular environmental compartments being assessed, and expected needs for current and future data comparison. In general, the major petroleum-specific target analytes that may be needed to be chemically characterised for oil source identification and environmental assessment include:

- Individual saturated hydrocarbons including *n*-alkanes (*n*-$C_8$ through *n*-$C_{44}$) and the selected isoprenoids pristane (2,6,10,14-tetramethyl-pentadecane) and phytane (2,6,10,14-tetramethyl-hexadecane). In some cases, another three highly abundant isoprenoid compounds—farnesane (2,6,10-trimethyl-$C_{12}$), 2,6,10-trimethyl-$C_{13}$, and norpristane (2,6,10-trimethyl-$C_{15}$)—are also included.

- Alkyl ($C_1$–$C_{14}$) cyclo-hexane homologous compound series. These homologous compounds exhibit a characteristic distribution pattern in mass-to-charge ratio (*m/z*) 83 mass chromatograms for different types of fuels, providing another useful fingerprint for characterising petroleum derivatives.
- The volatile aromatic hydrocarbons including BTEX (benzene, toluene, ethyl-benzene, and 3 xylene isomers) and alkylated ($C_3$- to $C_5$-) benzenes, naphthenes, and volatile paraffins and isoparaffins. Analysis of long-side-chain *n*-alkylbenzenes with the *n*-alkyl groups in the $C_7$–$C_{27}$ range for evaluation of fate of crude oil in the environment has been reported recently (Dutta and Harayama 2001).
- The 16 so-called EPA priority parent PAHs and, in particular, the petroleum-specific alkylated ($C_1$–$C_4$) homologues of selected PAHs (i.e., alkylated naphthalene, phenanthrene, dibenzothiophene, fluorene, and chrysene series). These alkylated PAH homologues are the backbone of chemical characterisation and identification of oil spill assessments (Table 3.2).
- Biomarker terpane and sterane compounds (Table 3.3). Analysis of selected ion peaks produced by these characteristic, environmentally persistent compounds generates information of great importance in determining sources, weathered state, and potential treatability.
- Measurements of bulk hydrocarbon groups. They include total petroleum hydrocarbons (TPHs), the unresolved complex mixtures (UCMs), the total saturates and total aromatics, and concentrations of asphaltenes and resins.
- Additives to petroleum products. They include alkyl lead additives (tetramethyl lead and trimethylethyl lead at *m/z* 253 and 223, dimethyldiethyl lead at *m/z* 267 and 223, methyltriethyl lead at *m/z* 281 and 223, tetraethyl lead at 295 and 237); oxygenates including substances such as ethanol, methanol, methyl tertiary butyl ether (MTBE), ethyl tertiary butyl ether (ETBE), and tertiary amyl methyl ether (TAME); fuel dyes used for differentiation among fuel grades; and antioxidant compounds or inhibitors (such as aromatic amines and alkyl-substituted phenols) added to fuels to retard auto-oxidation.
- Measurement of stable carbon isotope ratio ($\delta^{13}C$) of hydrocarbon groups and sometimes measurement of the isotopic composition of individual compounds by GC-IRMS for correlating spills with suspected sources are also included in many oil spill studies.

Another potentially valuable hydrocarbon group for oil spill identification is nitrogen and oxygen heterocyclic hydrocarbons. These heterocyclic hydrocarbons are generally only present in oils at relatively low concentrations compared to PAHs. However, they become enhanced with weathering

**Table 3.2 Source-Specific Alkylated Homologous PAHs and Other U.S. EPA Priority PAHs for Oil Spill Studies**

| Compounds | Code | Ring numbers | Target ions |
|---|---|---|---|
| Oil-characteristic alkylated PAHs | | | |
| Naphthalenes | | | |
| | $C_0N$ | 2 | 128 |
| | $C_1N$ | 2 | 142 |
| | $C_2N$ | 2 | 156 |
| | $C_3N$ | 2 | 170 |
| | $C_4N$ | 2 | 184 |
| Phenanthrenes | | | |
| | $C_0P$ | 3 | 178 |
| | $C_1P$ | 3 | 192 |
| | $C_2P$ | 3 | 206 |
| | $C_3P$ | 3 | 220 |
| | $C_4P$ | 3 | 234 |
| Dibenzothiophenes | | | |
| | $C_0D$ | 3 | 184 |
| | $C_1D$ | 3 | 198 |
| | $C_2D$ | 3 | 212 |
| | $C_3D$ | 3 | 226 |
| Fluorenes | | | |
| | $C_0F$ | 3 | 166 |
| | $C_1F$ | 3 | 180 |
| | $C_2F$ | 3 | 194 |
| | $C_3F$ | 3 | 208 |
| Chrysenes | | | |
| | $C_0C$ | 4 | 228 |
| | $C_1C$ | 4 | 242 |
| | $C_2C$ | 4 | 256 |
| | $C_3C$ | 4 | 270 |
| Other U.S. EPA priority PAHs | | | |
| Biphenyl | Bph | 2 | 154 |
| Acenaphthylene | Acl | 3 | 152 |
| Acenaphthene | Ace | 3 | 153 |
| Anthracene | An | 3 | 178 |
| Fluorancene | Fl | 4 | 202 |
| Pyrene | Py | 4 | 202 |
| Benz[a]anthracene | BaA | 4 | 228 |

**Table 3.2 Source-Specific Alkylated Homologous PAHs and Other U.S. EPA Priority PAHs for Oil Spill Studies (Continued)**

| Compounds | Code | Ring numbers | Target ions |
|---|---|---|---|
| Benzo[b]fluoranthene | BbF | 5 | 252 |
| Benzo[k]fluoranthene | BkF | 5 | 252 |
| Benzo[e]pyrene | BeP | 5 | 252 |
| Benzo[a]pyrene | BaP | 5 | 252 |
| Perylene | Pe | 5 | 252 |
| Indeno[1,2,3-cd]pyrene | IP | 6 | 276 |
| Dibenz[a,h]anthracene | DA | 5 | 278 |
| Dibenzo[ghi]perylene | BP | 6 | 276 |
| Internal standard and surrogates | | | |
| [$^2H_{14}$]Terphenyl | | | 244 |
| [$^2H_{10}$]Acenaphthene | | | 164 |
| [$^2H_{10}$]Phenanthrene | | | 188 |
| [$^2H_{12}$]Benz[a]anthracene | | | 240 |
| [$^2H_{14}$]Perylene | | | 264 |

because they are biorefractory and persistent in the environment. Most organic nitrogen hydrocarbons in crude oils are present as alkylated aromatic heterocyclics with a predominance of neutral pyrrolic structures over basic pyridine forms. They are chiefly associated with high boiling fractions with much of the nitrogen in petroleum being in asphaltenes. Individual and alkyl homologues of carbazole, quinoline, and pyridine have been identified in many crude oils (Meyer, Cartellieri, and Steinhart 2001; Bence and Burns 1995). These compounds may provide important clues for potential sources of hydrocarbons in the environment and for tracing petroleum molecules back to their biological precursors. Compared to the PAHs and biomarkers, the application of nitrogen- and oxygen-containing heterocyclic hydrocarbons in source identification is still in its infancy, and more research is clearly needed.

## Tiered Analytical Approach

The characterisation and identification of spilled oil and petroleum products can best be conducted using a tiered analytical approach (Uhler, Stout, and McCarthy 1998; Wang, Fingas, and Page 1999; Stout et al. 2002), by which sufficient details concerning the nature and origin, chemical composition changes, and weathering degrees of spilled oil under investigation can be gathered. Depending on the needs of the specific spill site investigation, the

**Table 3.3 Petroleum Biomarker Terpane and Sterane Compounds for Oil Spill Studies**

| Peak | Compound | Code | Empirical formula | Target ions |
|---|---|---|---|---|
| | | Sesquiterpanes (bicyclicterpanes) | | |
| | $C_{14}$ sesquiterpanes | | $C_{14}H_{26}$ | 123, 179 |
| | $C_{15}$ sesquiterpanes | | $C_{15}H_{28}$ | 123, 193 |
| | $C_{16}$ sesquiterpanes | | $C_{16}H_{30}$ | 123, 193, 207 |
| | | Diamondoids | | |
| | Adamantanes | | $C_{10}H_{16}$, alkyl-$C_{10}H_{15}$ | 136, 135, 149, 163, 177 |
| | Diamantanes | | $C_{14}H_{20}$, alkyl-$C_{14}H_{19}$ | 188, 187, 201, 215, 229 |
| | | Terpanes | | |
| 1 | $C_{19}$ tricyclic terpane | TR19 | $C_{19}H_{34}$ | 191 |
| 2 | $C_{20}$ tricyclic terpane | TR20 | $C_{20}H_{36}$ | 191 |
| 3 | $C_{21}$ tricyclic terpane | TR21 | $C_{21}H_{38}$ | 191 |
| 4 | $C_{22}$ tricyclic terpane | TR22 | $C_{22}H_{40}$ | 191 |
| 5 | $C_{23}$ tricyclic terpane | TR23 | $C_{23}H_{42}$ | 191 |
| 6 | $C_{24}$ tricyclic terpane | TR24 | $C_{24}H_{44}$ | 191 |
| 7 | $C_{25}$ tricyclic terpane (a) | TR25A | $C_{25}H_{46}$ | 191 |
| 8 | $C_{25}$ tricyclic terpane (b) | TR25B | $C_{25}H_{46}$ | 191 |
| 9 | Triplet: $C_{24}$ tetracyclic terpane + $C_{26}$ (S + R) tricyclic terpanes | TET24 + TR26A + TR26B | $C_{24}H_{42}$ + $C_{26}H_{48}$ | 191 |
| 10 | $C_{28}$ tricyclic terpane (a) | TR28A | $C_{28}H_{52}$ | 191 |
| 11 | $C_{28}$ tricyclic terpane (b) | TR28B | $C_{28}H_{52}$ | 191 |
| 12 | $C_{29}$ tricyclic terpane (a) | TR29A | $C_{29}H_{54}$ | 191 |
| 13 | $C_{29}$ tricyclic terpane (b) | TR29B | $C_{29}H_{54}$ | 191 |

| | | | | |
|---|---|---|---|---|
| 14 | Ts: 18α(H),21β(H)-22,29,30-trisnorhopane | Ts | $C_{27}H_{46}$ | 191 |
| 15 | 17α(H),18α(H),21β(H)-25,28,30-trisnorhopane | TH27 | $C_{27}H_{46}$ | 191, 177 |
| 16 | Tm: 17α(H),21β(H)-22,29,30-trisnorhopane | Tm | $C_{27}H_{46}$ | 191 |
| 17 | $C_{30}$ tricyclic terpane 1 | TR30A | $C_{30}H_{56}$ | 191 |
| 18 | $C_{30}$ tricyclic terpane 2 | TR30B | $C_{30}H_{56}$ | 191 |
| 19 | 17α(H),18α(H),21β(H)-28,30-bisnorhopane | H28 | $C_{28}H_{48}$ | 191 |
| 20 | 17α(H),21β(H)-25-norhopane | NOR25H | $C_{29}H_{50}$ | 191, 177 |
| 21 | 17α(H),21β(H)-30-norhopane | H29 | $C_{29}H_{50}$ | 191 |
| 22 | 18α(H),21β(H)-30-norneohopane ($C_{29}$Ts) | C29Ts | $C_{29}H_{50}$ | 191 |
| 23 | 17α(H)-diahopane | DH30 | $C_{30}H_{52}$ | 191, |
| 24 | 17α(H),21β(H)-30-norhopane (normoretane) | M29 | $C_{29}H_{50}$ | 191 |
| 25 | 18α(H) and 18β(H)-oleanane | OL | $C_{30}H_{52}$ | 191, 412 |
| 26 | 17α(H),21β(H)-hopane | H30 | $C_{30}H_{52}$ | 191 |
| 27 | 17α(H)-30-nor-29-homohopane | NOR30H | $C_{30}H_{52}$ | 191 |
| 28 | 17β(H),21α(H)-hopane (moretane) | M30 | $C_{30}H_{52}$ | 191 |
| 29 | 22*S*-17α(H),21β(H)-30-homohopane | H31S | $C_{31}H_{54}$ | 191 |
| 30 | 22*R*-17α(H),21β(H)-30-homohopane | H31R | $C_{31}H_{54}$ | 191 |
| 31 | Gammacerane | GAM | $C_{30}H_{52}$ | 191, 412 |
| 32 | 17β(H),21β(H)-hopane | (IS) | (Internal standard) | 191 |
| 33 | 22*S*-17α(H),21β(H)-30,31-bishomohopane | H32S | $C_{32}H_{56}$ | 191 |
| 34 | 22*R*-17α(H),21β(H)-30,31-bishomohopane | H32R | $C_{32}H_{56}$ | 191 |
| 35 | 22*S*-17α(H),21β(H)-30,31,32-trishomohopane | H33S | $C_{33}H_{58}$ | 191 |
| 36 | 22*R*-17α(H),21β(H)-30,31,32-trishomohopane | H33R | $C_{33}H_{58}$ | 191 |
| 37 | 22*S*-17α(H),21β(H)-30,31,32,33-tetrakishomohopane | H314S | $C_{34}H_{60}$ | 191 |

Continued

**Table 3.3 Petroleum Biomarker Terpane and Sterane Compounds for Oil Spill Studies (Continued)**

| Peak | Compound | Code | Empirical formula | Target ions |
|---|---|---|---|---|
| 38 | 22*R*-17α(H),21β(H)-30,31,32,33-tetrakishomohopane | H34R | $C_{34}H_{60}$ | 191 |
| 39 | 22S-17α(H),21β(H)-30,31,32,33,34-pentakishomohopane | H35S | $C_{35}H_{62}$ | 191 |
| 40 | 22*R*-17α(H),21β(H)-30,31,32,33,34-pentakishomohopane | H35R | $C_{35}H_{62}$ | 191 |
| | | Steranes | | |
| 41 | $C_{20}$ 5α(H),14α(H),17α(H)-sterane | S20 | $C_{20}H_{34}$ | 217 and 218 |
| 42 | $C_{21}$ 5α(H),14β(H),17β(H)-sterane | S21 | $C_{21}H_{36}$ | 217 and 218 |
| 43 | $C_{22}$ 5α(H),14β(H),17β(H)-sterane | S22 | $C_{22}H_{38}$ | 217 and 218 |
| 44 | $C_{27}$ 20*S*-13β(H),17α(H)-diasterane | DIA27S | $C_{27}H_{48}$ | 217 and 218, 259 |
| 45 | $C_{27}$ 20*R*-13β(H),17α(H)-diasterane | DIA27R | $C_{27}H_{48}$ | 217 and 218, 259 |
| 46 | $C_{27}$ 20*S*-13α(H),17β(H)-diasterane | DIA27S2 | $C_{27}H_{48}$ | 217 and 218, 259 |
| 47 | $C_{27}$ 20*R*-13α(H),17β(H)-diasterane | DIA27R2 | $C_{27}H_{48}$ | 217 and 218, 259 |
| 48 | $C_{28}$ 20*S*-13β(H),17α(H)-diasterane | DIA28S | $C_{28}H_{50}$ | 217 and 218, 259 |
| 49 | $C_{28}$ 20*R*-13β(H),17α(H)-diasterane | DIA28R | $C_{28}H_{50}$ | 217 and 218, 259 |
| 50 | $C_{29}$ 20*S*-13β(H),17α(H)-diasterane | DIA29S | $C_{29}H_{52}$ | 217 and 218, 259 |

| | | | | |
|---|---|---|---|---|
| 51 | $C_{29}$ 20*R*-13α(H),17β(H)-diasterane | DIA29R | $C_{29}H_{52}$ | 217 and 218, 259 |
| 52 | $C_{27}$ 20*S*-5α(H),14α(H),17α(H)-cholestane | C27S | $C_{27}H_{48}$ | 217 and 218 |
| 53 | $C_{27}$ 20*R*-5α(H),14β(H),17β(H)-cholestane | C27ββR | $C_{27}H_{48}$ | 217 and 218 |
| 54 | $C_{27}$ 20*S*-5α(H),14β(H),17β(H)-cholestane | C27ββS | $C_{27}H_{48}$ | 217 and 218 |
| 55 | $C_{27}$ 20*R*-5α(H),14α(H),17α(H)-cholestane | C27R | $C_{27}H_{48}$ | 217 and 218 |
| 56 | $C_{28}$ 20*S*-5α(H),14α(H),17α(H)-ergostane | C28S | $C_{28}H_{50}$ | 217 and 218 |
| 57 | $C_{28}$ 20*R*-5α(H),14β(H),17β(H)-ergostane | C28ββR | $C_{28}H_{50}$ | 217 and 218 |
| 58 | $C_{28}$ 20*S*-5α(H),14β(H),17β(H)-ergostane | C28ββS | $C_{28}H_{50}$ | 217 and 218 |
| 59 | $C_{28}$ 20*R*-5α(H),14α(H),17α(H)-ergostane | C28R | $C_{28}H_{50}$ | 217 and 218 |
| 60 | $C_{29}$ 20*S*-5α(H),14α(H),17α(H)-stigmastane | C29S | $C_{29}H_{52}$ | 217 and 218 |
| 61 | $C_{29}$ 20*R*-5α(H),14β(H),17β(H)-stigmastane | C29ββR | $C_{29}H_{52}$ | 217 and 218 |
| 62 | $C_{29}$ 20*S*-5α(H),14β(H),17β(H)-stigmastane | C29ββS | $C_{29}H_{52}$ | 217 and 218 |
| 63 | $C_{29}$ 20*R*-5α(H),14α(H),17α(H)-stigmastane | C29R | $C_{29}H_{52}$ | 217 and 218 |
| 64 | C30 steranes | C30S | $C_{30}H_{54}$ | 217 and 218 |
| | Monoaromatic steranes | | | 253 |
| | Triaromatic steranes | | | 231 |

tiered analytical approaches may vary. This gives the environmental forensic investigator the flexibility to determine how many tiers should be used and how much information is sufficient and necessary to address site- or incident-specific questions. The tiered approach used by the Environment Canada Oil Spill Research Program includes:

*tier 1*—determination of hydrocarbon groups and product type screening;
*tier 2*—determination of volatile hydrocarbons (e.g., BTEX and alkyl benzenes, low molecular weight alkyl-pentanes and alkyl-hexanes, smaller cyclo-pentanes and cyclo-hexanes, and various additives in lighter petroleum products);
*tier 3*—determination of the distribution pattern of target PAHs and biomarker components;
*tier 4*—determination and comparison of diagnostic ratios of the source-specific marker compounds (e.g., target biomarkers and PAHs) with the spill and suspected source oil samples and with the corresponding data from the database; and
*tier 5*—determination of weathered percentages of the residual oil and estimation of spill ages.

In this tiered analytical approach, high-resolution capillary GC-FID analysis is applied to determine hydrocarbon groups (e.g., TPH, UCM, the total saturates and total aromatics) and concentrations of total *n*-alkanes and major isoprenoid compounds (e.g., pristane and phytane) from *n*-$C_8$ to *n*-$C_{44}$, and to characterise the product types (e.g., crude oil, diesel, lube oil, or bunker C type fuel) in fresh to highly weathered oil samples. If needed, the thin layer chromatographic (TLC) or gravimetric methods are applied to determine the contents of asphaltenes and resins. The GC-MS analyses provide data on the source-specific marker compounds, including the target alkylated PAH homologues and other EPA priority PAHs, and biomarker terpane and sterane compounds. The MS detector is operated in the scan mode to obtain spectral data for identification of unknown components and in the selected ion mode (SIM) for quantitation of target compounds. An appropriate temperature program is selected to achieve near-baseline separation of all of the target components. Quantitation of the alkylated PAH homologues, other EPA priority PAHs, and biomarkers is performed using the internal standard method with the relative response factors (RRFs) for each compound determined during the instrument calibration. For analysis of BTEX and other alkyl benzenes, all oil samples are directly weighed and dissolved in *n*-pentane to an approximate concentration of 2 $mg.mL^{-1}$. Prior to analysis, the tightly capped oil solutions are put in a refrigerator for 30 min to precipitate the asphaltenes to the bottom of the vials in order to avoid performance deterioration of the column (Wang et al. 1995a).

Stout et al. (2002) proposed a similar tiered analytical approach to hydrocarbon fingerprinting in which the GC-FID and GC-MS methods used are modified from the standard U.S. EPA methods 8015 (GC-FID screening), 8260 (purge-and-trap GC-MS), and 8270 (semivolatile GC-MS) and other ancillary methods (including stable isotope measurements, alkyl lead fuel additive analysis, fuel oil dye analysis, simulated distillation curves, and density measurements). The development of this tiered approach is based on understanding the chemical composition of petroleum-, coal-, and combustion-derived hydrocarbons, and chromatographic behaviour of the major and minor constituents of each of these hydrocarbon assemblages. The progression of analytical techniques used at each step of the tiered approach focuses on the detailed analysis of particular hydrocarbon boiling point ranges (e.g., volatile-range hydrocarbons that comprise light distillates or semivolatile range hydrocarbons that comprise middle and residual distillates of petroleum or coal liquids) and particular classes of chemical compounds that comprise those fractions. This tiered approach has been successfully used for a number of spill case studies with site hydrocarbon contaminations known or unknown.

The quality and reliability of analytical measurements are dependent on the QA and quality control (QC) program employed. In order to support oil spill forensic investigations, quality management (including laboratory profile and mission, quality assurance and quality control system, updated standard operational procedures, personnel training program and record, up-to-date methodology, equipment management, sample management, data management, and workload management) must be strictly followed. The chemical measurements must be conducted within the framework of highly stringent, defensible, and reliable QC and QA programs (Page et al. 1995; Douglas et al. 1996, 2004; Wang, Fingas, and Page 1999; Wang et al. 2003; Stout et al. 2002; Faksness, Daling, and Hansen 2002; EPA 1997, 1998a, 1998b, 2001; ASTM 1997; ETC 2003). The QA/QC programs used by different laboratories may differ more or less in the course of sample handling and preparation, analysis, and reporting of analytical data, but the quality principles and practices are similar.

Quality assurance is a *definite plan* for laboratory operation that specifies standard procedures that help to produce data with defensible quality and reported results with a high level of confidence. The basic requirements of a QA program are to recognise possible errors, understand the measurement system used, and develop techniques and plans to minimise errors. The elements of quality assurance are quality control and quality assessment. Quality control includes good laboratory practices; updated standard operational procedures; sample collection, documentation, and calibration; standardisation; instrument maintenance; facilities maintenance; education and training; reporting of forensic analysis data, continuous improvement program; and inspection and validation. Quality assessment includes reference materials, replicates, splits, spikes, surrogates, collaborative tests, and statistical analysis.

Prior to sample analysis, a five-point response calibration curve should be established to demonstrate the linear range of the analysis. Check standards at the midpoint of the established calibration curves are analysed before and after each analytical batch of samples (7–10 samples) to validate the integrity of the initial calibration. The relative response factor (RRF) stability is a key factor in maintaining the quality of the analysis. A control chart for RRF values should be prepared and monitored. All samples and QC samples (procedural blank, matrix spike samples, duplicate and reference oil sample) are spiked with appropriate surrogates to measure individual sample matrix effects associated with sample preparation and analysis. PAH surrogate and matrix spike recoveries should be within 60–120%. Method detection limits (MDLs) of target compounds are performed according to the procedure described in the EPA protocol titled 'Definition and Procedure for the Determination of the Method Detection Limit' (Code of Federal Regulations 40CFR Part 136). Analysis and characterisation of forensic sample batches should be performed on the same instrument within the same analytical sequence by experienced chemists.

## Revised Nordtest Methodology for Oil Spill Identification

Nordtest is an institution under the Nordic Council of Ministers and acts as a joint Nordic body in the field of conformity assessment. In April 2000, Nordtest initiated the ongoing project of 'Revision of the Nordtest Methodology for Oil Spill Identification' in order to meet requirements of analytical methods for oil fingerprinting (being more *quantifiable*, *objective*, and *defensible*). The main objectives of this project are to (1) refine the existing Nordtest methodology into a technically more robust and more defensible oil spill identification methodology, and (2) adjust the revised Nordtest methodology into guidelines for the European Committee for Standardisation (CEN). The recommended analytical methodology (CEN 2002) has been tested and verified in a round robin study in which 12 different laboratories from 10 Nordic countries participated. The 'protocol/decision chart of the recommended methodology' includes three tiered levels of analyses and data treatment.

### *Level 1*

After sample preparation, the chemical fingerprinting starts with a GC-FID screening analysis on all samples. Results of this analysis form the basis for:

characterising the spill samples;

establishing selected isoprenoid indices/ratios (such as *n*-$C_{17}$/pristane, *n*-$C_{18}$/phytane, and pristane/phytane); and establishing a 'weathering check' (self-normalising to nonweathered or weathered-resistant compounds).

At this level of the investigation, the spill samples can be qualitatively and quantitatively compared to the suspected sources and obviously 'nonmatch' samples can be ruled out and eliminated from additional levels of analysis.

### *Level 2*

This level is analysis of spill and suspected source samples using GC-MS. The content and distribution of a suite of target petroleum biomarkers and PAHs are determined. The data produced from this analysis are used for generating:

a suite of calculated diagnostic ratios of PAHs;
a suite of calculated diagnostic ratios of target biomarkers; and
a weathering check from a suite of selected PAH groups.

### *Level 3*

In this level, the impact of weathering (based on weathering check data of *n*-alkanes from level 1 and PAH distribution from level 2) is assessed then the correlation studies are conducted. First, results from triplicate analyses are used to determine the analytical standard deviation, followed by the selection of the most robust diagnostic ratios using the Student's *t* statistical tool. Then, the results of spill and suspected source samples are compared, linear regressions are performed, and conclusions based on the 'fit' of the selected suite of robust diagnostic ratios between spill and suspected source samples can be made.

### *Conclusion*

The final assessment is concluded by the four operational and technical defensible identification terms: *positive match, probable match, inconclusive,* or *nonmatch.* These categories represent degrees of differences between the analyses of two oils according to the present criteria in ASTM method D3328:

*Positive match:* the chromatographic patterns of the samples submitted for comparison are virtually identical and the observed differences between the spill sample and suspected source are caused and can be explained by the acceptable analytical variance and/or weathering effects.
*Probable match*: The chromatographic patterns of the spill sample are similar to those of the samples submitted for comparison, except for (1) obvious changes that could be attributed to weathering, or (2) differences attributable to specific contamination.

*Inconclusive*: the chromatographic patterns of the spill sample are somewhat similar to that of the sample submitted for comparison, except for certain differences that are of such magnitude that it is impossible to ascertain whether the unknown is the same oil, heavily weathered, or a totally different oil.

*Nonmatch*: in the event of nonmatch, it is possible to stop the analytical procedures as soon as significant differences are confirmed because the sample is unlike the samples submitted for comparison. Otherwise, further levels of analyses will be performed to provide more specific, precise, and conclusive documentation.

## Two-Dimensional GC: An Emerging Technique for Fingerprinting Hydrocarbons

In recent years, a number of emerging instrumental techniques, such as GC-IRMS for isotopic composition of individual components in oil and petroleum products, ultrahigh-resolution Fourier transform ion cyclotron resonance mass spectrometry (FT-ICR MS), field desorption/ionisation FT-ICR MS and atmospheric pressure photoionisation pressure FT-ICR MS (Marshall and Rodgers 2004; Marshall et al. 2005), capillary GC with ICP-collision cell-MS detection (Bouyssiere et al. 2004), GC-field ionisation time-of-flight high-resolution MS for petroleum characterisation (Qian and Dechert 2002), and two-dimensional GC (GC × GC), have been applied for fingerprinting complex oil hydrocarbons, investigation of the low-concentration sulphur speciation in petroleum, and possibly ultimate characterisation of all of the chemical constituents of petroleum. Among these techniques, the GC × GC is the most studied and reported.

Since its invention in the 1990s, comprehensive two-dimensional gas chromatography (GC × GC) has continually evolved from an academic protocol to a fully integrated commercial system (Dimandja 2004; Dallüge, Beens, and Brinkman 2003). In this technique, two capillary GC columns are connected serially by a thermal modulator—the interface between the two separation dimensions. The thermal modulators can be further categorised into three types: heating, cryogenic, and jet-pulsed systems. Modulators periodically trap and then release smaller portions of a continuous stream of effluents. The first-dimension sample effluent is thus continuously transferred in smaller portions to the second-dimension column throughout the chromatographic run, and each transferred pulse generates a high-speed secondary gas chromatogram. Most often, the first-dimension separation uses a nonpolar phase to separate analytes by volatility difference, and the second dimension uses a more polar phase to separate first-dimension co-eluters by polarity difference. The resulting GC × GC chromatogram can be viewed in several formats, including surface, contour, and peak apex plots.

The key features of GC × GC are its greater resolving power and its ability to separate components into classes in samples. GC × GC chromatograms can have much higher peak capacities (exceeding 20,000) than conventional gas chromatograms (capacities rarely exceed 1,000). Another feature of GC × GC is that its separations can usually be done in a time comparable to the conventional GC because the significantly higher speed of the second dimensional GC allows increased peak capacity without increasing the length of the analysis. The separation of sample components into classes through structured chromatograms provides an additional means of identification and reduces the probability of peak overlap between numbers of different chemical classes (Dallüge et al. 2003).

This emerging technology has demonstrated a very promising perspective for the analyses of various complex organic mixtures such as food extracts, fatty acids, steroids, fly ash, sediments, and many others. In particular, the GC × GC technique has been successfully used to identify and quantify individual components and compound classes in crude oils and petroleum products. GC × GC chromatograms of crude oils and refined products showed distinct grouping of alkanes, cycloalkanes, alkylated PAHs, sulphur-containing PAHs, and hopane and sterane biomarkers (Phillips and Beens 1999; Frysinger and Gaines 2001; Dimandja 2004; Dallüge et al. 2003). Once chemical class locations and patterns are known, the two-dimensional chromatogram image can be rapidly inspected to detect variations in compound distribution and abundance. This procedure has been used to fingerprint an oil spill sample and match it to a potential source (Gaines et al., 1999), to characterise the chemical composition of a degraded no. 2 fuel (Reddy et al. 2002), and to investigate the chemical composition of the unresolved complex mixtures of hydrocarbons (UCM) in salt marsh sediments (Frysinger et al. 2003). Atomic emission detection (AED) and sulphur chemiluminescence detection (SCD) have been recently associated with GC × GC for analysis of petrochemical samples (van Stee et al. 2003; Hua et al. 2003). More importantly, with the introduction of the fast data-acquiring time-of-flight mass spectrometer (TOF-MS), the possibility of structure-related detection (i.e., identification of PAHs in petrochemical samples) has been created (Dallüge et al. 2003).

## Chemical Composition of Oil and Petroleum Products and Spill Identification

Generally, the chemical composition of fresh to mildly weathered oils and petroleum products can be readily revealed from their GC-FID traces, especially if the background hydrocarbon levels are low in the impacted environment. In addition, for measurements of TPH and other hydrocarbon groups in samples, GC-FID chromatograms provide a distribution pattern of petroleum hydrocarbons (e.g., boiling or carbon number range and profile of UCM), fingerprints of the major oil components (e.g., individual resolved *n*-alkanes and major isoprenoids), and information on the weathering extent of the spilled oil. Comparing biodegradation indicators (such as *n*-$C_{17}$/pristane and *n*-$C_{18}$/phytane) for the spilled oil with the source oil can also be used to monitor the effect of microbial degradation on the loss of hydrocarbons at the spill site. The GC-FID approach can be used to quickly screen the oil and refined product type (Figure 3.1). It is noted, however, that GC analyses alone may give limited oil diagnostic characteristics when the petroleum samples have been highly weathered. For defensible source identification, GC-MS analysis must be performed.

## Chemical Composition Features of Crude Oil

Crude oil compositions vary widely. Depending on the sources of carbon from which the oils are generated, the geologic environment in which they migrated, and from which reservoir (such as Middle East or North Sea), they may have

- dramatically varied compositions in the $C_5$–$C_{44}$ carbon range, such as relative amounts of paraffinic, aromatic, and asphaltenic compounds;
- large differences in the *n*-alkanes, isoprenoids, and cyclic-alkanes (such as alkyl cyclo-hexanes) concentrations and distribution patterns and UCM profiles;
- significantly different relative ratios of isoprenoids to normal alkanes; and
- large differences in distribution patterns and concentrations of oil-characteristic long side chain *n*-alkyl benzenes (the carbon number in the alkyl side chain can be up to $C_{27}$ for some oils), alkylated PAH homologues (many four- to six-ring unsubstituted PAHs are only minor components in oils), and biomarkers.

The most prominent aliphatics in most crude oil are the normal (straight chain) alkanes. In general, most crude oils exhibit an *n*-alkane distribution profile (GC-FID and GC/MS at *m/z* 85) of decreasing abundances with increasing carbon number. The maximum *n*-alkanes within the profile are variable from oil to oil. The smoothness of the *n*-alkane distribution profile in crude oil can be diagnostic. The carbon preference index (CPI) values of most oils are ~1. Oils with CPI values greater than 1 are often derived from source rock strata that contained relatively abundant land plant organic components including leaf waxes. CPI is defined as the total of *n*-alkanes

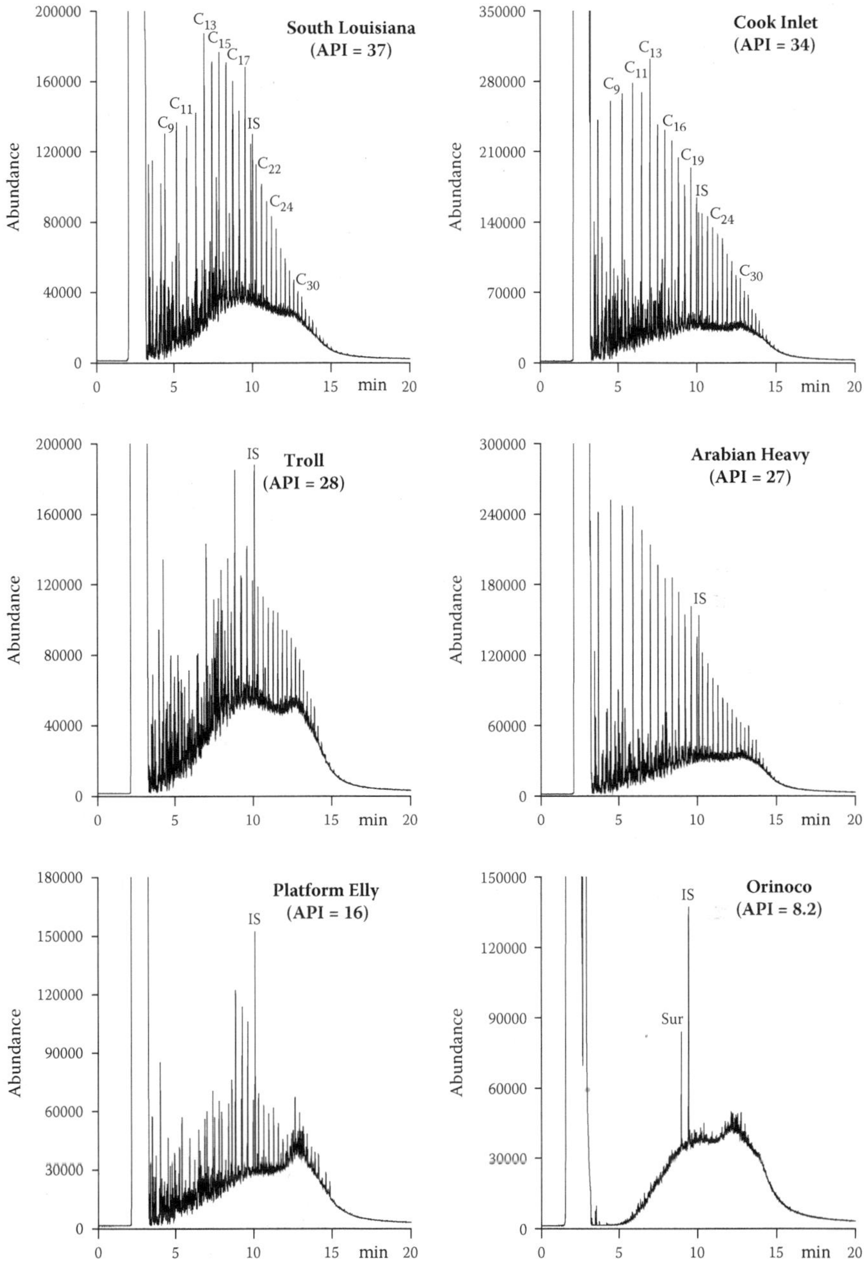


**Figure 3.1** GC-FID chromatograms of six oils. These six oils (South Louisiana, Cook Inlet, Troll, Arabian Heavy, Platform Elly, and Orinoco) are different as there are large differences in the distributions of resolved *n*-alkanes and isoprenoids and unresolved complex mixtures (UCMs) and also relative distribution patterns of UCMs. Note that the Orinoco sample has nearly no *n*-alkanes on its GC-FID chromatogram.

with odd carbon number divided by the total of $n$-alkanes with even carbon number in the carbon range of $C_8$–$C_{44}$:

$$\text{CPI} = (\text{the sum of odd } n\text{-alkanes})/(\text{the sum of even } n\text{-alkanes})$$

or, in the simplified formula,

$$\text{CPI} = (C_{23} + C_{25} + C_{27} + C_{29} + C_{31} + C_{33})/(C_{24} + C_{26} + C_{28} + C_{30} + C_{32} + C_{34})$$

The distributions of isoprenoids ($m/z$ 113) and alkyl ($C_0$- to $C_{15}$-) cyclohexane homologous series ($m/z$ 83) are also apparent in many crude oils. Biodegradation affects the straight-chain alkanes more than branched alkanes (isoprenoids). Therefore, determination and comparison of biodegradation indicators of $n$-$C_{17}$/pristane and $n$-$C_{18}$/phytane between the spilled oil and the source oil are often performed at this level to monitor the probable effects of microbial degradation at the spill site.

Figure 3.1 shows GC-FID chromatograms (by high-temperature program) for six different oils from different main production areas in the world. Table 3.4 summarises the hydrocarbon group analysis results for these six oils. They are different because there are not only large differences in the $n$-alkane distributions and UCM profiles, but also differences in hydrocarbon group composition and in relative ratios of isoprenoids to normal alkanes. Note that the Orinoco oil (a bitumen oil from Venezuela) has nearly no $n$-alkanes in its GC-FID chromatogram.

### *Background Hydrocarbons: Distinguishing Biogenic Hydrocarbons from Petrogenic Hydrocarbons in Oil-Contaminated Samples*

Differentiation of hydrocarbons from a range of sources is an essential part of any objective oil spill study. After oil spills, oil hydrocarbons often mix with other background hydrocarbons in the impacted area. One of the potential sources of hydrocarbons contributing to the background is biogenic hydrocarbons. Hydrocarbons from both anthropogenic and natural sources including biogenic sources are common in the marine and inland environments. Biogenic hydrocarbons are generated either by biological processes or in the early stages of diagenesis in recent marine sediments. Most soils and sediments contain some fraction of organic matter derived from biological sources including land plants, phytoplankton, animals, bacteria, macroalgae, and microalgae. It has been recognised (Cretney et al. 1987; Venkatesan 1988; Kolattukudy 1976; Page et al. 1995; Bence and Burns 1995) that the biogenic hydrocarbons have the following chemical composition characteristics:

**Table 3.4 Hydrocarbon Group Analysis Results for Six Example Crude Oils[a]**

| Hydrocarbon Groups | South Louisiana | Cook Inlet | Troll | Arabian Heavy | Platform Elly | Orinoco |
|---|---|---|---|---|---|---|
| Saturates (%) | 79.4 | 66.7 | 66.9 | 60.1 | 34.6 | 44.6 |
| Aromatics (%) | 16.9 | 25.2 | 26.6 | 24.6 | 32.4 | 27.3 |
| Resins (%) | 3.4 | 5.1 | 5.8 | 6.3 | 19.4 | 13.3 |
| Asphaltenes (%) | 0.4 | 3.1 | 0.7 | 9.0 | 13.6 | 14.8 |
| Resolved peaks/GC-TPHs (%) | 20.8 | 24.8 | 15.6 | 22.4 | 13.0 | 3.0 |
| GC-UCM/GC-TPH (%) | 79.2 | 75.2 | 84.4 | 77.6 | 87.0 | 97.0 |
| Total *n*-alkanes ($mg.g^{-1}$ oil) | 73.8 | 79.0 | 35.6 | 73.1 | 21.5 | — |
| $n$-$C_{17}$/pristane | 1.26 | 1.52 | 0.60 | 4.34 | 0.40 | — |
| $n$-$C_{18}$/phytane | 2.00 | 2.64 | 1.05 | 2.53 | 0.42 | — |
| Pristane/phytane | 1.95 | 2.22 | 1.96 | 0.65 | 1.05 | — |
| Total BTEX and $C_3$-benzenes[b] ($\mu g.g^{-1}$ oil) | 25.1 | 18.1 | 18.3 | 12.8 | 5.10 | 250 |
| Five alkylated PAH homologues ($\mu g.g^{-1}$ oil) | 12,844 | 12,088 | 16,670 | 8498 | 4226 | 3672 |
| Other U.S. EPA priority PAHs ($\mu g.g^{-1}$ oil) | 223 | 172 | 434 | 54.4 | 68.1 | 55.0 |

[a] GC-FID chromatograms shown in Figure 3.1.

[b] BTEX: benzene, toluene, ethyl-benzene and xylenes; $C_3$-benzenes include eight isomers.

Biogenic $n$-alkanes show a distribution pattern of odd carbon-numbered alkanes being much more abundant than even carbon-numbered alkanes in the range of $n$-$C_{21}$–$n$-$C_{33}$, resulting in unusually high CPI values. The presence of modern leaf plants can result in strong odd-carbon $n$-alkane dominance (CPI > 2).

There is a notable absence of the 'unresolved complex mixture (UCM)' hump in the chromatograms.

Pristane is often more abundant than phytane, suggesting a phytoplankton input and resulting in abnormally high pristane/phytane ratio values.

A 'biogenic cluster' (identified as olefinic hydrocarbons of biogenic origin) is present in the aromatic fractions.

Biogenic PAH perylene, an unsubstituted PAH produced in subtidal sediments, is widely distributed during early diagenesis.

Plant terpenoid biomarker compounds are present on occasion.

In some environmental forensic investigations, the CPI values were used to identify the contribution of modern hydrocarbons derived from modern plant leaf debris in soil and sediments. The presence of modern plant leaf waxes can impart a strong odd-carbon dominance (CPI > 2) that is unrelated to the petroleum contamination (Stout, Uhler, and McCarthy 2000). In the study of hydrocarbon biogeochemical changes of the Baffin Island oil spill (BIOS) experimental site, Cretney et al. (1987) found that the BIOS subtidal samples had high pristine/phytane ratios (5–15) and CPI values (3–11). High concentrations of pristane relative to phytane in most of beach and subtidal sediments indicate a biological hydrocarbon input from a marine source. In addition, the GC chromatograms of the aromatic fractions were typified by the olefinic hydrocarbon clusters. This cluster is a common feature of coastal marine subtidal sediments and is believed to be of marine planktonic or bacterial origin.

The possibility of in situ genesis of PAHs is indicated by the presence of perylene as a major PAH in almost all the beach and subtidal sediments. However, it should be noted that it cannot be used alone as a definitive source identification criterion because perylene is also produced in combustion processes. During the Nipisi spill study (Wang et al. 1998a), it was found that the background samples showed typical biogenic $n$-alkane distribution in the range of $C_{21}$–$C_{33}$; abundances of odd-carbon-number $n$-alkanes were much higher than those of even-carbon-number $n$-alkanes. The biogenic cluster was also obvious and no UCM was observed. No petrogenic hydrocarbons—in particular, no alkylated PAH homologues and petroleum-characteristic biomarker compounds such as pentacyclic hopanes and $C_{27}$–$C_{29}$ steranes—were detected. In addition, three plant terpenoid biomarker compounds eluted in the retention time window of 42–45 min with remarkable abundances. These were identified as 12-oleanene, 12-ursene, and 3-friedelene (all $C_{30}H_{50}$, MW = 410.7).

## General Chemical Composition Features of Refined Products

Refined petroleum products are obtained from crude oil through a variety of refining processes (Olah and Molnar 1995; Speight 2002), such as distillation, cracking, catalytic reforming, isomerisation, alkylation, and blending. Depending on the chemical composition of their 'parent' crude oil feedstocks, varying refining approaches and conditions, a wide range of applications, and regulatory and economic requirements, refined products can have a wide variety of chemical compositions. However, they can be still categorised in the following broad classes based on their general chemical composition features.

### *Light Distillates*

Light distillates are typically products in the $C_3$–$C_{13}$ carbon range. They include aviation gas (gasoline-type jet fuel, which has a wider boiling range than kerosene-type jet fuel and includes some gasoline fractions), naphtha (a liquid petroleum product that boils from about 30°C to approximately 200°C), and automotive gasoline. The GC traces of fresh light distillates are featured with a dominance of light-end, resolved hydrocarbons and a minimal UCM.

Gasoline is the generic term used to describe volatile, inflammable petroleum fuels used primarily for internal combustion engines. It is a complex mixture of hundreds of different hydrocarbons predominantly in the $C_4$–$C_{13}$ range, with the nominal boiling point range of 40–180°C or, at most, below 200°C. The composition of gasoline is best expressed in five major hydrocarbon classes: paraffins, isoparaffins (branched alkanes), aromatics, naphthenes (cyclo-alkanes), and olefins (PIANO). The bulk PIANO composition provides a useful cumulative parameter for fuel type (such as gasoline, aviation gasoline, or jet fuel) differentiation. Gasoline contains considerable BTEX and alkylated benzene compounds.

The properties of gasoline are quite diverse and the principal properties affecting the performance of gasoline are volatility and combustion characteristics. In order to improve some specific properties, such as the engine efficiency and antiknock properties, certain chemical compounds (*additives*) are often added to gasolines. They may include octane-boosting additives (such as methyl tertiary butyl ether, MTBE), oxidation inhibitors (such as aromatic amines and hindered phenols), corrosion inhibitors (such as carboxylic acids and carboxylates), anti-icing additives (such as alcohols, glycols, and surfactants), and antiknocking lead alkyls and dyes (oil-soluble solid and liquid dyes: red—alkyl derivatives of azobenzene-4-azo-2-naphthol; orange—benzene-azo-naphthol; yellow—para-diethylaminoazobenzene, and blue—1,4-

diisopropyl-aminoanthraquinone) for identification of different gasolines. Lead is a harmful pollutant and human exposure to all sources should be minimised. Phasing lead out of gasoline makes a very important contribution. Until approximately 1970, almost all gasoline used around the world contained lead—in many cases, at concentrations above 0.4 g per liter. Since the 1970s, the lead level in refined products in Canada, the United States, and many European countries has decreased substantially. Use of leaded gasoline in cars was completely banned in Canada, the United States, Germany, Denmark, and Sweden in 1993, 1996, 1996, 1995, and 1995, respectively.

### *Midrange Distillates*

Midrange distillates are typically products in a relatively broader carbon range ($C_6$–$C_{26}$) and include kerosene, aviation jet (turbine) fuels, and diesel products. Jet fuel is kerosene-based aviation fuel; it is used for aviation turbine power units and usually has the same distillation characteristics and flash point as kerosene. Jet fuels are manufactured predominantly from straight-run kerosene or kerosene–naphtha blends and are similar in gross composition; many of the differences in them are attributable to additives designed to control some fuel parameters, such as freeze and pour point characteristics. As Figure 3.2 shows, the chromatogram of a commercial jet A fuel is dominated by GC-resolved *n*-alkanes in a narrow range of $n$-$C_7$–$n$-$C_{18}$, with the maximum about $n$-$C_{11}$. The UCM is well defined.

Diesel fuels originally were straight-run products obtained from the distillation of crude oil. Currently, diesel fuel may also contain varying amounts of selected cracked distillates to increase the available volume. The boiling range of diesel fuel is approximately 125–380°C. One of the most widely used specifications (ASTM D-975) covers three grades of diesel fuel oils: diesel fuel no. 1, diesel fuel no. 2, and diesel fuel no. 4. Grades no. 1 and no. 2 are distillate fuels most commonly used in high-speed engines of the mobile type, in medium-speed stationary engines, and in railroad engines. Grade no. 4 diesel covers the class of more viscous distillates and, at times, blends of these distillates with residual fuel oils. The marine fuel specifications have four categories of distillate fuels and fifteen categories of fuels containing residual components (ASTM D-2069 method). Diesel consists of hydrocarbons in a carbon range of $C_8$–$C_{28}$ and has significantly high concentrations of *n*-alkanes, alkyl-cyclohexane, and PAHs. The properties of a given diesel are largely a function of the crude oil feedstock. The GC chromatogram of diesel fuel no. 2 is generally dominated by a nearly normal distribution of *n*-alkanes with maxima about $n$-$C_{11}$ to $n$-$C_{14}$. Also, a central UCM 'hump' is obvious (Figure 3.2). Once released to the water surface, midrange fuels spread very rapidly. Very large and thin films will often form, leading to quite rapid weathering of spilled fuel.

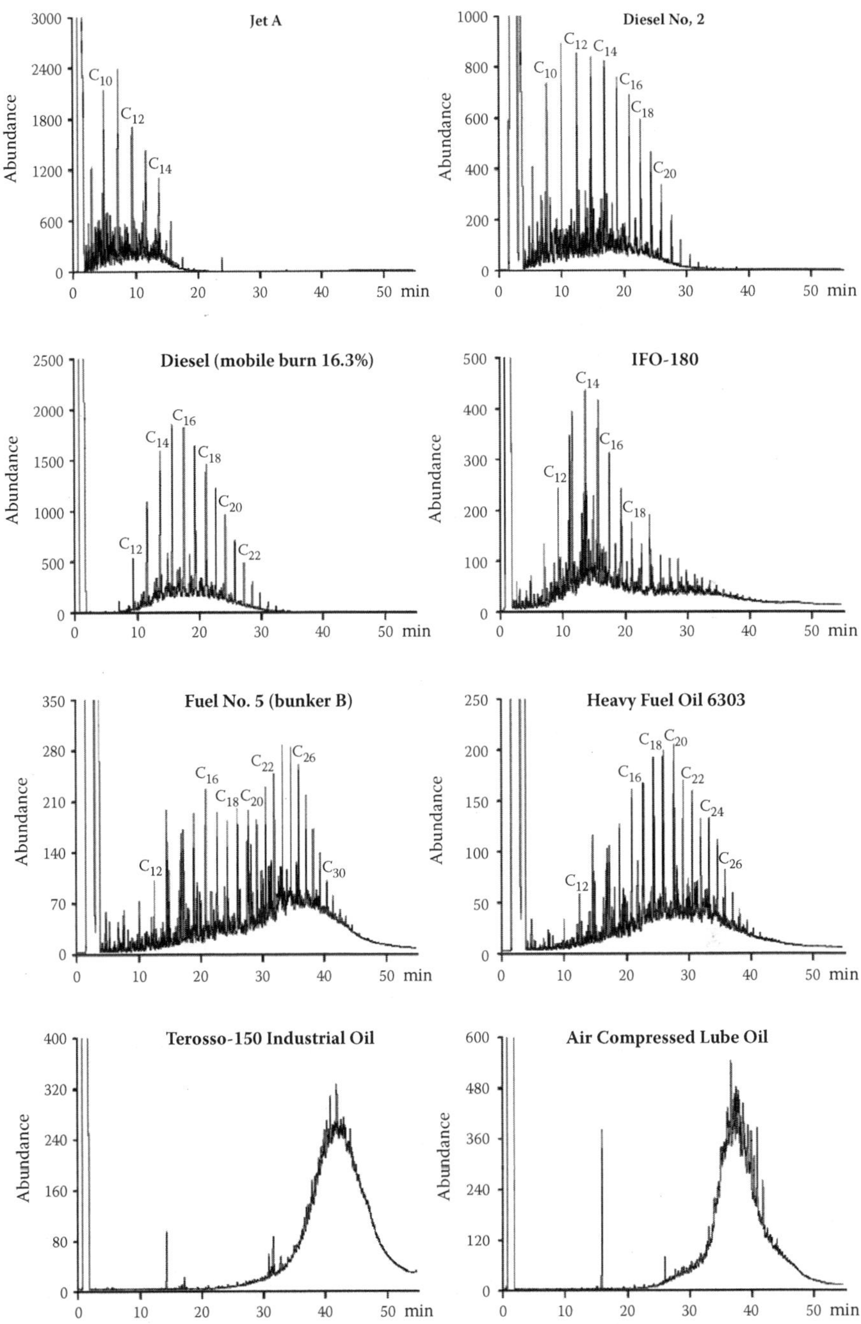


**Figure 3.2** GC-FID chromatograms of eight petroleum products (jet fuel, diesel, weathered diesel, IFO-180, fuel no. 5 (bunker B), heavy fuel oil, and two lube oils), illustrating differences of these products in the chromatographic profiles, carbon range, and UCM distribution patterns.

### *Classic Heavy Residual Fuel*

Heavy fuel oils (HFOs) are blended products manufactured from residues of various refinery distillation and cracking processes and are largely used in marine applications and industrial power generation. Classic heavy fuel types include fuel no. 5 and no. 6 (also known as bunker C). For many years the term 'bunker C fuel oil' has been widely used to designate the most viscous residual fuels for general land and marine use. Different grades of heavy fuel oils are expressed by the numbers of their kinetic viscosity in centistokes (cSt) at 50°C. The main grades are IFO 30, IFO 180, and IFO380. The chemical composition of bunker C (or IFO 380) can vary widely and remarkably, depending on production oilfields, production years, and processes that it has undergone. Currently, many bunker-type fuels are produced by blending residual oils with diesel fuels or other lighter fuels in various ratios to produce residual fuel oil of acceptable viscosity for marine or power plant use. The use of heavy fuel oils as bunker oil on ships has been found to be the main course of chronic oil pollution because of illegal discharge of residues and residual oil into the sea.

For comparison, the chromatograms of an IFO 180, a lighter residual fuel no. 5 (also called bunker B) and a heavy fuel oil 6303 (called bunker C or land bunker, from Imperial Oil Ltd., Nova Scotia, Canada) are also shown in Figure 3.2. The differences in the chromatographic profiles, carbon range, the shapes of UCM, distribution of *n*-alkanes and major isoprenoids, and diagnostic ratios of target alkanes (such as $n$-$C_{17}$/pristane and $n$-$C_{18}$/phytane) among these products are obviously considerable.

### *Lubricating Oil*

Petroleum-derived lubricating oil is a mixture produced by atmospheric and vacuum distillation of selected paraffinic and naphthenic crude oils. Solvent refining and/or hydrogen treatment is used to remove the nonhydrocarbon constituents and to increase the viscosity index, enhance the colour, and convert undesirable chemical structures (such as unsaturated hydrocarbons and aromatics) to less chemically reactive species. Solvent dewaxing is then used to remove the wax constituents and to improve the low-temperature properties. Finally, clay or hydrogen treatment is performed to prevent instability of the product. Lubricating oils may be divided into many categories according to the types of services and applications, such as motor oil, transmission oil, crankcase oil, hydraulic fluid, cutting oil, turbine oil, heat-transfer oil, electrical oil, and many others. However, there are two main groups: (1) oils used in intermittent service, such as motor and aviation oils, and (2) oils designed for continuous service, such as turbine oils. Chemical additives are often added to base oil to enhance the properties and to improve such characteristics as oxidation resistance and corrosion resistance of lubricating oil.

Small-scale lubricating oil spills and contaminations are quite common due to their wide application.

Figure 3.2 also includes the high-temperature (from 40 to 325°C) GC-FID chromatograms for two different lubricating oils. In general, lubricating oils have broad GC profiles in the carbon range of $C_{18}$–$C_{40}$ with boiling points greater than 340°C. Lubricating oil does not contain a lower boiling portion of petroleum hydrocarbons: They are largely composed of saturated hydrocarbons and their GC trace is often dominated by the UCM of hydrocarbons, with a very small amount of resolved peaks present. In lubricating oil such as hydraulic fluid, for example, the PAH concentrations can be very low, while the concentration of multiple condensed-ring biomarker compounds could be very high. Therefore, determination of these source-specific marker compounds often allows for successful identification and correlation between refined products from different sources.

### *Waste Oil*

Illegal discharges of oil from the machinery rooms of ships (e.g., bilge oil and sludge) have been found to be one of the major sources of oil pollution in areas of intensive shipping traffic (Dahlmann 2003). The bilge oils often consist of a mixture of light fuel oil, bunker oil, and waste lubricating oil. Bilge oil spills often involve different amounts of different products, which make identification of the spill sources more difficult. Bilge oils can have great variability in the final composition and therefore they can have significantly different GC-FID chromatograms. The final composition of spilled bilge oil is determined not only by the condition of the ship and ship's engine but also by the history of this type of oil on board (such as temperature, amount of water, and evaporation of light fuel portion). A mixture of light fuel oil and lubricating oil can be relatively more easily identified and distinguished because these two products have different carbon numbers and boiling ranges. A spill case study is presented later to illustrate how the source of an unknown waste oil spill is identified using the multicriterion approach.

## PAH Fingerprints of Oils and Petroleum Products

### *Distribution of Alkylated PAH Homologues and Other EPA Priority PAHs*

Crude oils and refined products from different sources can have very different PAH distributions. Also, many PAH compounds are more resistant to weathering than their saturated hydrocarbon counterparts (*n*-alkanes and isoprenoids) and volatile alkyl-benzene compounds, thus making PAHs one of the most valuable fingerprinting classes of hydrocarbons for oil identification. Examples of the PAH distribution of some oils and petroleum products (i.e., petrogenic PAHs) are illustrated in Figure 3.3. The oil products differ significantly in both the PAH concentrations and distribution patterns from the crude oils and from each other. Typically, in unweathered crude oils the alkylated naphthalenes and alkylated chrysenes are the most and least abundant PAHs among the five target alkylated PAH homologues, while many of four- to six-ring unsubstituted PAHs are only minor components or even absent in oils. The PAHs in each alkylated PAH homologous series, in general, exhibit distribution patterns where the $C_1$-, $C_2$-, and $C_3$-PAHs are more abundant than the parent ($C_0$-) and $C_4$-PAHs. This kind of characteristic PAH distribution profile has been termed as 'bell shaped'. By weathering or degradation, the bell-shaped distribution can be readily modified to the distribution profile of $C_0$- < $C_1$- < $C_2$- < $C_3$- (called *inverse sloped*) in most alkylated PAH homologous families.

As Figure 3.3 shows, jet A fuel has extremely high content of the naphthalene series (99%) among the five target alkylated PAH homologues, with the other four alkylated PAH series containing only 1% in total. In addition, no four- to six-ring PAHs were detected of the other 15 EPA priority PAHs. Diesel no. 2 has a high naphthalene content (86%), a low phenanthrene content (5%), and no chrysenes. In the no. 5 fuel and HFO 6303, the unusually high contents of the alkylated naphthalene and chrysene series are very pronounced. In the Orimulsion 400, the concentrations of the alkyl phenanthrenes and dibenzothiophenes are very high, accounting for approximately 38 and 22% of the total PAHs, respectively. In addition, the profile of each alkylated PAH family shows the distribution of $C_0 < C_1 < C_2 < C_3$ similar to the severely weathered oil, indicating that this oil was highly degraded during its geological formation.

### *Recommended Diagnostic Ratios of PAHs*

A number of diagnostic ratios of target alkylated PAH species have been developed and successfully used for source identification and differentiation, distinguishing inputs of pyrogenic hydrocarbons from petrogenic hydrocarbons and weathering indicators. These are briefly summarised in Table 3.5. Basic criteria that must be applied in selection of diagnostic ratios include:

variability (i.e., ability to discriminate between samples);
analytical precision of selected ratios; and
resistance to weathering.

A benefit of comparing diagnostic ratios of spilled oil and suspected source oils is that any concentration effects are minimised. In addition, the use of diagnostic ratios to correlate and differentiate oils tends to induce a self-normalising effect on the data since variations due to instrument operating conditions, operators, or matrix effects are minimised. Douglas et al.

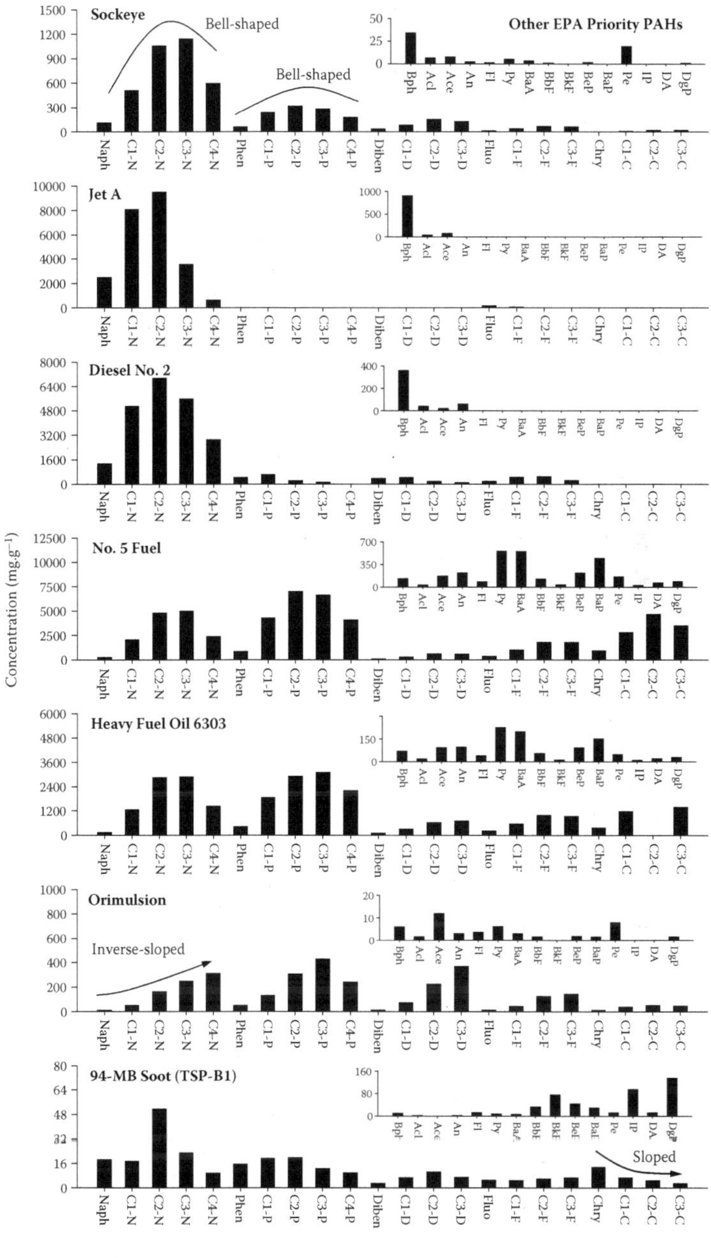


**Figure 3.3** Alkylated homologous PAH and other EPA priority PAH distributions for the Sockeye oil, jet A, diesel no. 2, no. 5 fuel, HFO 6303, Orimulsion, and 1994 Mobile burn soot sample, illustrating differences in PAH distribution features between different oils and oil products. Note that, for clarity, different scales are used for the Y-axis. Also, the PAH fingerprints and distinguishing features between petrogenic and pyrogenic PAHs are illustrated.

**Table 3.5 Diagnostic Ratios of PAHs Used in Oil Spill Fingerprinting Analysis**

| Diagnostic ratios | Ions monitored ($m/z$) |
|---|---|
| Double ratios | |
| $C_2D/C_2P$ vs. $C_3D/C_3P$ | 212, 206, 226, 220 |
| $C_3D/C_3P$ vs. $C_3D/C_3C$ | 226, 220, 270 |
| Pyrogenic index | Ions for target PAHs |
| $C_0C$:$C_1C$:$C_2C$:$C_3C$ | 228, 242, 256, 270 |
| Reten/C4-phen (reten:7-iospropyl-methyl-phen) | 270 |
| Ratios between alkylated PAH series | |
| Σphens/Σdibenzs, phen/Σphens | 128, 142, 156, 166, 170, 184 |
| Σnaphs/Σchrys, Σphens/Σchrys | 178, 192, 206, 220, 234 |
| Σdibenzs/Σchrys, Σfluos/Σchrys | 184, 198, 212, 226; 166, 180, 194, 208; 228, 242, 256, 270 |
| Ratios of isomer PAHs | |
| Methyl-dibenzothiophenes (4-:2-/3-:1:-m-DBT) | 198 |
| Methyl-phenanthrenes (3-+2-m-P)/(4-/9-m-+1-m-P) | 192 |
| 2-m-N/1-m-N and 2-m-N/((2-m-N+1-m-N) | 142 |
| An/Phen and An/(An+Phen) | 178 |
| Fluoranthene/Pyrene (Fl/Py) and FI/(FI+PY) | 202 |
| BaA/Chry and BaA/(BaA+Chry) | 228 |
| BeP/BaP and BeP/(BeP+BaP) | 252 |
| Indeno[1,2,3-cd]pyrene/benzo[ghi]perylene (IP/BP) and IP/(IP+BP) | 276 |
| Isomers in $C_3$-naphs and $C_4$-naphs | 156, 170 |
| Isomers in $C_2$-phens and $C_4$-phens | 206, 234 |
| Isomers in $C_1$-fluorenes | 180 |
| Cluster PAHs at $m/z$ 216 (six cluster PAHs are 2-m-fluoranthene, benzo(a)fluorene, benzo(b)fluorene, 1-m-pyrene, 4-m-pyrene, and 1-m-pyrene) | 216 |

(1996) determined double ratios ($C_2D/C_2P$ vs. $C_3D/C_3P$, the ratios of alkylated dibenzothiophenes to alkylated phenanthrenes) for more than 20 oils and refined products. They found that these ratios are very different among the studied oils and petroleum products from light jet fuel to heavy bunker C fuel. A method using the double ratio plots for identification and differentiation of petroleum product sources has been developed.

Because these ratios remain relatively stable over a wide weathering range (i.e., these PAH groups tend to weather at comparable rates), they were extensively used in the studies of the 1989 *Exxon Valdez* oil spill to distinguish Alaska North Slope (ANS) crude, its weathering products, and diesel refined from ANS feed stock from other petrogenic hydrocarbons including the sulphur-depleted Tertiary oil seeps in the region (Page et al. 1995; Bence, Kvenvolden, and Kennicutt 1996; Boehm et al. 1998, 2001). Douglas et al. (1996) also defined the $C_3D/C_3P$ and $C_3D/C_3C$ (the ratios of alkylated dibenzothiophenes to alkylated chrysenes) as 'source ratios' (the ratios that are almost constant because the compounds degraded at the same rate) and 'weathering ratios' (the ratios that change substantially with weathering and biodegradation), respectively. They were applied to describe oil depletion and to identify sources in subtidal sediment data from the *Exxon Valdez* spill and a North Sea oil spill. Hostettler, Rosenbauer, and Kvenvolden (1999) reported a method using the PAH refractory index ratio of two of the most refractory constituents of most oils (triaromatic steranes and methylchrysenes), as a source discriminant of hydrocarbon input for differentiation of three different oils (*Exxon Valdez* oil, Katalla oil, and PWS sediment hydrocarbons). Studies of characterisation of spilled oil residues and identification of unknown spill samples (Wang, Fingas, et al. 1997; Wang, Fingas, and Sergy 1994; Wang, Fingas, et al. 1998; Wang, Fingas, and Sigouin 2002) utilised a number of diagnostic ratios of selected source-specific alkylated PAHs in combination with determination of ratios of selected paired biomarkers for source identification and differentiation, determination of weathering extent and degree of surface and subsurface samples, and distinguishing between composition changes due to physical weathering and biodegradation.

### *PAH Isomer and Cluster PAH Analysis*

The use of the sum of the alkylated PAHs as multicomponent analytes in deriving diagnostic ratios for oil spill studies has made considerable advances, as described earlier. In recent years, research has been further expanded to use individual source-specific isomers within the same alkylation level and to determine the relative isomer-to-isomer distribution for oil spill source identification. As the alkylation levels increase, more isomers are detected (e.g., the $C_3$-dibenzothiophenes, as a group, contain more than 20 individual isomers with different relative abundances). The differences between the isomer distributions reflect the differences of the depositional environment during oil formation. Compared to PAH homologous groups at different alkylation levels, higher analytical accuracy and precision may be achieved due to the close match of physical and chemical properties of the isomers. Also, the relative distribution of isomers is subject to little interference from weathering in short-term or lightly weathered oils; hence, this approach can be positively used for oil spill identification.

On the other hand, it has been demonstrated that the position of the alkylation on the PAHs can influence the biodegradation rate of the isomers within an isomer group. This information can be used to sort out environmental factors such as the impact of biodegradation on the PAH distribution and to differentiate oil compositional changes due to physical weathering from those due to biodegradation. For example, the ratios among methyl dibenzothiophenes, methyl-phenanthrenes, and methyl and dimethyl naphthalenes have been studied and widely used for environmental forensic investigations.

### *Methyl Phenanthrenes*

All oils contain four methyl-phenanthrenes (in two pairs of doublet peaks: 3- and 2-, and 4-/9- and 1-m-P). Ratios among four methyl phenanthrene isomers have been shown to be related to the thermal history of crude oils and its source strata, and numerous methyl phenanthrene indices have been defined for monitoring the thermal maturities of oils (Radke, Welte, and Willsch 1986) and for spill oil source identification (Wang, Fingas, and Page 1999):

MPI 1 = 1.5(2-m-P + 3-m-P)/(P + 1-m-P + 9-m-P)
MPI 2 = 3(2-m-P)/(P + 1-m-P + 9-m-P)
2-m-P/1-m-P
(3- + 2-m-P)/(4-/9- + 1-m-P)

The 2-methyl phenanthrene was found to be more sensitive to biodegradation than the 1-methyl phenanthrene (Wang et al., 1998); therefore, it can be used as the indicator for biodegradation. It has also been reported (CEN 2002) that, in many crude oils, the first doublet peak is smaller than the second doublet peak, and the methyl-anthracene (the peak between the two doublet peaks) is often very small or insignificant. For many bunker fuels, the first doublet peak is higher than the second one and the methyl-anthracene is often more pronounced. However, it should be noted that bunker fuels can vary widely and remarkably in physical and chemical properties, depending on types of bunker fuels, blending ratios of residual oils to diesel fuels or other lighter fuels, and production processes used. Therefore, this parameter must be used very cautiously, as it may not always be valid.

### *Methyl Dibenzothiophenes*

Chromatographically well-resolved $C_1$-dibenzothiophene isomers (Fayad and Overton 1995; Wang and Fingas 1995b) are present in all oils at relatively high concentrations. Their relative abundance distributions vary significantly from different sources, which can be assessed by the following index:

$C_1$-dibenzothiophene distribution index = (4-:2-/3-:1-m-DBT)

A database of the relative ratios of the $C_1$-DBT isomers for several hundred crude, weathered, and biodegraded oils and petroleum products has been established and plots of 2-/3-methyldibenzothiophene versus 1-methyldibenzothiophene (both isomers are normalised relative to 4-methyldibenzothiophene) for these oils and oil products have been depicted. The plots show that the data points representing the various oils are very scattered. Another pronounced feature observed from the figure is that related oils produce tight clusters on the plot. The use of these ratios complements existing methods of oil characterisation, but it has its own distinct advantages for discrimination of different oils.

### *Other Relative Ratios of PAH Isomers*

Other selected PAH isomers (Boehm et al. 1997; Wang et al., 1998; Wang, Fingas, and Page 1999; Wang, Fingas, Shu, et al. 1999; Stout et al. 2002) used for oil fingerprinting studies include the ratio of retene (1-methyl-7-(1-methylethyl)-phenanthrene) to the total of C4-phenanthrenes; three isomers, each within $C_3$-naphthalenes (*m/z* 156) and $C_4$-naphthalenes (*m/z* 170); four isomers within $C_2$-phenanthrenes (*m/z* 206); two isomers within $C_4$-phenanthrenes (*m/z* 234); three isomers within $C_1$-fluorenes (*m/z* 180); 2-m-naphthalene/1-m-naphthalene (*m/z* 128); anthracene/phenanthrene (*m/z* 178); BaA/Chrysene (*m/z* 228); BeP/BaP (*m/z* 252); and indeno[1,2,3-cd]pyrene/benzo[ghi]perylene (*m/z* 276). Depending on the individual spill case and its degree of weathering, different diagnostic parameters may be selected and applied.

### *Cluster PAHs at* m/z *216*

As high-boiling biomarkers are rarely present in lighter fuels, it becomes increasingly difficult to compare two lighter fuel samples based on lower boiling compounds for source identification, especially for weathered fuel oils. It has been found (Dahlmann 2003) that the cluster PAH compounds at *m/z* 216 are relatively stable and can be used for comparing lighter fuel oil samples. Actually, not all compounds of this cluster are isomers. This cluster mainly represents six PAH compounds from different compound classes of aromatic hydrocarbons. It has, therefore, some advantages for discriminating between oils over an isomer cluster within a single compound class. These six cluster compounds have been identified to be 2-m-fluoranthene, benzo(a)-fluorene, benzo(b)-fluorene, 2-m-pyrene, 4-m-pyrene, and 1-m-pyrene. By normalising the peak abundances relative to 4-m-pyrene, which is often the most abundant in the cluster, a set of diagnostic ratios can be readily determined. This cluster ratio has been used for comparing three round robin spill fuel samples collected from a harbour spill in the Netherlands in 2004 (Wang et al. 2005).

### *Distinguishing Pyrogenic Hydrocarbons from Petrogenic Hydrocarbons*

PAH distributions are the most useful tool in distinguishing pyrogenic hydrocarbons from petrogenic hydrocarbons. The differences in PAH distribution between petrogenic and pyrogenic PAH sources were first recognised in modern sediment studies and then expanded to the environmental forensic interpretation of petrogenic, pyrogenic, and biogenic PAHs. As discussed before, petrogenic PAHs in most oils exhibit *bell-shaped* distribution patterns. In contrast, pyrogenic materials generally exhibit alkylated PAH homologue distribution patterns in which the parent PAHs are often the most abundant. The composition features of pyrogenic PAHs can be summarised as follows:

The dominance of the unsubstituted compounds over their corresponding alkylated homologues and this kind of PAH distribution profile of $C_0 \gg C_1 > C_2 > C_3 > C_4$ has been generically termed as *skewed* or *sloped* (Sauer and Uhler 1995).

The high molecular mass four- to six-ring PAHs dominate the low molecular mass two- to three-ring PAHs.

On the gross level PAH can comprise a much higher mass percentage in most pyrogenic source materials than in most petrogenic source materials.

As an example, Figure 3.3 also presents PAH fingerprints for the 1994 Mobile diesel burn soot sample, illustrating the distinguishing features of pyrogenic PAH distribution from the petrogenic PAH distribution.

Numerous quantitative diagnostic ratios have been defined to differentiate pyrogenic PAHs from other hydrocarbon sources (Blumer and Youngblood 1975; Bjøeseth 1985; Benlahcen et al. 1997; Sicre et al. 1987), including phenanthrene/anthracene (Ph/An), phenanthrene/methyl-phenanthrenes ($Ph/C_1$-Ph) and (Ph + An)/(Ph + An + $C_1$-Ph), fluoranthene/pyrene (Fl/Py) and Fl/(Fl + Py), benz[a]anthracene/chrysene (BaA/Ch), and BeP/(BeP + BaP). Wang et al. (1999) proposed a new pyrogenic index (PI) as a quantitative indicator for identification of pyrogenic PAHs and for differentiation of pyrogenic and petrogenic PAHs (Wang, Fingas, Shu, et al. 1999). The PI is defined as the ratios of the total of the other EPA priority unsubstituted three- to six-ring PAHs to the total of five target alkylated PAH homologues:

$$PI = \Sigma(\text{other three- to six-ring EPA PAHs})/\Sigma(\text{5 alkylated PAHs})$$

Compared to other diagnostic ratios obtained from individual compounds, this index ratio has its own distinct advantages:

Petrogenic and pyrogenic PAHs are characterised by dominance of five alkylated PAH homologous series and by dominance of unsubstituted high molecular weight PAHs respectively; therefore, determination of the changes in this ratio more truly reflects the difference in the PAH distribution between these two sets of hydrocarbons.

This ratio can offer better accuracy with less uncertainty than those relative ratios determined from individual PAH compounds.

This ratio shows great consistency from sample to sample and is subject to little interference from the concentration fluctuation of individual components within the PAH series.

Long-term natural weathering and biodegradation only slightly alter the values of this ratio, but the ratio will be dramatically altered by combustion.

Therefore, this index ratio can be used as a general and effective criterion to unambiguously differentiate pyrogenic PAHs and petrogenic PAHs.

The pyrogenic index values versus relative ratios of Ph/An for more than 100 oils and refined products have been determined. It is found that lighter refined products and most crude oils show PI ratios falling into a range of 0–0.01, while heavy oils and heavy fuels show significantly higher ratios in the range of 0.01–0.05. The ratio dramatically increases up to much higher values for pyrogenic materials (e.g., it increased to a range of 0.8–2.0 for the 1994 Mobile burn soot samples). The usefulness of the pyrogenic index in environmental forensic investigations for input of pyrogenic PAHs and spill source identification has been clearly demonstrated in several recent spill case studies (Meniconi et al. 2002; Tolosa et al. 2004; Wang, Fingas, and Lambert 2004).

## Biomarker Fingerprints of Oils and Petroleum Products

Biomarkers are useful in oil spill identification because they retain all or most of the original carbon skeleton of the original natural product and this structural similarity reveals more information about their origins and thermal history than other compounds (Peters and Moldowan 1993; Philp 1985). In comparison to *n*-alkanes and acyclic isoprenoids, many biomarkers are resistant to secondary processes, such as biodegradation. Therefore, chemical analysis of source-characteristic and environmentally persistent biomarkers generates information of great importance in determining the source of spilled oil, differentiating oils, and monitoring the degradation process and weathering state of oils under a wide variety of conditions. In the past decade, use of biomarker fingerprinting techniques to study spilled oils has greatly increased and biomarker parameters have been playing a prominent role in almost all oil spill work.

### *Distributions and Quantification of Biomarkers*

The cyclic terpane biomarkers in petroleum include sesqui- ($C_{15}$), di- ($C_{20}$), sester- ($C_{25}$), and triterpanes ($C_{30}$). The steranes are a class of biomarkers containing 21–30 carbons that are derived from sterols and include regular steranes, rearranged diasteranes and mono- and tri-aromatic steranes. Among them, the regular $C_{27}$-$C_{28}$-$C_{29}$ homologous sterane series are the most common (due to their abundance in the source organic matter) and useful steranes because they are highly specific for correlation (Peters and Moldowan 1993). Characterisation of these compounds is achieved by using GC-MS in the selected ion monitoring mode by the internal standard method: *m/z* 191 for tricyclic, tetracyclic, and pentacyclic terpanes; *m/z* 123 for bicyclic sesquiterpanes; *m/z* 217 and 218 for steranes; *m/z* 217 and 259 for diasteranes; *m/z* 253 for mono-aromatic steranes; and *m/z* 231 for tri-aromatic steranes.

Many oils show different composition and distribution patterns of biomarkers. The GC-MS chromatograms of terpanes (*m/z* 191) are often characterised by the terpane distribution in a wide range from $C_{20}$ to $C_{30}$, often with $C_{23}$ and $C_{24}$ tricyclic terpanes and $C_{29}$ αβ- and $C_{30}$ αβ-pentacyclic hopanes being prominent. As for steranes (at *m/z* 217 and 218), the dominance of $C_{27}$, $C_{28}$, and $C_{29}$ 20S/20R homologues among the $C_{20}$–$C_{30}$ steranes is often apparent. Figures 3.4 and 3.5 show GC-MS-SIM chromatograms at *m/z* 191 and 218 for Sockeye oil (California), Orimulsion-400 (Venezuela), HFO 6303, diesel no. 2 (Ontario), hydraulic oil (no. 1), and hydraulic oil (no. 3), respectively.

For Sockeye, $C_{28}$-bisnorhopane (a biomarker compound that is not a member of the regular hopane series), $C_{29}$ and $C_{30}$ αβ-hopane are the most abundant, with the concentration of $C_{28}$ even higher than $C_{29}$ and $C_{30}$ hopane (several other California heavy oils are found to be characterised by high concentrations of $C_{28}$-bisnorhopane too). For Orimulsion, $C_{23}$ terpane is the most abundant and the concentration of $C_{29}$ is lower than $C_{30}$ hopane. For HFO 6303, $C_{23}$ terpane is the most abundant, but nearly no homohopanes of $C_{31}$–$C_{35}$ were detected. Different from most bunker C type oils, the concentrations of both terpanes and steranes are quite low in HFO 6303. Only trace amounts of $C_{20}$–$C_{24}$ terpanes and $C_{20}$–$C_{22}$ steranes were detected in diesel no. 2 because the refining processes have removed or concentrated high molecular mass biomarkers from the corresponding crude oil feed stocks. In contrast, most lube oils contain high quantities of biomarkers. As Figures 3.4 and 3.5 show, hydraulic oil samples no. 1 and no. 3 had extremely high concentrations of target biomarkers (4701 and 5464 $\mu g.g^{-1}$ oil) in comparison with most crude oils (Wang et al. 2002). Most terpanes are in the high carbon range for both samples, but hydraulic oil no. 3 shows a markedly different distribution pattern of biomarkers from hydraulic oil no. 1. The abundances of $C_{23}$ and $C_{24}$ terpanes and $C_{31}$–$C_{35}$ homohopanes in hydraulic oil no. 3 are significantly lower and higher than the corresponding compounds in hydraulic oil no. 1, respectively.

In addition to their composition, the concentrations of biomarkers vary widely with the type of depositional environment (oxic/aoxic, freshwater/marine/hypersaline), type of organic matter (e.g., terrigenous origin or

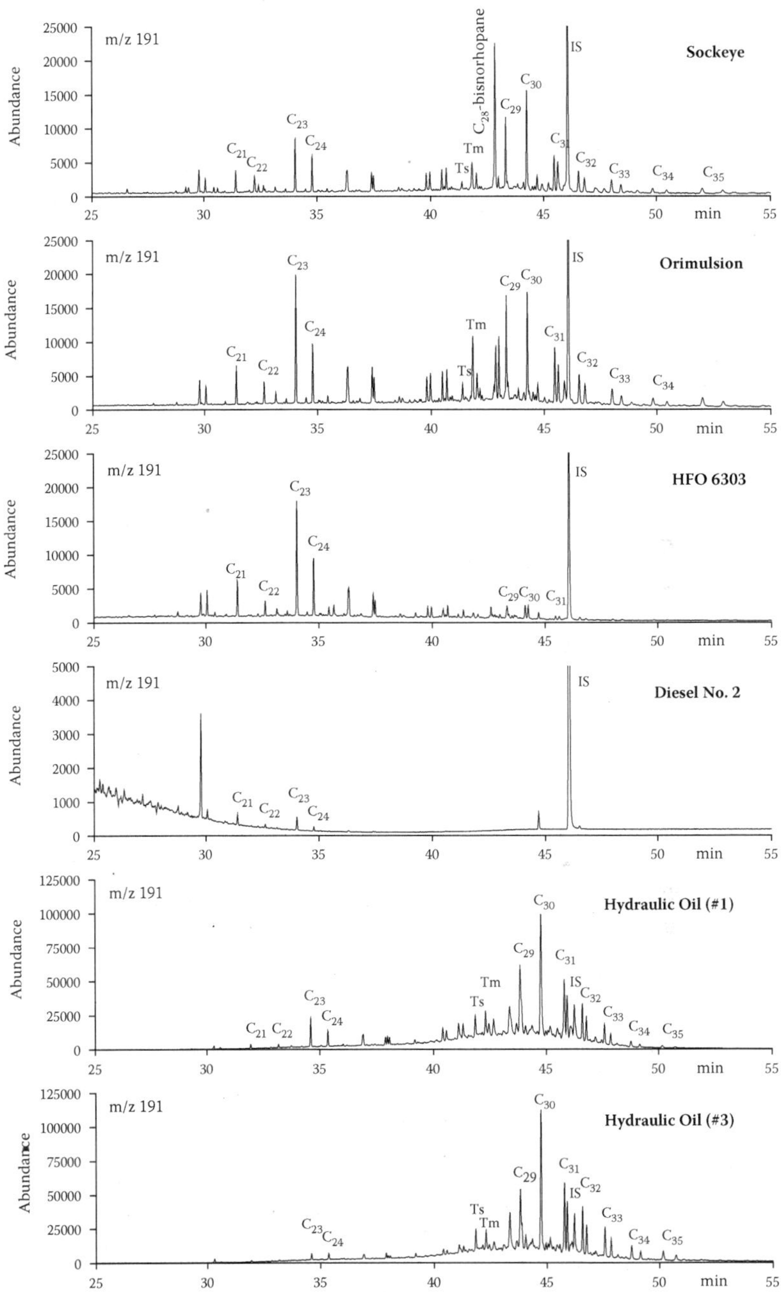


**Figure 3.4** Distribution of biomarker terpane compounds (at *m/z* 191) for Sockeye oil, Orimulsion, HFO 6303, diesel no. 2, and hydraulic oil no. 1 and no. 3 to illustrate the differences in the relative distribution of terpanes between oils and oil products.

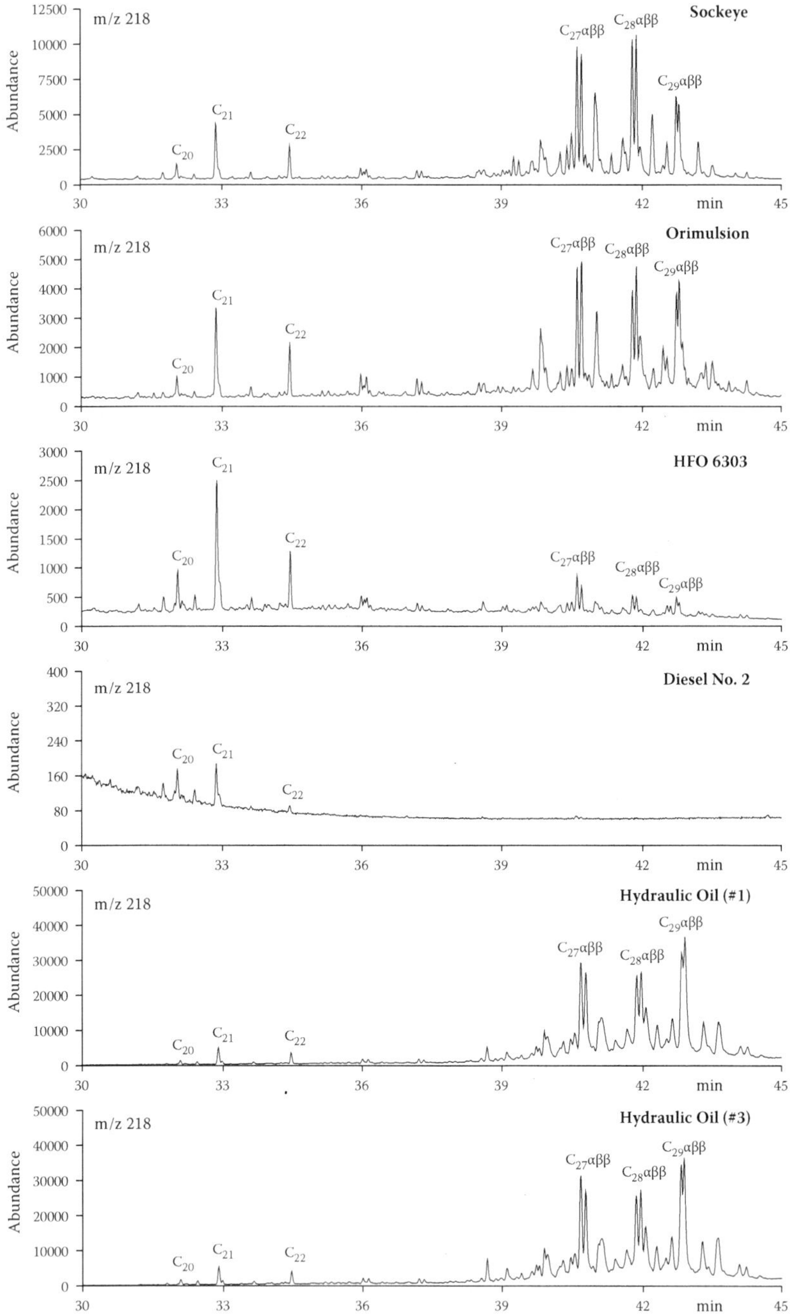


**Figure 3.5** Distribution of biomarker sterane compounds (at *m/z* 218) for Sockeye oil, Orimulsion, HFO 6303, diesel no. 2, and hydraulic oil no. 1 and no. 3 to illustrate the differences in the relative distribution of steranes between oils and oil products.

marine origin), maturity, and biodegradation. For a given type of organic material, the biomarker concentrations generally decrease with increasing thermal maturity. Very light oils or condensates (e.g., the Scotia Light) typically contain low biomarker concentrations. Characterisation of biomarkers should include determination of both concentrations and relative distributions and should not measure peak ratio alone. This is important because it is possible to have a situation where a source has a similar biomarker ratio but very different actual amounts of biomarkers. As an example, Table 3.6 presents the quantitation results of biomarkers in 10 common crude oil and refined products for comparison.

Severely weathered oils may exhibit completely different GC-FID chromatograms and *n*-alkane profiles or isoprenoid distributions from their source oil, but their biomarker distribution patterns may be unaltered. Characterisation of many long-term spilled oils (Wang, Fingas, and Sergy 1994, 1995; Wang, Fingas, et al. 1998; Wang et al. 2001) demonstrated that *n*-alkanes and isoprenoids in severely weathered samples could be completely lost, but the profiles of their GC-MS fingerprints at *m*/*z* 191 and 217/218 were nearly unaltered. Furthermore, the computed diagnostic ratios of a series of target pairs of biomarker compounds were also nearly identical. Thus, the fingerprinting of terpane and sterane biomarkers provides us with a powerful tool for tracking the source of the long-term weathered oil.

In addition to common biomarker terpanes and steranes, certain oils may also contain some specific biomarker compounds including several geologically rare acyclic alkanes, which can provide additional diagnostic information on the types of organic matter that give rise to the crude oil. For example, the geologically rare acyclic alkane botryococcane ($C_{34}H_{70}$) was used to identify a new class of Australian nonmarine crude oils (McKirdy et al. 1986). The presence of botryococcane indicates that the source rock contains remains of the algae *Botryococcus braunii*. The broad platform area of the northern North Sea, including Statford, Gullfacs, Brent, Oseberg, Troll, etc., seems to be characterised by relative high abundances of $C_{28}$-bisnorhopane (Dahlmann 2003). Thus, $C_{28}$-bisnorhopane can be regarded as a specific source parameter. Dahlmann (2003) also found that oils from the Niger Delta (Nigeria) and from Africa (in Angola Cabinda and Nemba crudes and in Kongo and Gabon crudes) are characterised by the presence of high abundant oleanane and gammacerane, respectively. The presence of 18α(H)-oleanane in benthic sediments in PWS, coupled with its absence in Alaska North Slope crude and specifically in *Exxon Valdez* oil and its residues, confirmed another petrogenic source (Bence et al. 1996). Other specific pentacyclic terpanes include $C_{30}$ 17α (H)-diahopane (potentially related to bacterial hopanoid precursors that have undergone oxidation and rearrangement by clay-mediated acidic catalysis), β-carotane ($C_{40}H_{78}$, highly specific for lacustrine deposition, highly abundant in Green River shale), gammacerane

**Table 3.6 Target Biomarkers in 10 Crude Oils and Refined Products**

| Oil samples | Arabian Light | Scotia Light | Cook Inlet | Orinoco bitumen | California (API = 11) | Diesel-02 | Korean diesel #1 (2002) | HFO 6303 | Used air compressor oil | Valvoline 10W-30 motor oil |
|---|---|---|---|---|---|---|---|---|---|---|
| | | | | Compounds ($\mu g.g^{-1}$ oil) | | | | | | |
| $C_{21}$ | 4.47 | 0.00 | 7.12 | 35.9 | 22.5 | 3.11 | 12.2 | 30.2 | 17.1 | 11.6 |
| $C_{22}$ | 4.73 | 0.00 | 2.77 | 21.1 | 8.86 | 1.42 | 6.68 | 12.0 | 14.3 | 15.2 |
| $C_{23}$ | 17.7 | 0.87 | 9.88 | 121 | 56.5 | 3.85 | 22.0 | 92.3 | 86.7 | 68.2 |
| $C_{24}$ | 6.60 | 0.61 | 6.16 | 56.8 | 39.3 | 1.39 | 9.74 | 45.9 | 45.0 | 25.5 |
| $C_{29}$ αβ | 152 | 3.32 | 45.0 | 79.2 | 69.3 | 0.00 | 22.5 | 14.4 | 190 | 864 |
| $C_{30}$ αβ | 125 | 5.79 | 125 | 83.6 | 109 | 0.00 | 24.4 | 11.5 | 414 | 718 |
| $C_{31}$ (S) | 79.9 | 1.74 | 45.0 | 42.7 | 46.1 | 0.00 | 5.52 | 3.70 | 180 | 385 |
| $C_{31}$ (R) | 65.7 | 1.24 | 35.6 | 30.4 | 32.7 | 0.00 | 3.77 | 3.30 | 148 | 305 |
| $C_{32}$ (S) | 48.1 | 0.95 | 29.0 | 27.2 | 32.5 | 0.00 | 3.85 | 3.20 | 142 | 238 |
| $C_{32}$ (R) | 29.8 | 0.79 | 20.1 | 16.2 | 22.0 | 0.00 | 2.12 | 2.00 | 96.1 | 164 |
| $C_{33}$ (S) | 27.0 | 0.00 | 16.6 | 17.4 | 25.1 | 0.00 | 1.37 | 1.50 | 104 | 140 |
| $C_{33}$ (R) | 17.8 | 0.00 | 12.3 | 10.5 | 17.6 | 0.00 | 1.00 | 1.20 | 69.5 | 91.7 |
| $C_{34}$ (S) | 14.4 | 0.00 | 10.5 | 13.0 | 17.9 | 0.00 | 0.00 | 1.10 | 78.3 | 77.6 |
| $C_{34}$ (R) | 8.80 | 0.00 | 7.48 | 5.96 | 11.6 | 0.00 | 0.00 | 0.90 | 43.1 | 51.6 |
| $C_{35}$ (S) | 14.7 | 0.00 | 5.98 | 13.0 | 23.0 | 0.00 | 0.00 | 0.78 | 72.5 | 85.7 |

| | | | | | | | | | | |
|---|---|---|---|---|---|---|---|---|---|---|
| $C_{35}$ (R) | 7.80 | 0.00 | 4.29 | 9.71 | 20.8 | 0.00 | 0.00 | 0.44 | 46.5 | 47.6 |
| Ts | 42.6 | 1.40 | 22.7 | 15.6 | 9.08 | 0.00 | 10.4 | 5.70 | 61.9 | 148 |
| Tm | 36.5 | 1.66 | 23.4 | 42.2 | 20.7 | 0.00 | 9.40 | 1.60 | 74.8 | 215 |
| $C_{27}$ αββ-steranes | 35.1 | 2.84 | 184 | 52.4 | 438 | 0.00 | 48.9 | 10.7 | 437 | 525 |
| $C_{28}$ αββ-steranes | 20.1 | 2.77 | 113 | 66.1 | 427 | 0.00 | 46.0 | 6.96 | 384 | 363 |
| $C_{29}$ αββ-steranes | 55.1 | 5.20 | 232 | 67.2 | 289 | 0.00 | 39.4 | 5.50 | 761 | 778 |
| **Total** | **814** | **29.2** | **958** | **827** | **1738** | **9.77** | **269** | **255** | **3466** | **5318** |
| | | | | | Diagnostic ratios | | | | | |
| $C_{23}/C_{24}$ | 2.68 | 1.42 | 1.60 | 2.13 | 1.44 | 2.78 | 2.25 | 2.01 | 1.93 | 2.68 |
| $C_{23}/C_{30}$ αβ | 0.14 | 0.15 | 0.08 | 1.45 | 0.52 | — | 0.90 | 8.03 | 0.21 | 0.09 |
| $C_{24}/C_{30}$ αβ | 0.05 | 0.11 | 0.05 | 0.68 | 0.36 | — | 0.40 | 3.99 | 0.11 | 0.04 |
| $C_{29}$ αβ/$C_{30}$ αββ | 1.22 | 0.57 | 0.36 | 0.95 | 0.64 | — | 0.92 | 1.25 | 0.46 | 1.20 |
| $C_{31}$(S)/$C_{31}$(S+R) | 1.22 | 1.40 | 1.26 | 1.40 | 1.41 | — | 1.47 | 1.12 | 1.22 | 1.26 |
| $C_{32}$(S)/$C_{32}$(S+R) | 1.61 | 1.20 | 1.44 | 1.67 | 1.48 | — | 1.81 | 1.60 | 1.48 | 1.45 |
| Ts/Tm | 1.17 | 0.84 | 0.97 | 0.37 | 0.44 | — | 1.10 | 3.56 | 0.83 | 0.69 |
| $C_{27}$ αββ-steranes/$C_{29}$ αββ-steranes | 0.64 | 0.55 | 0.79 | 0.78 | 1.52 | — | 1.24 | 1.95 | 0.57 | 0.67 |
| $C_{30}/(C_{31} + C_{32} + C_{33} + C_{34} + C_{35})$ | 0.40 | 1.23 | 0.67 | 0.45 | 0.44 | — | 1.39 | 0.63 | 0.42 | 0.45 |

($C_{30}H_{52}$, tentatively suggested as a marker for hypersaline episodes of source rock deposition), lupanes and bisnorlupanes (believed to indicate terrestrial organic matter input), and bicadinanes (highly specific for resinous input from higher plants).

It should be emphasised that if an oil shows any additional characteristic compositional features (such as extra biomarker peaks), these should, of course, always be included in the characterisation and considered in the identification and correlation.

### *Low-Boiling Sesquiterpanes in Oils and Lighter Petroleum Products*

The bicyclic biomarkers comprise one of the terpenoid classes. Sesquiterpanes with the drimane skeleton are ubiquitous components of crude oils and ancient sediments. Most bicyclic sesquiterpanes probably originate not only from higher plants but also from algae and bacteria (Alexander et al. 1984; Philp 1985; Fan, Qian, and Zhang 1991). Philp, Gilbert, and Riedrich (1981) have also suggested tricyclic diterpanes from higher plants may be a source of bicyclic terpanes through opening of the C-ring during maturation. The relative concentration of $C_{14}$ sesquiterpanes decreases with increasing maturation of the parent organic matter. The concentrations of $C_{14}$ bicyclic sesquiterpanes are higher at the immature stage, while those of $C_{15}$ drimane and $C_{16}$ homodrimane are relatively lower. As their higher molecular weight precursor dehydroxylates, the concentrations of drimane ($C_{15}$) and homodrimanes ($C_{16}$) gradually increase, and the concentrations of $C_{14}$ sesquiterpanes reduce (Cheng et al. 1991). For lighter petroleum products, the high-boiling-point pentacyclic triterpanes and steranes are generally absent or in very low abundance, while the low-boiling-point sesquiterpanes are more concentrated in these distillates. The sesquiterpanes elute between $n$-$C_{13}$ and $n$-$C_{16}$ in the GC-MS-SIM chromatograms and are monitored using $m/z$ 123, a base fragment ion common to all sesquiterpanes. Confirmation ions for the sesquiterpanes include $m/z$ 179 (the ion after sesquiterpane $C_{14}H_{26}$ loses $CH_3$), 193 (the ion after $C_{15}H_{28}$ loses $CH_3$ or after $C_{16}H_{30}$ loses $C_2H_5$), and 207 (the ion after $C_{16}H_{30}$ loses $CH_3$). Examination of GC-MS chromatograms for these characteristic ions of sesquiterpanes provides a highly diagnostic means of correlation, differentiation, and source identification for lighter petroleum products in comparison to the use of other hydrocarbon groups.

Figure 3.6 shows GC-MS chromatograms of sesquiterpanes at $m/z$ 123 for example crude oils and petroleum products. Peaks 1 and 2, 3–6, and 7–10 are $C_{14}$, $C_{15}$, and $C_{16}$ sesquiterpanes, respectively. Among these 10 compounds, peaks 5 and 10 are 8β(H)-drimane and 8β(H)-homodrimane, respectively. The distribution patterns of sesquiterpanes generally vary in crude oils and in refined petroleum products from different sources.

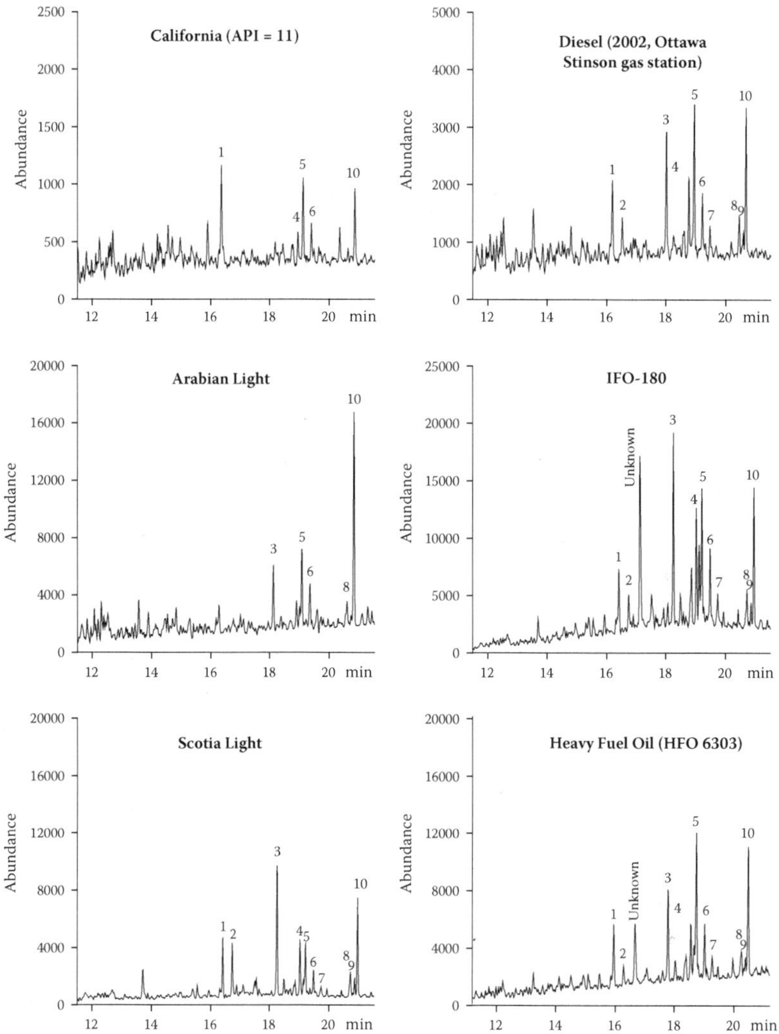


**Figure 3.6** GC-MS chromatograms at *m/z* 123 for sesquiterpane analysis of crude oils (California API 11, Arabian Light, and Scotia Light) and refined products (diesel no. 2, IFO-180, and heavy fuel oil). The different distributions of the sesquiterpanes demonstrate the differences between oils and refined products.

ANS, Arabian Light, and Scotia Light have high concentrations of sesquiterpanes with peak 10 ($C_{16}$) being the most abundant for the ANS and Arabian Light, and peak 3 ($C_{15}$) the most abundant for Scotia Light. The Arabian Light has the lowest concentration of $C_{14}$ sesquiterpanes (peaks 1 and 2), indicating that this oil is highly mature. By contrast, the heavy California API 11 oil has the highest concentration of $C_{14}$ sesquiterpane (peak 1), indicating that this oil may be relatively immature. Sesquiterpanes are absent in very light kerosene and heavy lubricating oils. However, IFO-180 and HFO-6303 (a bunker C fuel) have relatively high concentrations of sesquiterpanes. Jet A is characterised with peaks 3, 5, and 1 being the most abundant, while the diesel sample (2002, Ottawa) is characterised with peaks 5 and 10 being equally abundant and followed by peak 3. Fingerprinting a large number of middle distillate fuels like diesels demonstrates that sesquiterpanes are quite abundant in diesels. The differences in distribution patterns and concentrations of sesquiterpanes are often quite apparent between diesels (Stout et al. 2005; Wang et al. 2005).

Furthermore, diagnostic ratios of selected paired sesquiterpanes for a large number of oils and petroleum products have also been developed (Wang et al. 2005). In general, oils from different regions have ratios that cover quite a wide range. Cross-plots of diagnostic ratios of peak 4/peak 5 ($C_{15}$) versus the ratios of peak 3/peak 5 ($C_{15}$) for over 50 crude oils and refined products (both isomers 3 and 4 are normalised relative to isomer 5) demonstrate that different oils have different ratio values of peak 4/peak 5 and peak 3/peak 5, which fall in ranges of 0.2–1.2 and 0.1–2.1, respectively. Another feature is that related oils produce tight clusters on the plot (such as the cluster for Orimulsion samples from different batches and of the original Orinoco bitumen). This observation implies that the ratios of sesquiterpane isomers, in combination with other fingerprinting data, may be used to discriminate different oils and to identify the source of spill samples.

Oil spills were reported and sampled on March 17 and 23, 1998, at a sewer outlet flowing into the Lachine Canal in Quebec. Following the accident, a diesel fuel that was suspected to be the source of the spill was collected from a reservoir at a pumping station located in Lachine, Quebec. Biomarker fingerprinting of the samples revealed only trace amounts ($<10$ μg.g$^{-1}$ oil) of $C_{19}$–$C_{24}$ tricyclic terpanes, regular $C_{20}$–$C_{22}$ steranes, and diasteranes. However, the spill samples contained significant amount of sesquiterpanes. The GC-MS/SIM chromatogram at *m/z* 123 and diagnostic ratios of target sesquiterpanes of the spill samples were found to be nearly identical to that of the suspected source diesel. The only noticeable difference was that, compared to the suspected source diesel, the spilled sample had slightly higher abundances due to weathering. These similarities, in combination with other hydrocarbon results such as bulk hydrocarbon groups, *n*-alkane distribution, and fingerprints of alkylated PAHs, and diagnostic ratios of source-specific PAH compounds, argued strongly that the suspected diesel collected from the pumping station close to the spill site was the source of the spilled diesel. Stout et al. (2002) compared the chromatographic distributions of sesquiterpanes of two weathered diesel fuel samples from two adjacent petroleum terminal properties. The samples had been highly weathered, with *n*-alkanes completely lost. However, GC-MS (*m/z* 123) analysis results showed very different distribution profiles of sesquiterpanes between two samples, clearly indicating that two sources of diesel existed in the study area.

### *Diagnostic Ratios (Indices) of Biomarkers*

Biomarker diagnostic parameters have long been established and are widely used by geochemists (Peters and Moldowan 1993) for oil correlation, determination of organic input and depositional environment, assessment of thermal maturity, and evaluation of oil biodegradation. Many diagnostic biomarker parameters currently used in oil spill studies originated from geochemistry parameters. Table 3.7 lists some of the primary diagnostic ratios of biomarkers frequently used by environmental chemists for spilled oil identification, correlation, and differentiation. An important benefit of comparing diagnostic ratios of spilled oil and suspected source oils is that concentration effects are minimised. In addition, the use of ratios tends to induce a self-normalising effect on the data since variations due to the fluctuation of day-to-day instrument operating conditions, operator, and matrix effects are minimised. Therefore, comparison of diagnostic ratios more directly reflects differences of the target biomarker distribution between samples.

In a similar manner to the selection of the PAH diagnostic ratios, the criteria used for selection of the diagnostic ratios of biomarkers should also include:

specificity and diversity;
high resistance to weathering;
analytical precision; and
selection of peaks with relatively high abundance at the defined *m*/*z*.

More specifically, for the purpose of identification and correlation, one generally should choose diagnostic ratios that show a large variation between oils, reflect various types of geological processes, and are little altered by weathering. However, some flexibility in selecting ratios is also needed, as some ratios may not be applicable to some types of oils and oil products (e.g., diesel only contains trace or no high-boiling hopanes). Two types of frequently used diagnostic ratios are expressed in the following:

$$\text{Ratio} = a/b \tag{3.1}$$

or

$$\text{Ratio (\%)} = 100 \times a/(a + b) \tag{3.2}$$

where *a* and *b* could be the peak areas, peak heights, or concentrations. Ratios of the type defined in equation 3.1 range between 0 and infinity, whereas ratios defined in equation 3.2 range between 0 and 100. Ratios are defined from equation 3.1 for simplicity and easy comparison, but can readily be redefined using equation 3.2. The ratio $a/(a + b)$ increases linearly with *a*, and

**Table 3.7 Diagnostic Ratios of Biomarkers Frequently Used for Environmental Forensic Studies**

| Biomarker classes | Diagnostic ratios | Code |
|---|---|---|
| Acyclic isoprenoids | Pristane/phytane | pri/phy |
| | Pristane/*n*-$C_{17}$ | pri/$C_{17}$ |
| | Phytane/*n*-$C_{18}$ | phy/$C_{18}$ |
| Terpanes (*m/z* 191) | $C_{21}/C_{23}$ tricyclic terpane | TR21/TR23 |
| | $C_{23}/C_{24}$ tricyclic terpane | TR23/TR24 |
| | $C_{23}$ tricyclic terpane/$C_{30}$ αβ hopane | TR23/H30 |
| | $C_{24}$ tricyclic terpane/$C_{30}$ αβ hopane | TR24/H30 |
| | $C_{24}$ tertracyclic/$C_{26}$ tricyclic (S)/ $C_{26}$ tricyclic (R) terpane | Triplet ratio |
| | $C_{27}$ 18α,21β-trisnorhopane/$C_{27}$ 17α,21β-trisnorhopane | Ts/Tm |
| | $C_{28}$ bisnorhopane/$C_{30}$ αβ hopane | H28/H30 |
| | $C_{29}$ αβ-25-norhopane/$C_{30}$ αβ hopane | NOR25H/H30 |
| | $C_{29}$ αβ-30-norhopane/$C_{30}$ αβ hopane | H29/H30 |
| | Oleanane/$C_{30}$ αβ hopane | OL/H30 |
| | Moretane($C_{30}$ βα hopane)/$C_{30}$ αβ hopane | M30/H30 |
| | Gammacerane/$C_{30}$ αβ hopane | GAM/H30 |
| | Tricyclic terpanes ($C_{19}$-$C_{26}$)/$C_{30}$ αβ hopane | Σ(TR19–TR26)/H30 |
| | $C_{31}$ homohopane (22S)/$C_{31}$ homohopane (22R) | H31S/H31R |
| | $C_{32}$ bishomohopane (22S)/$C_{32}$ bishomohopane (22R) | H32S/H32R |
| | $C_{33}$ trishomohopane (22S)/$C_{33}$ trishomohopane (22R) | H33S/H33R |
| | Relative homohopane distribution | H31:H32:H33:H34:H35 |
| | Σ($C_{31}$–$C_{35}$)/$C_{30}$ αβ hopane | Σ(H31–H35)/H30 |
| | Homohopane index | H31/Σ(H31– H35) to H35/Σ(H31–H35) |
| Steranes and diasteranes (*m/z* 217 & 218) | $C_{27}$ 20S-13β(H), 17α(H)-diadterane/$C_{27}$ 20R-13β(H), 17α(H)-diadterane | DIA 27S/DIA 27R |
| | Relative distribution of regular $C_{27}$-$C_{28}$-$C_{29}$ steranes | C27:C28:C29 steranes |

**Table 3.7 Diagnostic Ratios of Biomarkers Frequently Used for Environmental Forensic Studies (Continued)**

| Biomarker classes | Diagnostic ratios | Code |
|---|---|---|
| | $C_{27}$ αββ/$C_{29}$ αββ steranes (at *m/z* 218) | C27ββ(S + R)/C29ββ(S + R) |
| | $C_{28}$ αββ/$C_{29}$ αββ steranes (at *m/z* 218) | C28ββ(S + R)/C29ββ(S + R) |
| | $C_{27}$ αββ/($C_{27}$ αββ + $C_{28}$ αββ + $C_{29}$ αββ) (at *m/z* 218) | C27ββ/(C27 + C28 + C29)ββ |
| | $C_{28}$ αββ/($C_{27}$ αββ + $C_{28}$ αββ + $C_{29}$ αββ) (at *m/z* 218) | C28ββ/(C27 + C28 + C29)ββ |
| | $C_{29}$ αββ/($C_{27}$ αββ + $C_{28}$ αββ + $C_{29}$ αββ) (at *m/z* 218) | C29ββ/(C27 + C28 + C29)ββ |
| | $C_{27}$, $C_{28}$, and $C_{29}$ ααα/αββ epimers (at *m/z* 217) | C27αα/C27ββ<br>C28αα/C28ββ<br>C29αα/C29ββ |
| | $C_{27}$, $C_{28}$, and $C_{29}$ 20S/(20S + 20R) steranes (at *m/z* 217) | C27 (20S)/C27 (20R)<br>C28 (20S)/C28 (20R)<br>C29 (20S)/C29 (20R) |
| | $C_{30}$ sterane index: $C_{30}$/($C_{27}$ to $C_{30}$) steranes | C30/(C27 to C30) steranes |
| | Selected diasteranes/regular steranes | |
| | Regular $C_{27}$-$C_{28}$-$C_{29}$ steranes/$C_{30}$ αβ-hopanes | C27-C28-C29 steranes/H30 |
| Sesquiterpanes (*m/z* 123) | Relative distribution of sesquiterpanes | |
| | $C_{14}$ group: peak 1/peak 2 | P1/P2 |
| | $C_{15}$ group: peak 3/peak 5, peak 4/peak 5, peak 6/peak 5 | P3/P5, P4/P5, P6/P5 |
| | $C_{16}$ group: peak 8/peak 10 | P8/P10 |
| | Intergroup: peak 1/peak 3, peak 1/peak 5, peak 3/peak 10, peak 5/peak 10 | P1/P3,P1/P5,P3/P10,P5/P10 |
| Adamantanes (*m/z* 135, 149, 163, 177) | Methyl adamantane index: 1-MA/(1- + 2-MA) | MAI |
| | 1,4-DMA, cis/1,4-DMA, trans | |
| | dimethyl admantane index: | DMAI |
| | 1,3-DMA/(1,3- + 1,4- + 1,2-DMA) | |
| | 1,3,4-TMA, cis/1,3,4-TMA, trans | |
| | trimethyl adamantane index: | TMAI |

Continued

**Table 3.7 Diagnostic Ratios of Biomarkers Frequently Used for Environmental Forensic Studies (Continued)**

| Biomarker classes | Diagnostic ratios | Code |
|---|---|---|
| | 1,3,4-DMA, cis/(1,3,4-DMA, cis + 1,3,4-DMA, trans) | |
| | Ethyl adamantane index: 1-EA/(1- + 2-EA) | EAI |
| Diamantanes (*m/z* 187, 201, 215) | Methyl-diamantane index: 4-MD/(1- + 3- + 4-MD) | MDI |
| | Relative distribution of diamantanes: $C_0$-D:$C_1$-D:$C_2$-D:$C_3$-D | |
| Triaromatic steranes (*m/z* 231) | $C_{20}$ TA/($C_{20}$ TA + $C_{21}$ TA) | |
| | $C_{26}$ TA (20S)/sum of $C_{26}$ TA (20S) through $C_{28}$ TA (20R) | |
| | $C_{27}$ TA (20R)/$C_{28}$ TA (20R) | |
| | $C_{28}$ TA (20R)/$C_{28}$ TA (20S) | |
| | $C_{26}$ TA (20S)/[$C_{26}$ TA (20S) + $C_{28}$ TA (20S)] | |
| | $C_{28}$ TA (20S)/[$C_{26}$ TA (20S) + $C_{28}$ TA (20S)] | |
| Monoaromatic steranes (*m/z* 253) | $C_{27}$-$C_{28}$-$C_{29}$ monoaromatic steranes (MA) distribution | |

*Notes:* Ratios are defined from equation 3.1 for simplicity, but can be readily redefined using equation 3.2. For example, in accordance with equation 3.2, the ratio of $C_{29}$ αβ-30-norhopane/$C_{30}$ αβ hopane (H29/H30) can be readily redefined as H29/(H29 + H30) × 100%.

therefore it can be plotted on a linear scale. Reporting of decimals for ratio values greater than 10% is unnecessary as it would suggest a higher precision than the one that can be achieved in reality. For ratio values smaller than 10% two significant figures are recommended (e.g., 2.4 and 2.6% instead of 2 and 3%).

The second type of the ratio is preferably used in the CEN methodology (CEN 2002) because it is considered to avoid the potential problems (such as division by zero or very large values) connected with the ratios of the type *a/b*. In the CEN method, a suite of 18 diagnostic biomarker ratios (nine hopane ratios at *m/z* 191, six ratios of regular steranes and diasteranes at *m/z* 217 and 218, and three ratios of triaromatic steranes) has been selected as defensible indices to differentiate among qualitatively similar oils from spills and available candidate sources. This methodology has been successfully used for a round robin study (Faksness et al. 2002). Seven oil samples (two artificially weathered 'spill' samples and five possible sources) were analysed following the recommended CEN analytical protocol. The round robin study was a difficult case because two spill samples and three of the suspected sources were highly correlated to one another. These samples were from the same oil field in the North Sea but from different production wells. This round robin study demonstrates the potential of this methodology as a strong, technically defensible tool in oil spill identification and differentiation.

During the *Arrow* oil spill, the ratio of the most abundant $C_{29}$–$C_{30}$ hopane was defined and used as a reliable source indicator (Wang, Fingas, and Sergy 1994). Zakaria et al. (2000) studied oil pollution in the Straits of Malacca. Various samples including Malaysia oil, Middle East crude oils, Southeast Asian crude oils, tarballs, sediments, and mussels were collected and analysed. The analytical results in this study demonstrated the utility of the $C_{29}/C_{30}$ ratio, $\Sigma(C_{31} - C_{35})/C_{30}$, and the homohopane index as molecular tools to distinguish the source of petroleum in the straits. Barakat et al. (1997) studied biomarker properties of five crude oils from the Gulf of Suez, Egypt. The results revealed significant differences in biomarker distribution and diagnostic ratios within the oils that suggested two oil types and one mixed type. The triplet ratio, in general, varies in oils from different sources and is dependent upon sources, depositional environment, and maturity. The triplet ratio was first used in a chemistry study of North Slope crude by Kvenvolden, Rapp, and Bourell (1985) in which the ratio was ~2. *Exxon Valdez* oil (an Alaska North Slope crude) and its residues also have triplet ratios of ~2; in contrast, many tarballs collected from the shorelines of Prince William Sound have triplet ratios of ~5.

During January and February 1996, a significant number of tarball incidents occurred along the coasts of Vancouver Island of British Columbia (BC), Washington (WA), Oregon (OR), and California (CA). The diagnostic values of source-specific biomarkers and PAH isomer compounds of representative tarball samples and the suspected source Alaska North Slope (ANS) oil were determined and compared (Wang, Fingas, Landriault, et al. 1998). The results clearly revealed the following:

- Almost all diagnostic ratios for the ANS oil differ significantly from those of the tarball samples, indicating the ANS oil was not the source oil of the tarball samples.
- All the relative ratios are almost identical for samples BC-1 and BC-2, indicating they were from the same source.
- The tarball sample from CA was very similar in concentrations and diagnostic ratios of target biomarkers with the samples BC-1 and BC-2, but it had markedly different PAH isomeric ratios from samples BC-1 and BC-2, indicating CA tarballs may have another source different from the BC samples.

Cross-plots (double ratios), which are more sensitive to maturity and depositional environment than a single diagnostic ratio, are frequently used in oil geochemistry for oil–oil correlation and determination of oil source and depositional environment. Zakaria et al. (2000, 2001) proposed utility of the cross-plots of the $C_{29}$ $\alpha\beta/C_{30}$ $\alpha\beta$ hopane ratio versus the homohopane index $\Sigma(C_{31} - C_{35})/C_{30}$ hopane as key biomarker indicators; they successfully distinguished a large number of tarball samples that originated from Southeast Asian crude oil sources from those of Middle East sources.

The biomarker fingerprinting results described previously strongly suggest a basic rule in environmental forensic investigations: A negative correlation of biomarkers is strong evidence for a lack of relationship between samples, but a positive correlation of certain biomarkers is not necessarily proof that samples are related because some oils from different sources can show similar characteristics of biomarkers. It must be emphasised that there is no single parameter that can unambiguously identify the source of unknown spills. Therefore, in order to correlate or differentiate samples reliably and defensively, the multiparameter approach must be initiated in many cases. That is, analyses of more than one suite of analytes must be performed, and numerous indices with different individual specificities within one suite of analytes must be selected and compared.

## Weathering Check Using Hydrocarbon Fingerprints

### *Oil Weathering*

When crude oil or petroleum products are accidentally released to the environment, whether on water or land, they are immediately subject to a wide variety of changes in physical and chemical properties that in combination are termed 'weathering'. The most important weathering processes include evaporation, dissolution, dispersion, and microbial degradation. In the short term, evaporation is the single most important and dominant weathering process. The rates of weathering of an oil can be very different and are controlled by a number of spill conditions and natural processes such as the type of oil spilled, the local environmental conditions, and natural population of indigenous microbial and microbiological activities during and after the spill. In the first few days following a spill, the weathering is largely caused by evaporation and the loss can be up to 70 and 40% of the volume of light and medium crudes, respectively. For heavy or residual oils the losses are only a few percent by volume. For lighter petroleum products, evaporation has a great effect on the amount remaining on water or land.

Major chemical compositional changes due to weathering are summarised as the following:

For lightly weathered oils and refined products (e.g., <15% weathered), the abundances of low-end $n$-alkanes are significantly reduced. However, the ratios of $n$-$C_{17}$/pristane and $n$-$C_{18}$/phytane are found to be virtually unaltered. The losses of BTEX and $C_3$-benzene compounds are obvious, and the most abundant two-ring alkylated naphthalene series appear slightly enriched.

For moderately weathered oils and refined products (e.g., ~15–30% weathered), significant losses occur in $n$-alkanes and relatively low molecular weight isoprenoid compounds. Rapid loss of volatile aromatic compounds is clear. The loss of $C_0$- and $C_1$-naphthalenes can be significant. The ratio of GC-resolved peaks to UCM can be considerably decreased due to the preferential loss of resolved hydrocarbons over the unresolved complex hydrocarbons. The biomarker compounds are enriched.

For severely weathered oils and refined products, not only $n$-alkanes but also branched and cyclo-alkanes are heavily or completely lost, and the UCM becomes extremely pronounced, resulting in a significant increase in the relative ratios of UCM/GC-TPH and in the substantial decrease in a relative ratios of resolved peaks to GC-TPH. The BTEX and alkyl benzene compounds are completely lost. Pronounced decreases in the alkylated naphthalene series relative to other alkylated PAH homologous series are clearly observed. A profile in each alkylated PAH family showing the distribution of $C_0$- < $C_1$- < $C_2$- < $C_3$- is clearly developed. The alkylated chrysene series is significantly enhanced relative to other PAH series. Biomarker compounds are concentrated because of their refractory nature and high resistance to biodegradation.

## *Weathering Check Using n-Alkanes and GC Traces*

Generally speaking, GC-FID traces (including the quantitative $n$-alkane distribution and the relative size of UCM in the GC chromatograms) are useful only for estimation of general weathering trends and identification of the carbon range of remaining hydrocarbons, but they are not source specific. Figure 3.7 shows quantitative $n$-alkane (mg.g$^{-1}$ oil) distribution of the Alberta Sweet Mixed Blend at weathered percentages of 0 and 45%, respectively, to illustrate the effects of evaporative weathering on the $n$-alkane distributions. Another chromatographic feature useful for indication of the weathering degree is the increase in the relative size of UCM and the decrease in the ratio of GC-resolved peaks to UCM due to the preferential loss of resolved hydrocarbons over the unresolved complex hydrocarbons by weathering.

The weathering degree can be readily checked by comparing the weathering index (WI), defined as the sum of $n$-$C_8$, $n$-$C_{10}$, $n$-$C_{12}$, and $n$-$C_{14}$ divided by the sum of $n$-$C_{22}$, $n$-$C_{24}$, $n$-$C_{26}$, $n$-$C_{28}$ (Wang and Fingas 1995a):

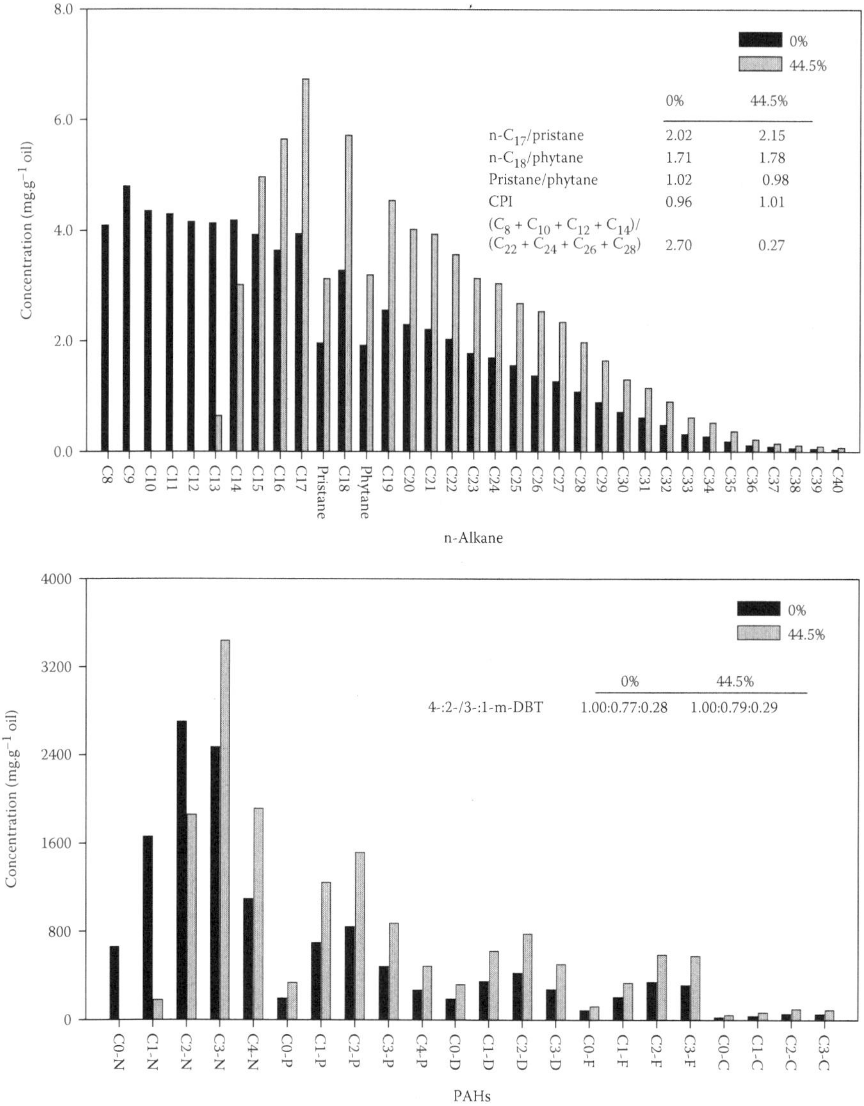


**Figure 3.7** Comparison of *n*-alkane (top panel) and PAH (bottom panel) distribution for the Alberta Sweet Mixed Blend at evaporative weathered percentages of 0 and 45%, to illustrate the effect of physical weathering on the oil chemical composition changes.

$$WI = (n\text{-}C_8 + n\text{-}C_{10} + n\text{-}C_{12} + n\text{-}C_{14})/(n\text{-}C_{22} + n\text{-}C_{24} + n\text{-}C_{26} + n\text{-}C_{28})$$

Alkane susceptibility to weathering is clearly correlated with the carbon number or chain length; that is, the shorter the chain length is (e.g., the lower the boiling points), the more easily it can be weathered. Therefore, as the weathered percentages increase, the value of the numerator significantly decreases and the denominator, in contrast, grows larger, resulting in a continuous decrease of the WI values. As for $n$-$C_{17}$/pristane and $n$-$C_{18}$/phytane, they are traditionally used as biodegradation indicators to monitor the relative degree of biodegradation in spill samples, based on the fact that branched alkanes (isoprenoids) are less susceptible to biodegradation than $n$-alkanes. However, it must be noted that, for severely weathered samples, not only $n$-alkanes but also branched and cyclo-alkanes can be heavily or completely lost, and the UCM can become very pronounced with only very few saturated hydrocarbon peaks not degraded. Therefore, one must be cautious in using the ratios of $n$-$C_{17}$/pristane and $n$-$C_{18}$/phytane to estimate the degree of oil biodegradation because the use of these ratios may underestimate the magnitude of oil biodegradation in many cases. Under such circumstances, GC-FID analysis is of little value for source identification.

### *Weathering Check Using PAHs*

Generally, the major compositional changes of alkylated PAH compounds due to weathering can be summarised as the following:

The BTEX compounds are rapidly lost as the weathering degree increases. When oils are weathered to a certain degree (approximately in the range of 20–25% weathered for most oils), the BTEX and $C_3$-benzenes could be completely lost (Wang, Fingas, Landriault, et al. 1995).

The naphthalene series is pronouncedly decreased relative to other alkylated PAH series; in particular, the losses of naphthalene and methylnaphthalenes can be very apparent.

An increase in alkylation level within a given PAH family decreases susceptibility to biodegradation, leading to the development of a profile in each alkylated PAH family from a 'bell-shaped' distribution (Sauer and Uhler 1995) to the 'inverse-sloped' distribution of $C_0$- < $C_1$- < $C_2$- <$C_3$-.

For other EPA priority unsubstituted PAHs, high molecular weigh PAHs are less susceptible to lower molecular weight ones.

No noticeable loss of the chrysene series is observed in many spill cases.

The quantitative results from PAH analysis (bottom panel, Figure 3.7) can be used as a weathering check as well. As Figure 3.7 shows, in comparison to the unweathered Alberta Sweet Mix Blend (ASMB) oil, the naphthalene was nearly completely lost and the concentrations of $C_1$- and $C_2$-naphthalenes were significantly reduced. In contrast, the concentrations of the other four target alkylated PAH series were increased due to the oil volume reduction. The CEN methodology (2002) suggests that all PAH data are normalised relative to $C_{30}$-hopane and presented graphically. For middle distillate fuels (e.g., diesels) in which no high boiling biomarkers are present, some diagnostic ratios of alkylated PAHs (such as $C_3$-dibenzothiophenes/$C_3$-chrysenes) have been used as a weathering indicator for estimation of their weathering percentages (Douglas et al. 1996). Boehm et al. (1997) reported that after the *Exxon Valdez* spill, one of the key parameters to source identification was the presence or absence of alkylated chrysenes to differentiate between crude and refined products.

The PAH compositional changes can be also used to distinguish abiotic and biotic weathering. The abiotic (physical) weathering is more predictable, especially for the PAH isomers. For example, the isomeric distributions within $C_1$-phenanthrenes (four isomers), $C_1$-dibenzothiophenes (three isomers), $C_1$-fluorenes (three isomers), and $C_1$-naphthalenes (2-methyl- and 1-methyl-naphthalene) exhibited great consistency in their relative ratios as the physical weathering percentages of the ASMB oil increased from 0 to 45% (Wang and Fingas 1995a). As for biodegradation or biotic weathering, it is generally a long-term process. It affects straight-chain *n*-alkanes more than branched alkanes, and alkanes more than other hydrocarbon classes; GC-resolved compounds more than GC-unresolved complex hydrocarbons; and small aromatics more than large aromatic compounds. More importantly, it is usually more PAH isomer specific. The study of biodegradation of various crude oils (Wang et al. 1998) revealed that bacteria preferentially degraded 2-/3-methyl dibenzothiophene among $C_1$-dibenzothiophenes, 2-methylphenanthrene among $C_1$-phenanthrenes, the first-eluted methyl-fluorene among 3 $C_1$-fluorenes, and 1,3- and 1,6-dimethyl-naphthalene among $C_2$-naphthalenes. In sharp contrast, no such compositional changes of PAH isomers were observed for physical and/or short-term weathered oils. It should be noted that the alterations in chemical composition of naturally weathered spilled oils generally result from the combined effects of abiotic and biotic weathering, rather from any single weathering effect.

### *Weathering Check Using Biomarkers*

It has been recognised that terpane and sterane compounds are very resistant to weathering. In laboratory studies of biodegradation of nine Alaska oils and oil products (Wang, Blenkinsopp, et al. 1997) and eight Canadian oils (Wang et al. 1998) by a defined bacterial consortium incubated under freshwater and cold/marine conditions, it was found that the fingerprint patterns of triterpanes and steranes showed no changes after incubation, despite extensive saturate and aromatic losses; the ratios of selected paired biomarkers also remained constant. Therefore, biomarkers can, in many cases, be and have been used as conserved internal references for estimation of oil weathering percentages. However, it should not be translated that biomarkers are not biodegradable. Biodegradation can be observed in severely weathered samples. Based on geochemical studies, Peters and Moldowan (1993) have created a 'quasi-stepwise' sequence for assessing the extent to which biomarkers have been degraded. Munoz et al. (1997) found that isoprenoids were severely degraded and biomarkers were more or less altered 8 years after an oil spill in a peaty mangrove in a tropical ecosystem. They also found that norhopanes were the most biodegradation resistant among the studied terpane and sterane groups and the $C_{30}$ αβ-hopane appeared more sensitive to weathering than its higher homologues.

In a recent study on long-term fate and persistence of the spilled Metula oil in a marine marsh environment, Wang et al. (2001) found that, in highly degraded asphalt pavement samples, even the most refractory cyclic biomarker compounds showed some degree of biodegradation. The degree of biodegradation of biomarkers was not only molecular mass and size dependent, but also stereo-isomer dependent. The biomarkers were generally degraded in the declining order of importance as: diasterane > $C_{27}$ steranes > tricyclic terpanes > pentacyclic terpanes > norhopanes ~ $C_{29}$ αββ-steranes. The degradation of steranes was in the order of $C_{27} > C_{28} > C_{29}$ with the stereochemical degradation sequence 20R ααα steranes > (20R + 20S) αββ steranes > 20S ααα steranes. For the pentacyclic homohopanes, degradation of $C_{35} > C_{34} > C_{33} > C_{32} > C_{31}$ was apparent with significantly preferential degradation of the 22R epimers over 22S epimers. $C_{30}$ αβ hopane appeared more degradable than the 22S epimers of $C_{31}$ and $C_{32}$ homohopanes, but had roughly the same biodegradation rate as the 22R epimers of $C_{31}$ and $C_{32}$ homohopanes. $C_{29}$-18α(H), 21β(H)-30 norneohopane, and $C_{29}$ αββ 20R and 20S stigmastanes appeared to be the most biodegradation-resistant terpane and sterane compounds, respectively, among the studied target biomarkers.

## A Case Study: Using a Multicriterion Approach for Source Identification of Unknown Spill Samples

An oil spill to the Rouge River and Detroit River was discovered and reported in the second week of April (April 8–13) 2002. Several thousand gallons of oil (by estimation) spilled into the Rouge River and travelled about 2 miles to the Detroit River. It then floated in several small patches down the river into northern Lake Erie. Several thousand gallons more spilled into the Rouge River during that weekend. The two spills were related and heavy rains flushed the additional oil out of the sewer and into the river. Environment Canada (EC) Ontario Region conducted an aerial survey of the Detroit River. They also surveyed the majority of the areas by vessel. The spill impacted approximately 43 km of U.S. and Canadian shorelines. The presence of a sheen over the majority of the impacted river area was observed. On the shore it appeared as a black coat and was typically 0.2–1.0 mm thick. EC Ontario Region collected a number of spill samples from various spots and sent 11 samples to the Oil Research Laboratory for analysis. The integrated multicriterion analytical approach was applied for this case study to defensibly identify the spilled oil (Wang et al. 2004).

After the sample extractions, appropriate volumes of the concentrated extracts containing approximately 30–40 mg of total solvent extractable material (TSEM) were spiked with appropriate amounts of surrogates and then quantitatively transferred into chromatographic columns for sample cleanup and fractionation. Hexane (12 mL) and 50% benzene in hexane (v/v, 15 mL) were used to elute the saturated and aromatic hydrocarbons, respectively. For each sample, half of the hexane fraction (F1) was used for analysis of the total GC-detectable saturates, *n*-alkanes and isoprenoids, and biomarker compounds; half of the 50% benzene fraction (F2) was used for analysis of alkylated homologous PAHs and other EPA priority unsubstituted PAHs; and the remaining halves of F1 and F2 were combined into a fraction (F3) and used for the determination of the TPH and UCM.

### *Product Type Screen and Determination of Hydrocarbon Groups*

Figure 3.8 shows the GC-FID chromatograms of fraction 3 of three representative spill samples for TPH and *n*-alkane analysis. The saturated fractions F1 demonstrated very similar GC-FID chromatogram profiles to their corresponding fraction 3. Table 3.8 summarises the hydrocarbon group analysis results of the spill samples.

The major chemical composition features of TPH and saturate hydrocarbons in the samples are summarised as follows:

The GC traces of both the F1 and F3 of the spill samples were clearly dominated by large UCM (located in the $n$-$C_{18}$–$n$-$C_{36}$ range) with almost no *n*-alkane being detected after $n$-$C_{20}$. The ratios of all the GC-resolved peaks to the total GC area were determined to be only 0.06 for three samples (Table 3.8). The GC chromatographic profile and shape of the UCM humps are significantly different from those of crude oils and most refined products. In addition, the ratios of the total saturates to the GC-TPH were determined to be about 90%, much higher than that for most crude oils. All the GC trace features (Figure 3.8) suggest that the major portion of the spilled oil might be a lubricating oil.

The resolved *n*-alkanes mainly distributed in the diesel carbon range ($C_8$–$C_{27}$), suggesting the minor portion of the spill oil was a diesel. No *n*-alkane with a carbon number smaller than $C_{10}$ and greater than $C_{24}$ was detected. The total *n*-alkanes including pristane and phytane were

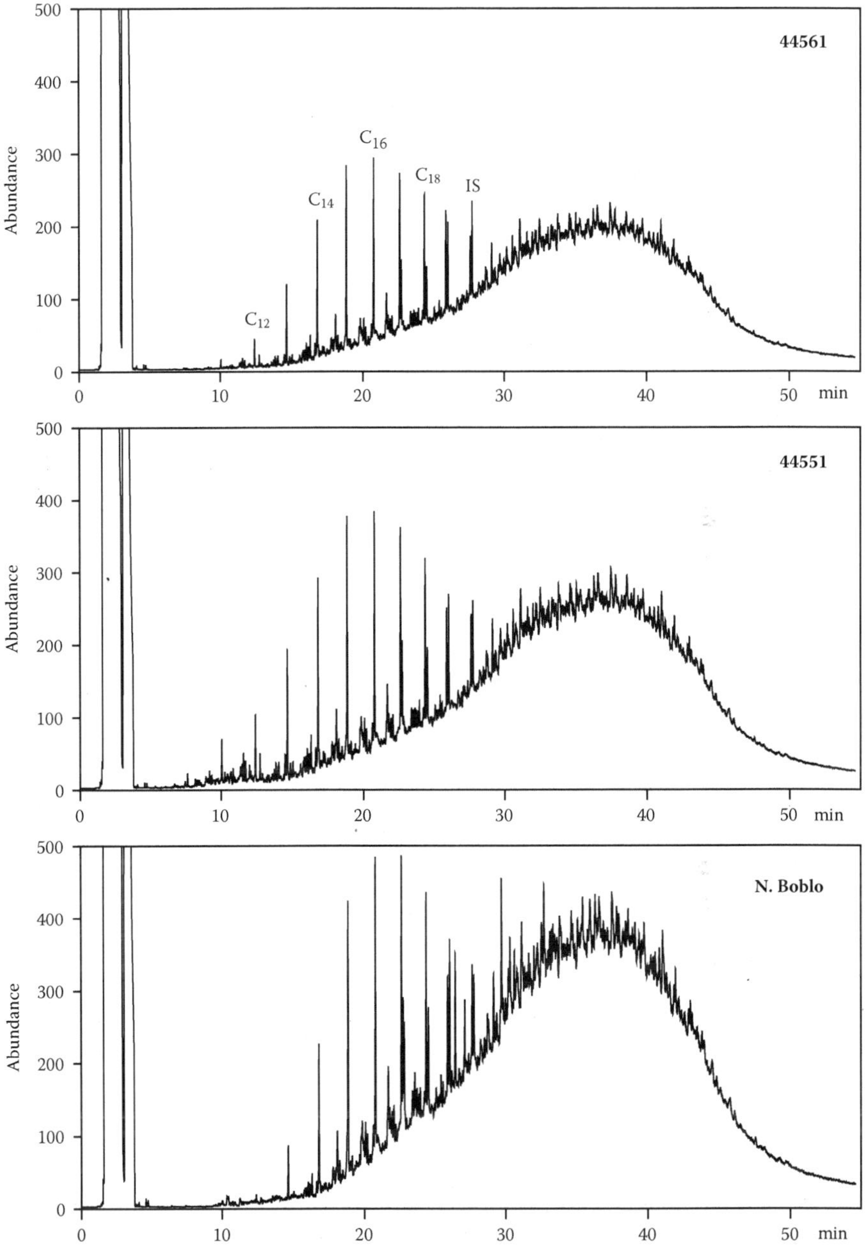


**Figure 3.8** GC-FID chromatograms of three Detroit River spill samples. The GC traces are featured by dominance of large UCM with small amounts of resolved peaks being detected in the lubricating oil carbon range (retention time: 24–50 min).

**Table 3.8 Hydrocarbon Group Analysis Results of the Detroit River Spill Samples**

| Samples | 44561 | 44551 | N. Boblo |
|---|---|---|---|
| Sample weight (g) | 19.8 | 122 | 5.30 |
| Final volume of extract (mL) | 25.0 | 100 | 5.0 |
| Total TSEM (g) | 7.59 | 40.9 | 0.46 |
| TSEM conc. ($mg.g^{-1}$ sample) | 384 | 335 | 87.5 |
| GC-TPH ($mg.g^{-1}$ TSEM) | 449 | 494 | 513 |
| GC-saturates ($mg.g^{-1}$ TSEM) | 398 | 442 | 452 |
| GC-saturates/GC-TPH (%) | 89 | 90 | 88 |
| GC-aromatics/GC-TPH (%) | 11 | 10 | 12 |
| Resolved peaks/total GC area (F3) | 0.06 | 0.06 | 0.06 |
| UCM/GC-TPH (F3) | 0.94 | 0.94 | 0.94 |
| Total *n*-alkanes ($mg.g^{-1}$ TSEM) | 9.30 | 10.4 | 8.58 |
| $n$-$C_{17}$/pristane | 1.93 | 2.00 | 2.09 |
| $n$-$C_{18}$/phytane | 1.78 | 1.79 | 1.83 |
| Pristane/phytane | 1.19 | 1.14 | 1.09 |

determined to be only 9.3, 10.4, and 8.6 $mg.g^{-1}$ of TSEM for samples 1, 2, and 3, respectively. Using the estimated value of 120 mg *n*-alkanes per gram of diesel and in consideration of weathering effect, the percentage of diesel in the spill samples may be estimated not to exceed 20% of the total hydrocarbons detected.

Three samples showed nearly identical GC chromatographic profiles and *n*-alkane distribution patterns, as well as the nearly identical diagnostic ratios (Table 3.8) of $n$-$C_{17}$/pristane, $n$-$C_{18}$/phytane, and pristane/phytane. This implies that they were most likely the same oil and from the same source, and some small differences were likely caused by weathering.

All quantitative GC results implied that the spill samples were largely composed of lubricating oil mixed with a smaller portion of diesel fuel, the diesel in the samples had been weathered and degraded, and the diesel portion in sample 3 had been more weathered (most probably by more evaporation and water washing in its longer journey from the spill source to the destination) than samples 1 and 2.

### *Determination of Oil-Characteristic Alkylated PAH Homologues and Their Diagnostic Ratios*

PAH analysis results (Figure 3.9) demonstrate the following:

The relative distribution patterns and profiles of alkylated PAHs are very much the same for the spilled samples, in particular for samples 1 and 2, further implying they were from the same source.

The five target alkylated PAH homologous series and other EPA priority PAHs were determined to be 1404, 1479, and 1028 μg.g$^{-1}$ TSEM, and 250, 257, and 167 μg.g$^{-1}$ TSEM for samples 1, 2, and 3, respectively. Compared to crude oils and most refined products such as jet fuel and diesel (>10,000 μg.g$^{-1}$ for most oils), the PAH concentrations in these spill samples are relatively low. The dominance of alkylated naphthalene and phenanthrenes among five target alkylated PAH homologous series is pronounced for all three samples.

Sample 2 still contained small amounts of BTEX and $C_3$-benzene compounds. In comparison, almost no BTEX and other alkyl benzene compounds were detected in samples 1 and 3. This fact further demonstrates that sample 2 was least weathered. The loss of lighter molecular weight naphthalene and $C_1$- and $C_2$-naphthalenes was obvious for all three samples, resulting in development of the relative distribution of $C_0$-N < $C_1$-N < $C_2$-N < $C_3$-N. This relative distribution pattern is particularly obvious for the more weathered sample 3. For other EPA priority PAHs, the more weathered sample 3 also demonstrated lower concentrations of lighter two- and three-ring PAHs (biphenyl, acenaphthylene, and acenaphthene).

Analysis of the diagnostic ratios of source-specific PAH isomers clearly revealed that the relative distribution of PAH isomers 4-, 2-/3-, and 1-methyl dibenzothiophene at *m/z* 198 and (3- + 2-methyl-phenanthrene) to (4-/9- + 1-methyl-phenanthrene) at *m/z* 192 were found to be very closely matching; the double ratios (C2D/C2P:C3D/C3P) were also nearly identical (0.22:0.31, 0.22:0.30, and 0.23:0.30 for samples 1, 2, and 3, respectively).

It has been well demonstrated that, in general, lubricating oils only contain small quantities of PAH compounds, while PAH concentrations are high in diesel. Obviously, detected PAHs in these spill samples were largely contributed by the small portion of diesel in spill samples.

### *Input of Pyrogenic PAHs to the Spill Samples*

Another pronounced PAH compositional feature (Figure 3.9) is that among the alkylated phenanthrene, fluorene and chrysene series, the parent PAHs are most abundant and their concentrations are even higher than their corresponding alkylated homologous constituents. In particular, the highest abundance of parent chrysene over its alkyl-substituted homologues and the decrease in relative abundances with increasing level of alkylation (i.e., in the order of $C_0$-C > $C_1$-C > $C_2$-C > $C_3$-C) was very pronounced. This kind of PAH distribution profile has been generically termed as 'skewed' or 'sloped'. The pyrogenic index was determined to be as high as 0.16 for three samples, far higher than the corresponding values for crude oils and refined products

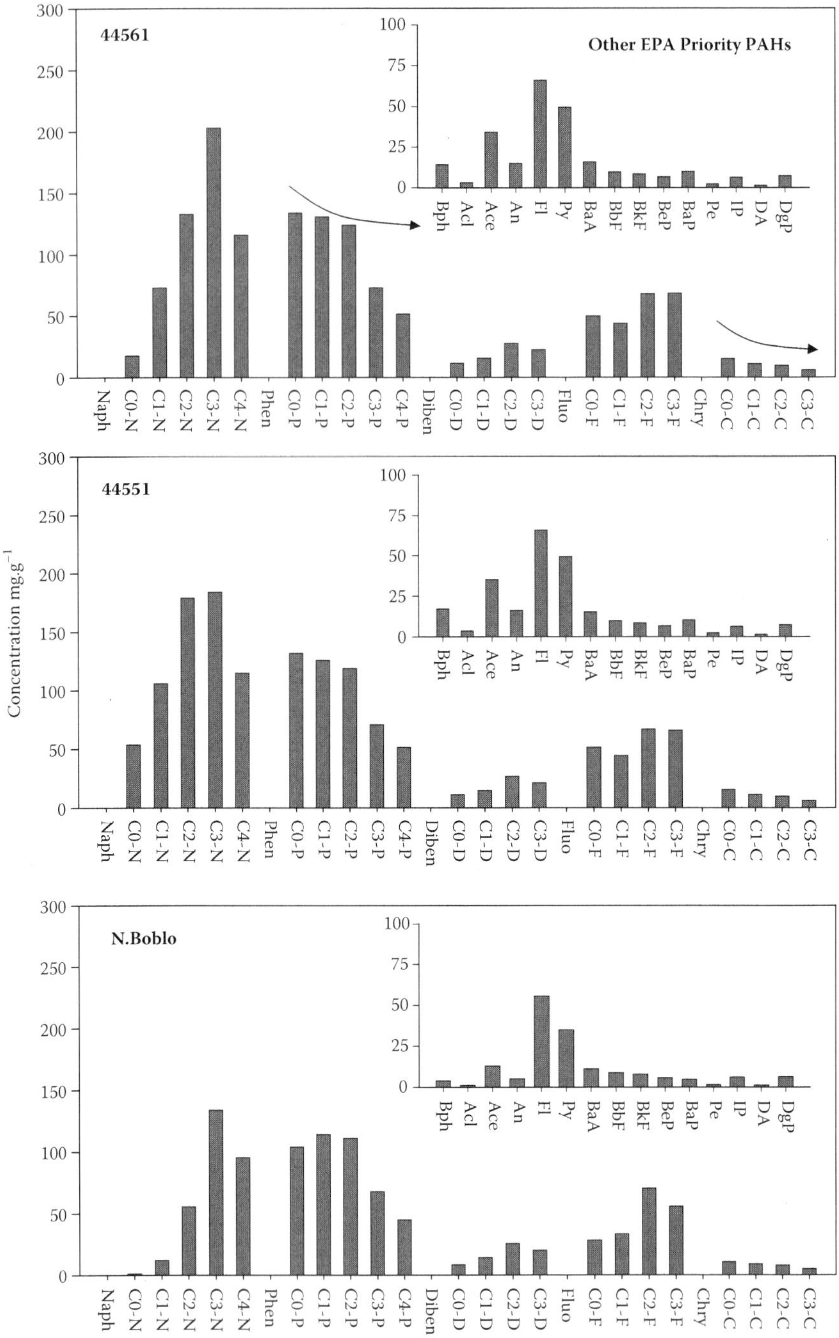


**Figure 3.9** Distribution of alkylated PAHs in the Detroit River spill samples. The distributions of other EPA priority PAHs are shown in the inserts. The input of pyrogenic PAHs is clearly demonstrated.

(exclusively smaller than 0.06), defensively indicating the formation and presence of pyrogenic PAHs in the spill samples. In addition, the relative ratios of chrysene to benz[a]anthracene were determined to be very close to 1.0, also far higher than the same ratios for crude oils and refined products. All these features indicate the input of pyrogenic PAHs.

The most likely source of pyrogenic PAHs in used motor oils is combustion 'blow-by' past the piston rings of exhaust gasses directly into the crankshaft cavity. Excessive heat in the motor lubrication process can also increase the concentration of PAHs, in particular the high molecular weight PAHs, in used lubricating oils. Therefore, it can be reasonably concluded that the pyrogenic PAHs found in the spilled oil were most probably produced from combustion and motor lubrication processes, and the oil in these spill samples was waste lubricating oil.

### *Characterisation of Biomarker Compounds*

Biomarker characterisation results reveal the following:

The samples show nearly identical distribution patterns of biomarkers and these biomarkers were mostly from the lubricating oil portion of the spill samples. It has been demonstrated that diesels do not contain high molecular weight biomarkers and only contain traces of low molecular weight biomarker compounds ($C_{20}$–$C_{24}$).

The totals of the target biomarkers were determined to be 1103, 941, and 941 $\mu g.g^{-1}$ TSEM for samples 1, 2, and 3, respectively.

The diagnostic ratios of target biomarker compounds $C_{23}/C_{24}$, $C_{29}$ αβ-hopane/$C_{30}$ αβ-hopane, Ts/Tm, $C_{31}$(22S)/$C_{31}$(22S+22R), $C_{32}$(22S)/$C_{32}$(22S+22R), $C_{33}$(22S)/$C_{33}$(22S+22R), $C_{34}$(22S)/$C_{34}$(22S+22R), $C_{35}$(22S)/$C_{35}$(22S+22R), and $C_{31}$/($C_{31}$ to $C_{35}$) were very similar.

All this evidence, in combination with the TPH and PAH analysis results, unambiguously points to the conclusion that the three spill samples came from the same source.

It is important to note that the fingerprinting results described previously highlight the necessity to analyse for more than one suite of analytes in source identification. Characterisation of PAH and biomarker compounds must include determination of both concentrations and relative distributions and should not measure peak ratios alone. This is important because it is possible to have a situation where a source might have a similar biomarker ratio but very different actual amounts of biomarkers. In summary, the fingerprinting results described earlier highlight the necessity to analyse for more than one suite of analytes in forensic investigation and spill source identification:

The spill samples were largely composed of used lubricating oil mixed with a smaller portion of diesel fuel.
The diesel in the samples had been weathered and degraded.
The diesel portion in sample 3 collected from N. Boblo Island was more weathered (most probably by more evaporation and water washing) than samples 1 and 2.
Three samples were from the same source.
Most PAH compounds were from the diesel portion in the spill samples, while the biomarker compounds were largely from the lube oil portion.
Input of pyrogenic PAHs, most probably produced from combustion and motor lubrication processes, was apparent.

## Conclusions

The advances in petroleum hydrocarbon fingerprinting and data interpretation methods and approaches in the last two decades have now allowed for detailed qualitative and quantitative characterisation of spilled oils. Chemical fingerprinting is a powerful tool for hydrocarbon source identification and differentiation, when it is applied properly. However, in many cases, particularly for complex hydrocarbon mixtures or extensively weathered and degraded oil residues, no single fingerprinting analysis can meet the objectives of forensic investigation and quantitatively allocate hydrocarbons to their respective sources. Under such circumstances, integrated multiple-parameter approaches are always needed and used, more than one suite of analytes must be performed, and other independent techniques such as isotope analysis may be applied to support correlations. If a large number of spill and source candidate samples are involved, statistical and numerical analysis techniques (such as principal component analysis) for data analysis are always performed. Development in hydrocarbon fingerprinting techniques will continue as analytical and statistical techniques evolve. It can be anticipated that these developments will further enhance the utility and defensibility of oil hydrocarbon fingerprinting.

## Acknowledgements

We thank Dr. Chun Yang and Mr. Mike Landriault of Emergencies Science and Technology Division for performing some laboratory work and working on some graphics.

## References

Alexander, R., R. I. Kagi, R. A. Noble, and J. K. Volkman. (1984) Identification of some bicyclic alkanes in petroleum. In *Advances in organic geochemistry,* vol. 6, ed. P. A. Schenck, J. W. De Leeuw, and G. W. M. Lijmbach. Oxford: Pergamon Press, 1983.

ASTM Method 3328-90. (1997) In *Annual book of ASTM standards, water (II),* vol. 11.02. Philadelphia, PA: American Society for Testing and Materials.

Barakat, A. O., A. Mostafa, M. S. El-Gayar, and J. Rullkotter. (1997) Source-dependent biomarker properties of five crude oils from the Gulf of Suez, Egypt. *Organic Geochemistry,* 26: 441–450.

Bence, A. E., and W. A. Burns. (1995) Fingerprinting hydrocarbons in the biological resources of the *Exxon Valdez* spill area. In Exxon Valdez *oil spill: Fate and effects in Alaska waters,* ed. P. G. Wells, J. N. Butler, and J. S. Hughes, 84–140. Philadelphia, PA: ASTM.

Bence, A. E., K. A. Kvenvolden, and M. C. Kennicutt, II. (1996) Organic geochemistry applied to environmental assessments of Prince William Sound, Alaska, after the *Exxon Valdez* oil spill—A review. *Organic Geochemistry,* 24: 7–42.

Benlahcen, K. T., A. Chaoui, H. Budzinski, J. Bellocq, and P. Garrigues. (1997) Distribution and sources of polycylic aromatic hydrocarbons in some Mediterranean coastal sediments. *Marine Pollution Bulletin,* 34: 298–305.

Bjøeseth, A., and T. Ramdahl. (1985) Chapter 1: Source and emissions of PAH. In *Handbook of polycyclic aromatic hydrocarbons, volume 2,* ed. A. Bjøeseth and T. Ramdahl, 1–20. New York: Marcel Dekker.

Blumer M., and W. W. Youngblood. (1975) Polycyclic aromatic hydrocarbons in soils and recent sediments. *Science,* 188: 53–55.

Boehm, P. D., G. S. Douglas, W. A. Burns, P. J. Mankiewicz, D. S. Page, and A. E. Bence. (1997) Application of petroleum hydrocarbon chemical fingerprinting and allocation techniques after the *Exxon Valdez* oil spill. *Marine Pollution Bulletin,* 34: 599–613.

Boehm, P. D., D. S. Page, W. A. Burns, A. E. Bence, P. J. Mankiewicz, and J. S. Brown. (2001) Resolving the origin of the petrogenic hydrocarbon background in Prince William Sound, Alaska. *Environmental Science & Technology,* 35: 471–479.

Boehm, P. D., D. S. Page, E. S. Gilfillan, A. E. Bence, W. A. Burns, and P. J. Mankiewicz. (1998) Study of the fate and effects of the *Exxon Valdez* oil spill on benthic sediments in two bays in Prince William Sound, Alaska. 1. Study design, chemistry and source fingerprinting. *Environmental Science & Technology,* 32: 567–576.

Bouyssiere, B., P. Leonhard, D. Profrock, F. Baco, C. L. Garcia, S. Wibur, and A. Pronge. (2004) Investigation of the sulfur speciation in petroleum products by capillary GC with ICP-collision cell-MS detection. *Journal of Analytical Atomic Spectrometry,* 19: 700–702.

CEN. (2002) Revision of the Nordtest methodology for oil spill identification for the European Committee for Standardization (CEN), SINTEF Report STF66 A02027, Norway.

Cheng, K., W. Jin, Z. He, and J. Chen. (1991) Application of sesquiterpanes to the study of oil–gas source: The gas–rock correlation in the Qingdongnan Basin. *Journal of Southeast Asian Earth Science,* 5: 189–195.

Cretney, W. J., D. R. Green, B. R. Fowler, B. Humphrey, D. L. Fiest, and P. D. Boehm. (1987) Hydrocarbon biogeochemical setting of the Baffin Island oil spill experimental site. I, Sediments. *Arctic,* 40: 51–55.

Dahlmann, G. (2003) *Characteristic features of different oil types in oil spill identification.* Germany: Berichte des BSH 31, ISSN 0946-6010, 48 pp.

Dallüge, J., J. Beens, and U. Brinkman. (2003) Comprehensive two-dimensional GC: A powerful and versatile analytical tool. *Journal of Chromatography, A,* 1000: 69–108.

Dimandja, J. D. (2004) GC × GC. *Analytical Chemistry,* 76: 167A–174A.

DOE. (2004) Energy Information Administration Web site, United States Department of Energy, http://eia.doe.gov/.

Douglas, G. S., A. E. Bence, R. C. Prince, S. J. McMillen, and E. L. Butler. (1996) Environmental stability of selected petroleum hydrocarbon source and weathering ratios. *Environmental Science & Technology,* 30: 2332–2339.

Douglas, G. S., W. A. Burns, A. E. Bence, D. S. Page, and P. Boehm. (2004) Optimizing detection limits for the analysis of petroleum hydrocarbons in complex environmental samples. *Environmental Science & Technology,* 38: 3958–3964.

Dutta, T. K., and S. Harayama. (2001) Analysis of long-side-chain alkylaromatics in crude oil for evaluation of their fate in the environment. *Environmental Science & Technology,* 35: 102–107.

EPA. (1997) *EPA test methods for evaluating solid waste* (SW-846), update III. U.S. EPA, Office of Solid Waste and Emergency Response, Washington, D.C.

EPA. (1998a) *EPA guidance for quality assurance project plans,* EPA QA/G-5. U.S. EPA, Washington, D.C.

EPA. (1998b) *Guidance for data quality assessment, practical method for data analysis,* EPA AQ/G-9 QA 97. Updated, U.S. EPA, Washington, D.C.

EPA. (2001) *EPA requirements for quality assurance project plans,* EPA QA/R-5. U.S. EPA, Washington, D.C.

ETC (Environmental Technology Center). (2003) Quality assurance Web site: http://qa.etc.ec.gc.ca/.

ETC Method. (2002) *Analytical methods for determination of oil components,* ETC Method No.: 5.3/1.3/M, 2002. Environmental Technology Center, Environment Canada, Ottawa, Ontario.

Faksness, L. G., P. S. Daling, and A. B. Hansen. (2002) Round robin study—Oil spill identification. *Environmental Forensics,* 3: 279–292.

Fan, P., Y. Qian, and B. Zhang. (1991) Characteristics of biomarkers in the recent sediments from Qinghai Lake, Northwest China. *Journal of Southeast Asian Earth and Science,* 5: 113–128.

Fayad, N. M., and E. Overton. (1995) A unique biodegradation pattern of the oil spilled during the 1991 Gulf War. *Marine Pollution Bulletin,* 30: 239–246.

Fingas, M. (2001) *The basics of oil spill cleanup,* 2nd ed. New York: Lewis Publishers.

Frysinger, G., R. B. Gaines, L. Xu, and C. M. Reddy. (2003) Resolving the unresolved complex mixture in petroleum-contaminated sediments. *Environmental Science & Technology,* 37: 1653–1662.

Frysinger, G. S., and R. B. Gaines. (2001) Separation and identification of petroleum biomarkers by comprehensive two-dimensional gas chromatography. *Journal of Separation Science,* 24: 87–96.

Gaines, R. B., G. S. Frysinger, M. A. Hendrick-Smith, and J. Stuart. (1999) Oil spill source identification by comprehensive two-dimensional gas chromatography. *Environmental Science & Technology,* 33: 2106–2112.

Hostettler, F. D., R. J. Rosenbauer, and K. A. Kvenvolden. (1999) PAH refractory index as a source discriminant of hydrocarbon input from crude oil and coal in Prince William Sound, Alaska. *Organic Geochemistry,* 30: 873–879.

Hua, R., Y. Li, W. Liu, J. Zheng, and H. Wei. (2003) Determination of sulphur-containing compounds in diesel oils by comprehensive two-dimensional GC with a sulphur chemiluminescence detector. *Journal of Chromatography, A,* 1019: 101–109.

Kolattukudy, P. E. (1976) *Chemistry and biochemistry of natural waxes.* New York: Elsevier.

Kvenvolden, K. A., J. B. Rapp, and J. H. Bourell. (1985) In *Alaska North Slope oil/rock correlation study,* ed. L. B. Magoon and G. E. Claypool. American Association of Petroleum Geologists Studies in Geology, No. 20, 593–617.

Marshall, A. G., S. Kim, T. M. Schaub, J. M. Purcell, D. F. Smith, and R. P. Rodgers. (2005) Characterisation of petroleum by high resolution field desorption/ionisation and atmospheric pressure photoionisation FT-ICR mass spectrometry. *Preprints, ACS Fuel Chemistry Division,* 50(1): 221–222.

Marshall, A. G., and R. Rodgers. (2004) Petroleomics: The next grand challenge for chemical analysis. *Accounts of Chemical Research,* 37: 53–59.

McKirdy, D. M., R. E. Cox, J. K. Volkman, and V. J. Howell. (1986) Botryococcane in a new class of Australian nonmarine crude oils. *Nature,* 320: 57–59.

Meniconi, M. G., I. T. Gabardo, M. E. Carneiro, S. M. Barbanti, G. C. Silva, and C. G. Massone. (2002) Brazilian oil spills chemical characterization—Case studies. *Environmental Forensics,* 3: 303–322.

Meyer, S., S. Cartellieri, and H. Steinhart. (2001) Simultaneous determination of PAHs hetero-PAHs, and their degradation products in creosote-contaminated soils. *Analytical Chemistry,* 71: 4023–4029.

Munoz, D., M. Guiliano, P. Doumenq, F. Jacquot, P. Scherrer, and G. Mille. (1997) Long-term evolution of petroleum biomarkers in mangrove soil. *Marine Pollution Bulletin,* 34: 868–874.

NRC (National Research Council). (2002) *Oil in the sea III: Inputs, fates, and effects.* Washington, D.C.: The National Academies Press.

Olah, G. A., and A. Molnar. (1995) *Hydrocarbon chemistry.* New York: Wiley-Interscience.

Page, D. S., P. D. Boehm, G. S. Douglas, and A. E. Bence. (1995) Identification of hydrocarbon sources in the benthic sediments of Prince William Sound and the Gulf of Alaska following the Exxon Valdez spill. In: *Exxon Valdez oil spill: Fate and effects in Alaska waters,* ed. P. G. Wells, J. N. Butler, and J. S. Hughes, 41–83, ASTM STP 1219. Philadelphia, PA: ASTM.

Peters, K. E., and J. W. Moldowan. (1993) *The biomarker guide: Interpreting molecular fossils in petroleum and ancient sediments.* Englewoood Cliffs, NJ: Prentice Hall.

Phillips, J. B., and Beens, J. (1999) Comprehensive two-dimensional GC: A hyphenated method with strong coupling between the two dimensions. *Journal of Chromatography, A,* 856: 331–347.

Philp, R. P. (1985) *Fossil fuel biomarkers, application and spectra.* New York: Elsevier.

Philp, R. P., T. D. Gilbert, and J. Riedrich. (1981) Bicyclic sesquiterpanoids and triterpenoids in Australia crude oils. *Geochimica Cosmochimica Acta,* 45: 1173–1180.

Qian, K., and G. Dechert. (2002) Recent advance in petroleum characterization by GC field time-of-flight high resolution mass spectrometry. *Analytical Chemistry,* 74: 3977–3983.

Radke, M. D., H. Welte, and H. Willsch. (1986) Maturity parameters based on aromatic hydrocarbons: Influence of the organic matter type. *Organic Geochemistry,* 10: 51–63.

Reddy, C. M., T. I. Eglinton, A. Hounshell, H. K. White, L. Xu, R. B. Gaines, and G. S. Frysinger. (2002) The West Falmouth oil spill after 30 years: The persistence of petroleum hydrocarbons in salt marsh sediments. *Environmental Science & Technology,* 36: 4754–4760.

Sauer, T. C., and A. D. Uhler. (1995) Pollutant source identification and allocation: Advances in hydrocarbon fingerprinting. *Remediation,* 25–50.

Scott, A. S., D. U. Allen, and J. M. Kevin. (2005) Middle distillate fuel fingerprinting using drimane-based bicyclic sesquiterpanes. *Environmental Forensics,* 6: 241–252.

Sicre, M. A., J. C. Marty, A. Salion, X. Aparicio, J. Grimalt, and J. Albaiges. (1987) Aliphatic and aromatic compounds in aerosols. *Atmospheric Environment,* 21: 2247–2259.

Speight, J. G. (2002) *Handbook of petroleum product analysis.* Hoboken, NJ: Wiley-Interscience.

Stout, S. A., A. D. Uhler, and K. J. McCarthy. (2000) Recognizing the confounding influences of 'background' contamination in 'fingerprinting' investigation. *Soil, Sediment and Groundwater,* 35–38.

Stout, S. A., A. D. Uhler, K. J. McCarthy, and S. Emsbo-Mattingly. (2002) In *Introduction to environmental forensics,* ed. B. L. Murphy and R. D. Morrison, 137–260. London: Academic Press.

Stout, S. A., A. D. Uhler, and K. J. McCarthy. (2005) Middle distillate fuel fingerprinting using drimane-based bicyclic sesquiterpanes. *Environmental Forensics* 6(3): 241–251.

Tolosa, I., S. de Mora, M. Sheikholeslami, J. Villeneuve, J. Bartocci, and C. Cattini. (2004) Aliphatic and aromatic hydrocarbons in coastal Caspian Sea sediments. *Marine Pollution Bulletin,* 48: 4–60.

Uhler, A. D., S. A. Stout, and K. J. McCarthy. (1998) Increased success of assessments at petroleum sites in 5 steps. *Soil and Groundwater Cleanup,* Dec/Jan: 13–19.

Van Stee, L. L. P., J. Beens, R. J. J. Vreuls, and U. Brinkman. (2003) Comprehensive two-dimensional GC: A powerful and versatile analytical tool. *Journal of Chromatography, A,* 1000: 69–108.

Venkatesan, M. I. (1988) Occurrence and possible sources of perylene in marine sediments—A review. *Marine Chemistry,* 25: 1–27.

Wang, Z. D., S. Blenkinsopp, M. Fingas, G. Sergy, M. Landriault, L. Sigouin, J. Foght, K. Semple, and D. W. S. Westlake. (1997) Chemical composition changes and biodegradation potentials of nine Alaska oil under freshwater incubation conditions. *Preprints of Symposia, American Chemical Society,* 43: 828–835.

Wang, Z. D., and M. Fingas. (1995a) Study of the effects of weathering on the chemical composition of a light crude oil using GC/MS and GC/FID. *Journal of Microcolumn,* September, 7: 617–639.

______. (1995b) Use of methyldibenzothiophenes as markers for differentiation and source identification of crude and weathered oils. *Environmental Science & Technology*, 29: 2841–2849.

Wang, Z. D., M. Fingas, S. Blenkinsopp, G. Sergy, M. Landriault, and L. Sigouin. (1998a) Study of the 25-year-old Nipisi oil spill: Persistence of oil residues and comparisons between surface and subsurface sediments. *Environmental Science & Technology*, 32: 2222–2232.

Wang, Z. D., M. Fingas, S. Blenkinsopp, G. Sergy, M. Landriault, L. Sigouin, J. Foght, K. Semple, and D. W. S. Westlake. (1998) Comparison of oil composition changes due to biodegradation and physical weathering in different oils. *Journal of Chromatography*, 809: 89–107.

Wang, Z. D., M. Fingas, M. Landriault, L. Sigouin, B. Castel, D. Hostetter, D. Zhang, and B. Spencer. (1998) Identification and linkage of tarballs from the coasts of Vancouver Island and northern California using GC/MS and isotopic techniques. *Journal of High Resolution Chromatography*, 21: 383–395.

Wang, Z. D., M. Fingas, M. Landriault, L. Sigouin, Y. Feng, and J. Mullin. (1997) Using systematic and comparative analytical data to identify the source of an unknown oil on contaminated birds. *Journal of Chromatography*, 775: 251–265.

Wang, Z. D., M. Fingas, M. Landriault, L. Sigouin, and N. Xu. (1995a) Identification of alkylbenzenes and direct determination of BTEX and (BTEX + $C_3$-benzenes) in oils by GC/MS. *Analytical Chemistry*, 67: 3491–3500.

Wang, Z. D., M. Fingas, and P. Lambert. (2004) Characterization and identification of Detroit River mystery oil spill (2002). *Journal of Chromatography*, 1038: 201–214.

Wang, Z. D., M. Fingas, and K. Li. (1994) Fractionation of ASMB oil, identification and quantitation of aliphatic, aromatic and biomarker compounds by GC/FID and GC/MSD. *Journal of Chromatographic Science*, 32: 361–366; 367–382.

Wang, Z. D., M. Fingas, E. H. Owens, L. Sigouin, and C. E. Brown. (2001) Long-term fate and persistence of the spilled Metula oil in a marine salt marsh environment: Degradation of petroleum biomarkers. *Journal of Chromatography*, 926: 275-190.

Wang, Z. D., M. Fingas, and D. Page. (1999) Oil spill identification. *Journal of Chromatography*, 843: 369–411.

Wang, Z. D., M. Fingas, and G. Sergy. (1994) Study of 22-year-old Arrow oil samples using biomarker compounds by GC/MS. *Environmental Science & Technology*, 28: 1733–1746.

______. (1995) Chemical characterization of crude oil residues from an arctic beach by GC/MS and GC/FID. *Environmental Science & Technology*, 29: 2622–2631.

Wang, Z. D., M. Fingas, Y. Y. Shu, L. Sigouin, M. Landriault, and P. Lambert. (1999) Quantitative characterization of PAHs in burn residue and soot samples and differentiation of pyrogenic PAHs from petrogenic PAHs—The 1994 Mobile burn study. *Environmental Science & Technology*, 33: 3100–3109.

Wang, Z. D., M. Fingas, and L. Sigouin. (2002) Using multiple criteria for fingerprinting unknown oil samples having very similar chemical composition. *Environmental Forensics*, 3: 251–262.

Wang, Z. D., B. P. Hollebone, M. Fingas, B. Fieldhouse, and J. Weaver. (2003) Development of a physical and chemical property database for ten US EPA-selected oils. In *Proceedings of the 26th Arctic and Marine Oil Spill Program (AMOP) Technical Seminar*, Environment Canada, Ottawa, pp. 111–142.

Wang, Z. D., C. Yang, M. Fingas, B. Hollebone, A. S. Hansen, and J. H. Christensen. (2005) Characterization, weathering, and application of sesquiterpanes to source identification of spilled lighter petroleum products. *Environmental Science & Technology,* 39: 8700–8707.

Zakaria, M. P., A. Horinouchi, S. Tsutsumi, H. Takada, S. Tanabe, and A. Ismail. (2000) Oil pollution in the Straits of Malacca, Malaysia: Application of molecular markers for source identification. *Environmental Science & Technology,* 34: 1189–1196.

Zakaria, M. P., T. Okuda, and H. Takada. (2001) PAHs and hopanes in stranded tarballs on the coast of Peninsular Malaysia: applications of biomarkers for identifying source of oil pollution. *Marine Pollution Bulletin,* 12: 1357–1366.

# Biomarkers and Stable Isotopes in Environmental Forensic Studies

# 4

R. PAUL PHILP
TOMASZ KUDER

## Contents

Introduction....114
Biomarkers in Environmental Forensics....115
Biomarkers in Environmental Forensics of Crude Oil and Refined Products....120
Other Forensic Applications of Biomarkers....131
Stable Isotopes in Environmental Forensics....132
Isotope Geochemistry—Principles....134
Isotope Fractionation....135
Stable Isotopes in Tracking Contaminant Sources....137
Data Interpretation—Different Contaminant Sources or Diagenetic Changes?....139
When CSIA Works for Identification of Source Signatures....139
Field Applications....142
Stable Isotopes in Contaminant Attenuation Studies....145
Qualitative Evidence of Biodegradation....146
Quantitative Interpretation of CSIA....147
Interpreting the Biodegradation of a Sequence of Intermediates....152
Identification of the Mechanism of Biodegradation....154
Limitations of CSIA....154
Applications of CSIA....156
Dating of Contaminant Spills....157
Summary....160
References....161

## Introduction

The past decade has seen an exponential increase in the number of forensic geochemical applications discussed in the literature and utilised in litigation. The reasons for this are numerous and some of these are summarised here. First, petroleum geochemistry is now a mature science (Peters, Walters, and Moldowan 2005); many of the techniques developed for correlation of crude oils and suspected source rocks or other oils based on the biomarker concept are used routinely in forensic applications for correlation of spilled products, both crude oils and refined products (see chapter 3). Second, the past decade has seen a tremendous increase in the number of isotope applications to forensic problems. Determination of stable isotopes is not new and the capability has been there for more than 50 years. However, the ability to determine the isotopic composition of individual compounds in complex mixtures is relatively new and came about with the development and commercial availability of the combined gas chromatograph–isotope ratio mass spectrometer (GC-IRMS) in the late 1980s and early 1990s. Initially, only carbon isotopes could be determined with this approach, but now it is possible to determine hydrogen and nitrogen isotopes in the same manner.

The coming together of the mature techniques with the developments in isotope geochemistry has led to significant advances in the ability to undertake source determinations at contaminated sites, unravel commingled plumes of contaminants, and evaluate the onset or state of natural attenuation at a contaminated site. It is the purpose of this chapter to briefly review the major developments in these two important areas and illustrate them with examples from the recent literature. For the most part, comments will be limited to hydrocarbon-related products in the environment, along with chlorinated compounds and the omnipresent MTBE (methyl *t*-butyl alcohol). There are numerous other applications, but these are currently the areas receiving the most attention in the forensic geochemistry arena. Any forensic geochemistry study is basically directed toward answering four major questions:

What is the product?
Where did it come from?
How long has it been there?
Is it going away or degrading naturally?

As you may well imagine, all four of these questions to some degree or other are geared to answering the ultimate question: Who is going to pay for the cleanup?

It is our intent to review this topic by addressing the first four questions; the ultimate question does not need to be addressed here since it depends upon the data obtained in answering the first four questions. Each question will be evaluated and advances made in these areas discussed along with examples. When appropriate, reference will be made to earlier reviews on similar topics for the sake of brevity. It is also important to emphasise that, wherever possible, both of these approaches—the isotopes and the GC-MS analyses—should be used to complement each other. This may not always be possible (e.g., in the case of single components), but more complex mixtures will certainly benefit from both techniques being used together.

## Biomarkers in Environmental Forensics

The concept of biomarkers was introduced in the late 1960s and early 1970s by Eglinton and Calvin (1967) at the same time that significant developments were taking place in analytical instrumentation, particularly the combination of gas chromatography and mass spectrometry (GC-MS) and associated ancillary techniques such as single ion monitoring (SIM) or multiple ion detection (MID). It is very important to clarify the history of biomarker development in the context of forensic geochemistry since a cursory examination of the contemporary geochemical literature does not give significant credit to this early pioneering work or simply ignores it. There have been many papers published in the past 2 or 3 years, particularly in the environmental field, that give the impression that the authors were totally responsible for discovering and developing the concept of biomarkers. This is very unfortunate and short sighted, and provides a distorted view of the history and development of this approach. Blumer (1972) and co-workers (Blumer, Souza, and Sass 1970) were among the earliest scientists who pioneered fingerprinting of marine pollutant hydrocarbons, and the significance and utility of biomarkers in pollution studies have been thoroughly reviewed in Volkman, Revill, and Murray (1997). Thereafter, numerous workers have extensively used molecular markers for identification of the sources of the pollutants (Albaiges and Albrecht 1979; Farran et al. 1987; Kvenvolden et al. 1993; Hosteller and Kvenvolden 1994; Prince et al. 1994; Wang, Fingas, and Sergy 1994; Bence, Kvenvolden, and Kennicutt 1996).

The concept of biomarkers had its origin in early studies on the origin of life and the search for earliest life forms in Precambrian rocks and organic compounds in the returned lunar samples of the 1970s (Eglinton and Calvin 1967; Calvin 1969). The Precambrian studies identified compounds in these extremely old rocks that could be associated with blue green algae and other early life forms (Burlingame et al. 1965). Many of these compounds were relatively simple molecules such as 7- and 8-methylheptadecane (Han and Calvin 1969), which are still used today as indicators of a blue green algal contribution to recent sediments. In the early 1970s attention switched to the characterisation of organic matter in very recent lake sediments, following the old adage that the key to the past lies in the present (Brooks et al. 1976 and 1977). These studies led to the identification of numerous compounds that could be associated with specific inputs of organic matter to lacustrine environments.

All of these studies at the extremes of the geologic record made extensive use of the biomarker concept. In brief, a biomarker can be defined as a compound in the geologic record with a carbon skeleton essentially identical to the carbon skeleton of its functional precursor molecule (Eglinton and Calvin 1967). An example would be cholesterol, a widely occurring natural product that, upon being deposited into a natural environment over an extended period of geologic time, will be converted into cholestane. The carbon skeleton of cholestane is identical to that of cholesterol, so the presence of cholestane in a crude oil tells you the ancient source material for that oil included cholesterol. The presence of that biomarker and many others helps the petroleum geochemist reconstruct the origin and history of the oil, providing useful information in the search for new accumulations of crude oil.

Two key episodes in the development of this work were based on the Ph.D. thesis of Brenda Kimble at the University of Bristol in 1972, who, for the first time, recognised that many of these complex biomarkers occurred in rocks and sediments with stereochemistries that varied as a result of maturity. The second important landmark was the recognition by Wolfgang Seifert (Seifert and Moldowan 1978, 1981; Seifert et al. 1978), a geochemist with Chevron in the mid-1970s, that the stereochemistry of the molecules observed by Kimble could be used to evaluate changes in the maturity of source rocks responsible for crude oil generation. Seifert was primarily responsible for getting the concept of biomarkers utilised in petroleum exploration studies despite stiff resistance from upper-level management.

Today it is a totally different picture in which the biomarker approach is a routine, mature science used by major and minor oil companies alike. It is used in exploration, reservoir, and production chemistry. By quantifying a wide range of biomarkers in a crude oil or source rock extract, it is possible to build up a picture of the type of organic material responsible for a crude oil and the environment in which it was deposited. Recently, there has been significant effort to utilise specific compounds as age-diagnostic biomarkers to estimate the geological age of the oil (Moldowan et al. 1993). Biomarkers are typically present in crude oils in relatively low concentrations but readily determined by GC-MS and MID. Therefore, if a crude oil is analysed by GC-MS to determine the distribution of steranes and terpanes, two fingerprints are obtained, one for each family of biomarkers (Figure 4.1). From an exploration point of view it is fairly important that we have a good idea of the identity of as many of these compounds as possible since they will be used to reconstruct depositional environments, the nature of source material, maturity levels, etc. The fingerprints themselves can also be used as correlation

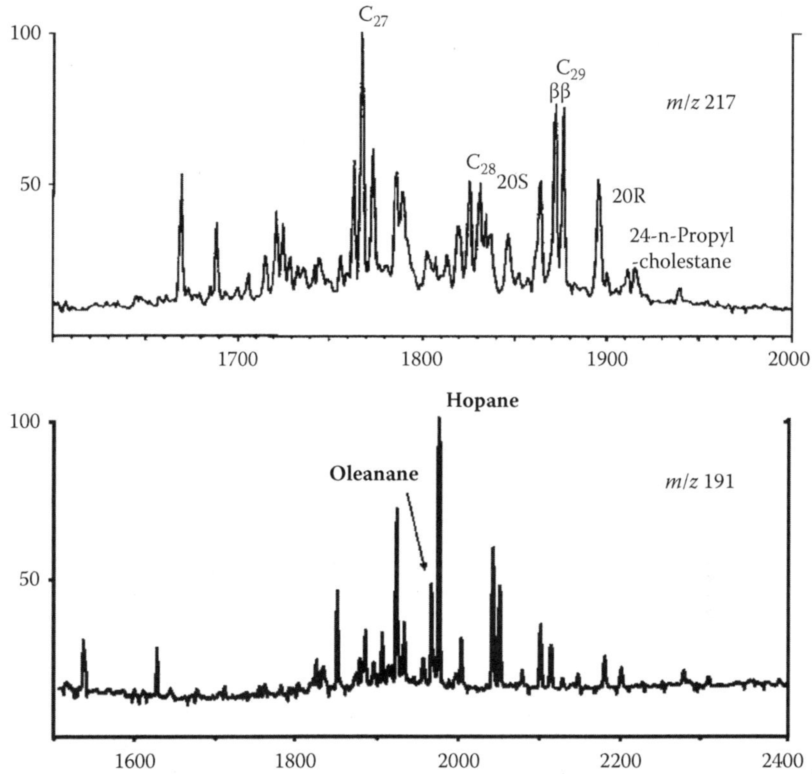


**Figure 4.1** Biomarker distributions for crude oils are determined through the use of GC-MS and multiple ion detection. Sterane and terpane distributions are obtained through monitoring the ions at mass 217 and 191, respectively, as shown here.

tools to determine whether two oils are derived from the same source rock or whether an oil is derived from a specific source rock. All of these applications are used extensively in the petroleum industry and are well documented in the literature. For additional information, the reader is referred to the most recent monograph on this topic by Peters et al. (2005).

There is a wide range of hydrocarbon biomarkers that have been identified in the past three or four decades for use in petroleum exploration (Philp 1985). These compounds range from the simple *n*-alkanes to the more complex polycyclic compounds such as the hopanes. Other classes include the isoprenoids, sesquiterpanes, tricyclic terpanes, tetracyclic terpanes, and steranes (Figure 4.2). For the most part, all of the commonly used biomarkers are saturated compounds and many of the precursor molecules have been identified. There are also a number of compounds that have been utilised where the identity of the compound is not known. However, if that same compound, as identified by GC retention time and mass spectral data, is present in several samples, it is often possible to infer information on its origin

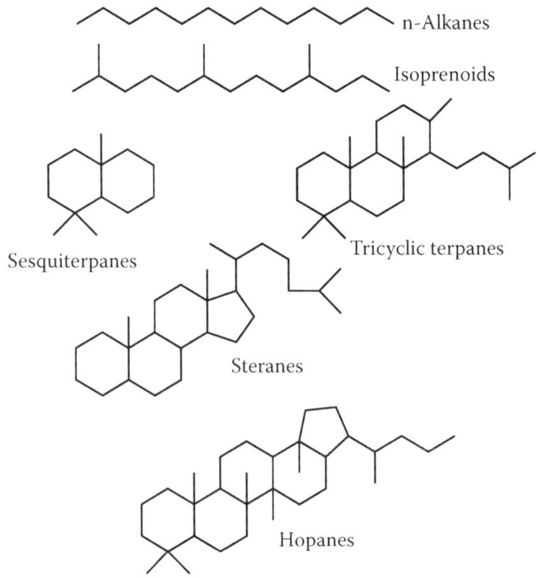


**Figure 4.2** Structures of the common classes of biomarkers used in environmental forensics and petroleum exploration studies.

and significance. Such compounds may be in use for several years before a synthetic chemist is able to synthesise the material and unambiguously prove its structure.

It is also very important to reiterate the fact that none of these compounds would have been discovered if it were not for the development of the combined GC-MS system and all the ancillary techniques associated with this development. Crude oils and refined products are extremely complex mixtures of hydrocarbons; for the most part, with the exception of the *n*-alkanes, most of the biomarkers are hidden in the baseline of a conventional gas chromatogram. Gas chromatography will provide a fingerprint showing the distribution of the major components in the sample (Figure 4.3). However, it does not unambiguously identify these compounds or provide any information on the minor compounds in the oils such as the biomarkers that are hidden in the baseline of the chromatogram. All of the commonly used families of biomarkers have characteristic ions associated with them that can be used for MID analyses (Table 4.1 summarises the most commonly used ions). For crude oils, the most commonly used biomarker fingerprints are the steranes and terpanes at mass to charge ratio (*m*/*z*) 217 and 191, respectively. There is a tremendous volume of information in the geochemical literature on the significance of the distribution of these compounds. Some of this information is equally applicable to the environmental samples, but in many cases it is not necessary to interpret the fingerprints in such detail since the primary use of the fingerprints in the environmental studies is for correlation purposes.

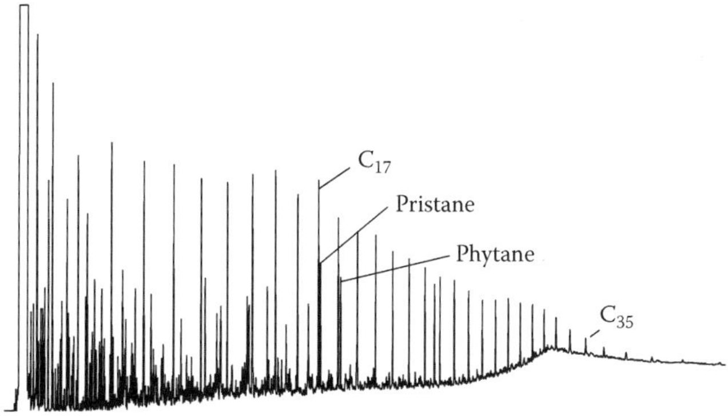


**Figure 4.3** Gas chromatography will provide a fingerprint showing the distribution of the major components in a crude oil or refined product, but it does not provide information on the identity of the individual compounds or the biomarkers.

**Table 4.1 Ions Commonly Used for Biomarker Monitoring in Crude Oils and Refined Products**

| Biomarker | Ion (*m/z*) |
|---|---|
| *n*-Alkanes | 85 |
| Isoprenoids | 183 |
| Bicyclic sesquiterpanes | 109,123 |
| Diamondoids | 135, 187 |
| Tricyclic terpanes | 191 |
| Tetracyclic terpanes | 191 |
| Pentacyclic terpanes | 191 |
| Steranes | 217, 218 |
| Methyl steranes | 231, 232, 245 |
| Monoaromatic steranes | 253 |
| Triaromatic steranes | 231 |

Examples of sterane and terpane chromatograms are shown in Figure 4.1. In the sterane chromatogram certain components are labeled. A petroleum geochemist would be particularly interested in such features as the relative proportions of the $C_{27}$:$C_{28}$:$C_{29}$ steranes since this gives an indication of the type of source material in the original source rock. Ratios such as 20S/20S + 20R for the $C_{29}$ steranes or the $\alpha\alpha/(\alpha\alpha + \beta\beta)$ $C_{29}$ steranes provide an indication of the relative maturity of the sample—important information from an exploration point of view. Similarly, in the terpane fingerprint, various ratios can be calculated for maturity determinations; the presence of an abundant $C_{29}$ norhopane relative to the $C_{30}$ hopane is typically indicative of an oil derived from a carbonate source rock. The presence of certain triterpanes, for example oleanane, can indicate a specific input of source material. Oleanane is a classic biomarker indicative of a higher plant source, specifically angiosperms; furthermore, this limits the age of the source material to Late Cretaceous/Early Tertiary.

Bicadinanes are another group of terpanes that are used as unique source indicators for higher plant resins and are predominantly observed in oils from Southeast Asia and Indonesia. These specific marker compounds are very important for reconstructing the origin or history of an oil and are equally important for environmental studies in many cases. For example, if an oil slick or residue is observed in the ocean and no obvious source is nearby, the presence of bicadinanes in the oil will indicate that it came from a tanker carrying oil from Southeast Asia. A search of the shipping traffic in that area may then provide information on the passage of tankers carrying crudes from Southeast Asia that had passed through that area, hence assisting in tracking down the culprits responsible for the spill.

Another potential environmental application of parameters initially derived for exploration purposes is the maturity parameters. For example, in a crude oil the ratio of the 22S/22S + 22R for the $C_{31}$ homohopanes will be approximately 0.65. However, in a very immature sample the ratio will be 0.1 or less since immature samples are dominated by the 22R epimer (Peters et al. 2005). In the situation where the extent of an oil well blow out or leak is being determined, soil samples collected in the vicinity of the well where the oil concentration is highest will be dominated by the mature stereochemistry, and samples will have ratios around 0.65. Moving away from the well to areas where the mature signature has been mixed with the microbial background and immature signature, the values will decrease until you reach the point where you only see the microbial background, which will delineate the extent the plume has spread from the well.

## Biomarkers in Environmental Forensics of Crude Oil and Refined Products

The use of biomarkers in environmental problems is widespread and powerful; in most applications, the biomarker distributions are being used as fingerprints in the same way that fingerprints are used at a crime scene. If the fingerprint of the spilled product matches the fingerprint of the suspected source in the pipeline, storage tank, or tanker, then that is taken as a very strong piece of evidence that the two samples are related. A significant advantage of the biomarkers is that these compounds are relatively resistant to biodegradation. In the short time between most spills and sample collection, few, if any, changes will be expected to compounds such as the steranes and terpanes.

Any crude oil is a very complex mixture of hydrocarbons, and compounds containing N, S, and O atoms, with an overall carbon number distribution ranging from $C_1$ to $C_{70}$, or higher in some cases. The hydrocarbons comprise a variety of families including saturates and aromatics and then, within each of these two main groups, many smaller families of compounds. All of these compound classes have been studied extensively by petroleum geochemists over the years and it has been well documented that the *n*-alkanes are the most susceptible to biodegradation even over a relatively short period of time—weeks to months. However, branched hydrocarbons, such as the isoprenoids, or cyclic hydrocarbons such as steranes and terpanes are very resistant to biodegradation, as manifested by their presence in oils that have been undergoing extensive biodegradation for several millions of years.

To illustrate the effects of biodegradation, an oil that has been biodegraded under laboratory conditions is shown in Figure 4.4. Note the relatively rapid degradation of the *n*-alkanes, but then in Figure 4.5 the terpanes

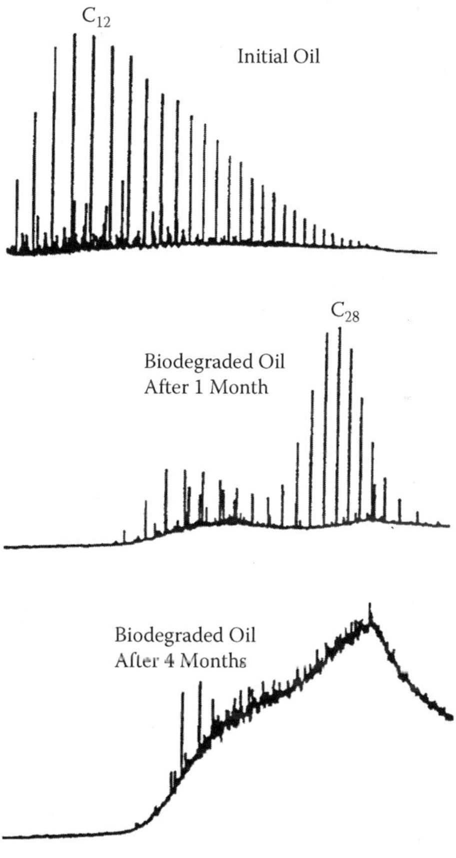


**Figure 4.4** The effects of biodegradation are illustrated in the chromatograms shown in this figure obtained under laboratory conditions. The *n*-alkanes are removed in a relatively short period of time, ultimately producing a large hump of unresolved compounds.

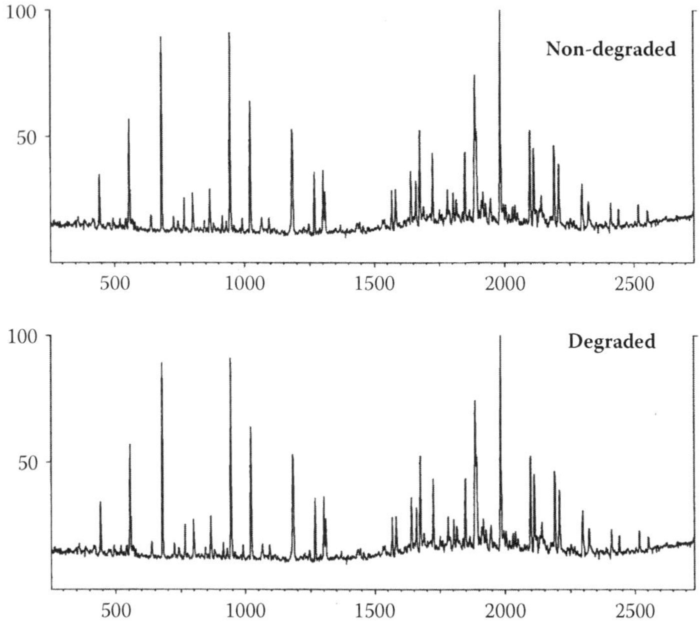


**Figure 4.5** Short-term degradation of an oil may remove the *n*-alkanes, but the biomarker distributions, terpanes, and steranes present in the very heavily biodegraded residue and the original oil will be virtually identical. The terpanes as shown here were determined by GC-MS and single ion monitoring at *m/z* 191.

and steranes observed in the original oil and the very heavily biodegraded residue are virtually identical. In any environmental forensic study, the goal is to get as many pieces of evidence as possible to correlate the spilled sample with the suspected source or sources. Whilst the steranes and terpanes are the most commonly used groups of biomarkers, there are a number of additional biomarker fingerprints that can be used, such as the isoprenoids, diamondoids, and sesquiterpanes. In addition, the isotopic composition of the samples can be used as supporting data for the correlations.

In view of the vast amounts of oil being transported around the globe on a daily basis, there have been many reported major oil spills since the late 1960s and probably even more unreported spills from tankers and pipelines. It would be impossible to report on all of these incidents but, without a doubt, biomarkers have played a significant role in many of these cases by establishing the origin of the spill along with monitoring the fate of the spilled product. The first supertanker incident was the *Torrey Canyon,* which ran aground on Seven Stones Reef off the southwest coast of England in 1967, contaminating significant areas of the coast in both England and France. Since that time the busiest shipping lanes in Europe, Southeast Asia, and the Middle East have seen a steady number of incidents on a regular basis:

In the first Gulf War in the early 1990s, an estimated 460 million gallons of crude oil were released into the Persian Gulf.
In June 1979 the Ixtoc 1 in the Gulf of Mexico blew out, spilling an estimated 140 million gallons of crude oil into the open sea.
In January 1993 the *Braer* went aground off the Shetland Islands, releasing 26 million gallons of oil.
In March 1989 the *Exxon Valdez* ran aground in Prince William Sound, Alaska, spilling 10.92 million gallons of crude oil.
In December 1999 the *Erika* broke apart and sank off Brittany on the French Atlantic coast, spilling 3 million gallons of oil.
In November 2002 the *Prestige* sank off the coast of Spain and spilled 77,000 tons of oil.

Table 4.2, taken from the Web site of the International Tanker Owners Pollution Federation Limited, provides a summary of the 20 major tanker incidents over the past 40 years.

**Table 4.2 Summary of Major Oil Spills from Tankers Worldwide over the Past 40 Years**

| Position | Ship name | Year | Location | Spill size (tonnes) |
|---|---|---|---|---|
| 1 | *Atlantic Empress* | 1979 | Off Tobago, West Indies | 287,000 |
| 2 | *ABT Summer* | 1991 | 700 nautical miles off Angola | 260,000 |
| 3 | *Castillo de Bellver* | 1983 | Off Saldanha Bay, South Africa | 252,000 |
| 4 | *Amoco Cadiz* | 1978 | Off Brittany, France | 223,000 |
| 5 | *Haven* | 1991 | Genoa, Italy | 144,000 |
| 6 | *Odyssey* | 1988 | 700 nautical miles off Nova Scotia, Canada | 132,000 |
| 7 | *Torrey Canyon* | 1967 | Scilly Isles, United Kingdom | 119,000 |
| 8 | *Sea Star* | 1972 | Gulf of Oman | 115,000 |
| 9 | *Irenes Serenade* | 1980 | Navarino Bay, Greece | 100,000 |
| 10 | *Urquiola* | 1976 | La Coruna, Spain | 100,000 |
| 11 | *Hawaiian Patriot* | 1977 | 300 nautical miles off Honolulu, Hawaii | 95,000 |
| 12 | *Independenta* | 1979 | Bosphorus, Turkey | 95,000 |
| 13 | *Jakob Maersk* | 1975 | Oporto, Portugal | 88,000 |
| 14 | *Braer* | 1993 | Shetland Islands, United Kingdom | 85,000 |
| 15 | *Khark 5* | 1989 | 120 nautical miles off Atlantic coast of Morocco | 80,000 |
| 16 | *Aegean Sea* | 1992 | La Coruna, Spain | 74,000 |
| 17 | *Sea Empress* | 1996 | Milford Haven, United Kingdom | 72,000 |
| 18 | *Katina P* | 1992 | Off Maputo, Mozambique | 72,000 |
| 19 | *Nova* | 1985 | Off Kharg Island, Gulf of Iran | 70,000 |
| 20 | *Prestige* | 2002 | Off Galicia, Spain | 63,000 |

While it might appear that the source for most of these spills is obvious and one could question the need for a forensic investigation, there is another side to many of these investigations. For example, in the case of the *Exxon Valdez* spill, tar ball residues from the beaches of Prince William Sound were collected several years after the *Exxon Valdez* incident and the gas chromatograms of these samples all showed signs of extensive degradation (Kvenvolden et al. 1995). On the basis of the gas chromatograms alone, it was totally impossible to obtain any information on the possible relationships between these residues and the *Exxon Valdez* oil. The bulk isotope data provided the initial information to indicate different sources for some of these residues. As can be seen from Figure 4.6, two of the residues were significantly isotopically heavier. The biomarker data provided additional evidence to show that the two residues labeled A and B in Figure 4.6 were derived from a source different from the *Exxon Valdez* (Figure 4.6c). The presence of the bisnorhopane in the samples shown in Figures 4.6a and 4.6b is characteristic of Monterey (CA)-derived crude oils (Moldowan et al. 1984), which are also known to be isotopically heavier with values around –24‰. Therefore, the samples showed that the two residues in Figures 4.6a and 4.6b were derived from Monterey crude and the residue in Figure 4.6d from oil of a similar source to that of the *Exxon Valdez*.

The reason Californian crude was in Alaska is a relatively simple question to answer. Prior to the discovery of commercial quantities of oil in Alaska, oil had to be imported from California and temporarily stored in tanks on the edge of Prince William Sound. In 1964 a massive earthquake struck the region of Valdez, the tanks ruptured, oil spilled into the sound, and, in view of the overall magnitude of the disaster, the oil was left in the sound to degrade naturally with very little cleanup. The biomarkers in crude oils, apart from the *n*-alkanes, are very resistant to biodegradation, and over a time period of 50 years, little change will occur to the biomarker distributions.

The Straits of Malacca is another waterway through which a large number of tankers pass on a daily basis; they originate from the Middle East, transporting crude oil and petroleum products to Singapore (currently Asia's largest oil refinery centre), Japan, Hong Kong, Korea, and China (Zakaria et al. 2000). In late 1997, the Chinese tanker *An Tai* carrying Middle East oil ran aground in the straits and spilled 235 tons of crude oil. The oil threatened to destroy the remaining mangrove forest that lines the central coast of the western part of peninsular Malaysia and to have a significant effect on the local aquaculture fisheries in the area. However, in this area there are many other possible sources of contamination resulting from the rapid urbanisation and industrialisation coupled with domestic petroleum production in Malaysia. In order to get an indication of contributions from these various sources, the spilled oil from the tanker plus numerous tar balls on the beaches were collected and characterised in detail using a wide range of bio-

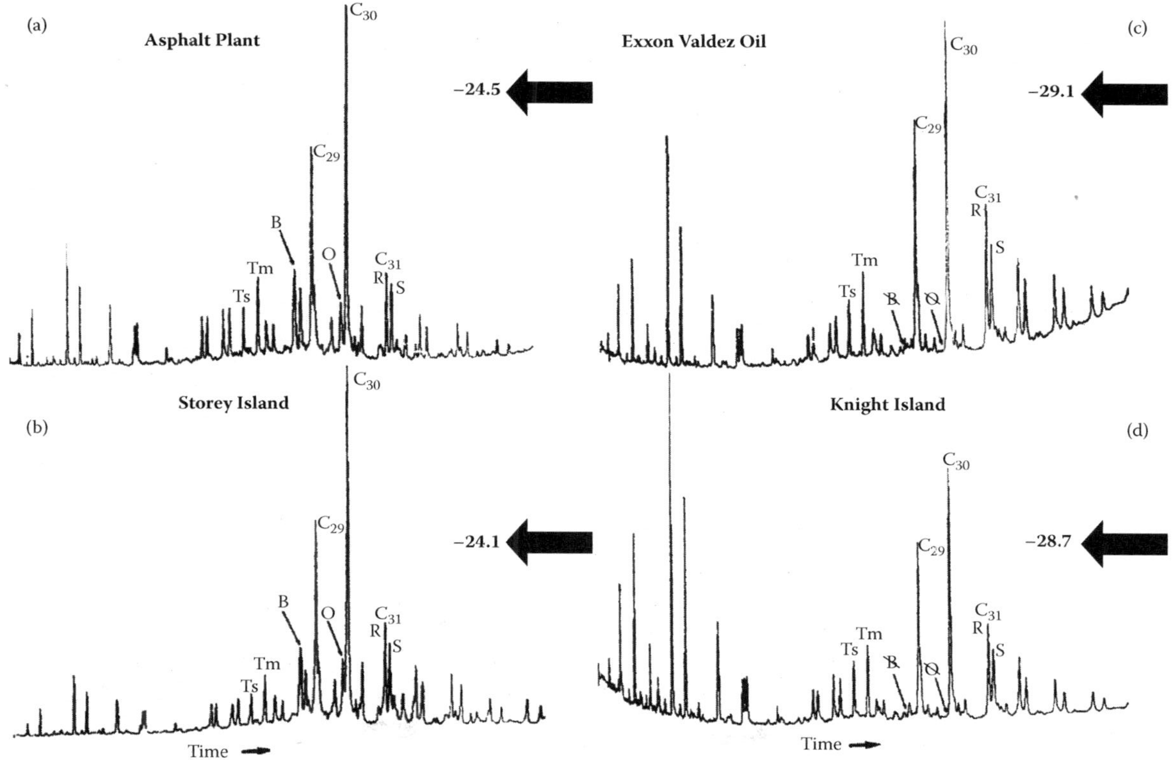


**Figure 4.6** Crude oil residues from Prince William Sound several years after the *Exxon Valdez* incident were collected and analysed by GC-MS and carbon isotope were data (‰) determined. Four of the residues illustrated in this diagram could be differentiated on the basis of their isotopic data along with the terpane distributions as determined from the *m/z* 191 chromatograms. The B and O biomarkers are absent from samples (c) and (d).

marker compounds such as isoprenoid alkanes, steranes, hopanes, and polycyclic aromatic hydrocarbons (PAHs). In this particular case, the problem was somewhat simplified since Kuwait crude has a very characteristic fingerprint, dominated by the $C_{29}$ hopane. The locally produced oils are dominated by the presence of oleanane and the hopanes maximise at $C_{30}$ and not the $C_{29}$. Based on the differences in these two fingerprints, it was relatively easy to differentiate the two sources and obtain an indication of their contributions to the contaminated areas.

A third example is from the case of the *Erika* tanker that broke up off the coast of France in 1999 and released significant quantities of heavy fuel oil, which subsequently drifted ashore (Mazeas and Budzinski 2002a). Oil residues and oiled bird feathers were collected all along the Atlantic shoreline of France after the wreck. The aim of the study was to differentiate oil residues and oiled bird feathers related to the *Erika* oil spill from those resulting from numerous tar ball incidents that occurred after the *Erika* oil spill. A combination of the alkane and PAH distributions was used along with the isotopic compositions of individual PAH components to differentiate those residues from the *Erika* from the other sources. All the oiled birds appeared to have been contaminated by the *Erika* oil, but samples from the southern regions of the Atlantic Coast had a different molecular fingerprint compared to the *Erika* oil, indicating that they were not related to the *Erika* oil spill. Molecular isotopic composition of saturate hydrocarbons and phenanthrene compounds permitted unambiguous differentiation of samples related to the *Erika* oil spill from those due to other unrelated tar ball incidents. The stability of the PAH components to biodegradation will also be of use for the long-term monitoring in this region. Over the longer term, as the more viable compounds are removed or altered, the molecular isotopic composition of the PAH compounds should continue to be of use in differentiating the *Erika* residues from those of other incidents.

There is evidence and papers have been published to demonstrate that hopane does indeed degrade and, if that is the case, then the amount of oil that has degraded will be grossly overestimated. For example, a study of Venezuelan crude showed that over a relatively short 5-week period, regular hopanes, including the $C_{30}$ hopane, were removed under aerobic conditions. In these experiments, no 25-norhopanes were formed (Bost et al. 2001). Active oil seeps in the Santa Barbara channel, for example, have extensively altered hopane fingerprints, dominated by the $C_{35}$ extended hopanes, indicating relatively rapid degradation of the hopanes (Requejo and Halpern 1995). Therefore, it would appear that there is a need to exercise caution prior to utilisation of the $C_{30}$ hopane as an internal standard.

Crude oils contain a wide range of biomarkers, but the situation is different for refined products. The majority of the conventional biomarkers elute in the region of the chromatogram above $C_{25}$. Most refined products are dominated by lower boiling components and do not contain steranes and terpanes. One exception is because motor oils and lubricating oils, since these are typically in the range $C_{27}$–$C_{35}$. The chromatograms for these products are typically a large unresolved hump of branched and cyclic compounds as seen in Figure 4.7. However, motor oils derived from different crude oils can be differentiated isotopically, although the range of values for motor oils derived from different sources is relatively small, covering a range of about 2‰. If the motor oils are derived from different crude oils, the biomarkers will provide another tool to differentiate the source of the motor oils, as illustrated in Figure 4.8.

In recent years, primarily as a result of the work in petroleum geochemistry, a number of biomarkers have become available to differentiate refined products, such as diesel, from different sources. These include adamantanes, sesquiterpanes, and partially degraded steranes (Figure 4.2), all of which elute in the diesel boiling range and are easily detectable by GC-MS. Adamantanes have been investigated in great detail by Dahl et al. (1999) and have been used to obtain information on the extent of thermal cracking of crude oils. These compounds are extremely stable and very resistant to biodegradation, which makes them extremely valuable for correlation of environmental samples. Sesquiterpanes were among the earliest of low molecular weight biomarkers discovered and sources for many of these compounds have been proposed (Philp, Gilbert, and Friedrich 1981; Alexander et al. 1984). Degraded steranes were also discovered in a petroleum exploration investigation (Jiang, Philp, and Lewis 1988); although these compounds do not appear to have been used in environmental studies, they could be used to support information from the adamantanes and sesquiterpanes.

Diesel products, which are a lower boiling point fraction of crude oil than motor oils, are for the most part below the range for the presence of common steranes and terpanes. However, compounds that do occur in that boiling point range include admantanes and sesquiterpanes. Again, these compounds have been thoroughly investigated in numerous crude oils for the purposes of petroleum exploration and production. An example illustrating the combined use of isotopes and adamantanes is shown in Figures 4.8–4.10. Samples from the two monitoring wells were at the centre of a dispute as to whether or not they were related and hence part of the same plume of hydrocarbons. Both samples were identified as a refined product (namely, diesel); however, the sample in monitoring well 1 appeared different, although these differences are simply due to biodegradation, as can be observed from the much smaller *n*-$C_{17}$/Pr value in this sample compared to monitoring well 6. The samples were analysed by GC-IRMS and carbon values were determined for the isoprenoids; these values are shown in Figure 4.9, suggesting a similarity in origins for the two samples. However, one should not depend solely upon one piece of data to suggest a relationship between two samples. Hence, the two samples were analysed by GC-MS; in particular, the distribution of the adamantanes was

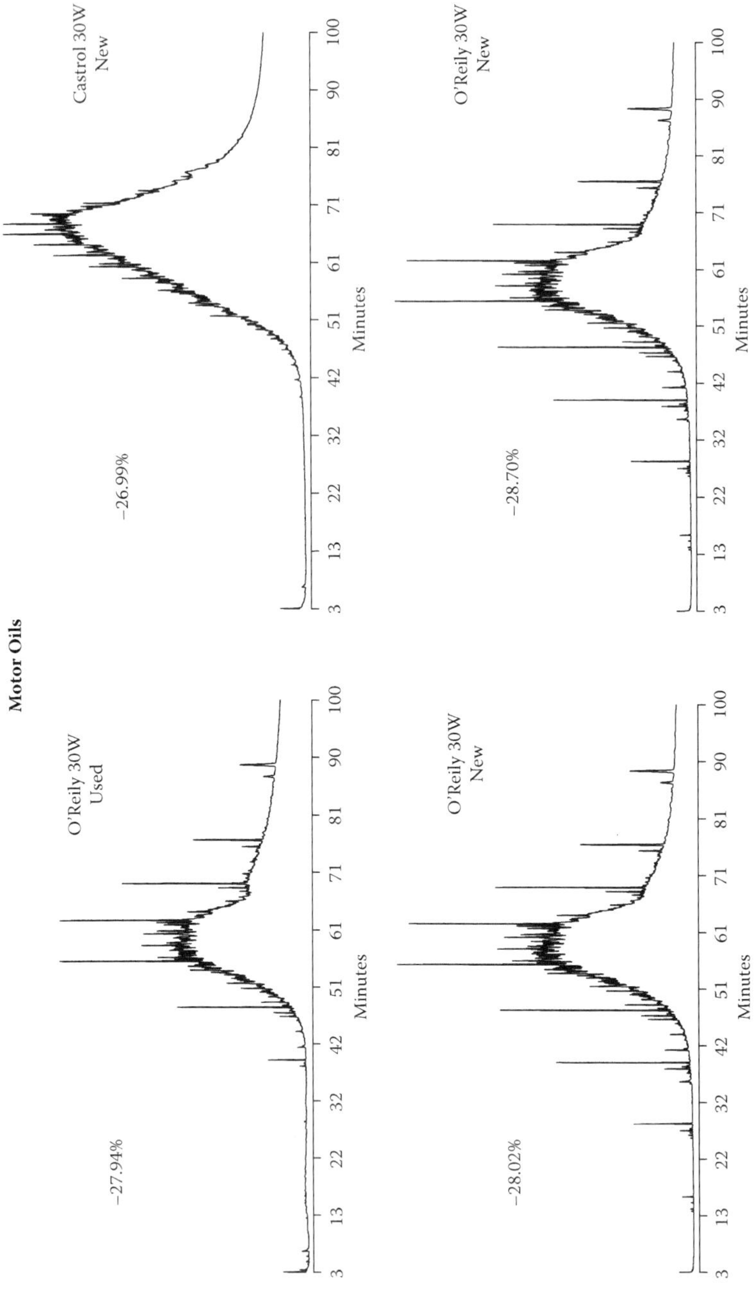


**Figure 4.7** Chromatograms of motor oils are dominated by a large unresolved hump of branched and cyclic compounds as illustrated here for four different oil samples. The isotope values cover a small range.

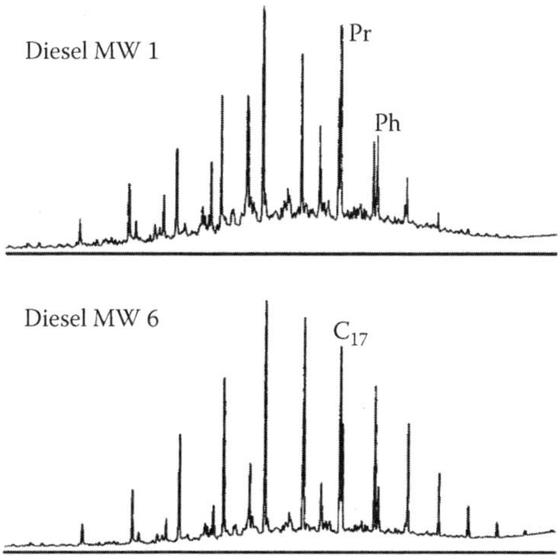


**Figure 4.8** Gas chromatograms of free product from two wells at a contaminated site. The upper sample is showing signs of biodegradation and it is unclear from these chromatograms as to whether or not the samples are related.

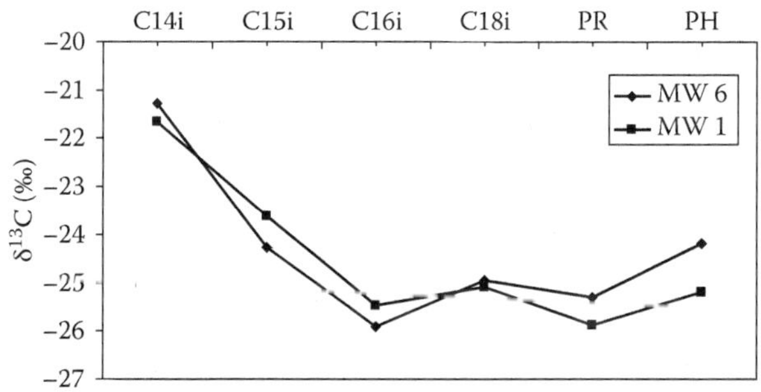


**Figure 4.9** The isotope data were complemented by the GC-MS analyses to determine the adamantane distributions as shown in the chromatograms.

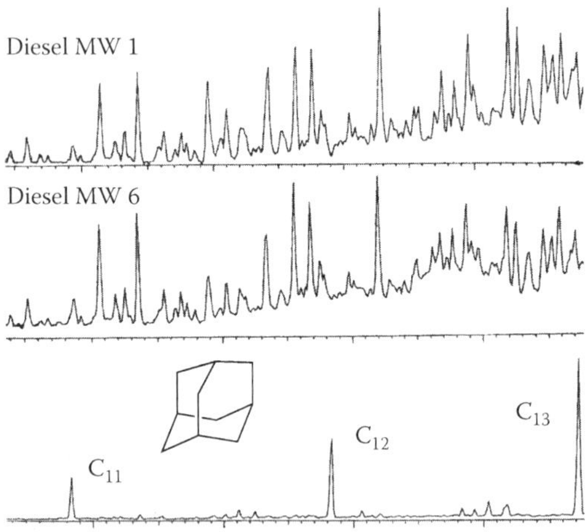


**Figure 4.10** Carbon isotope values for the individual isoprenoids in these two samples suggested that they are probably related to each other and derived from a common source.

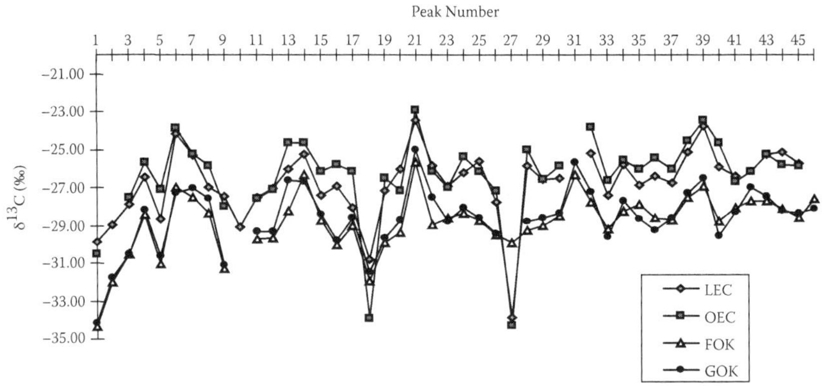


**Figure 4.11** The isotopic composition of individual compounds in four gasoline samples determined by GC-IRMS has resolved the four gasolines into two groups of two, derived from different crude oils.

determined and is shown in Figure 4.10. The similarity in these distributions along with the isotope data and other biomarker fingerprints not shown here clearly established a relationship between these two samples.

Gasolines, an even more volatile distillation fraction of crude oils, are totally devoid of any conventional biomarkers. For the most part, the majority of components in a gasoline are known and have been identified. Characterisation and correlation of gasolines by GC and GC-MS are, in many cases, of limited use since many gasolines from different sources may have similar chromatograms. However, this is a situation where, in the absence of biomarkers, the utilisation of GC-IRMS may be used to differentiate gasolines if they are derived from different crude oils. The crude oils themselves may be isotopically distinct and this distinction carries over into the refined gasoline. To illustrate this, Figure 4.11 shows four gasolines analysed by GC-IRMS; as can be seen from this figure, the isotopic composition of the individual compounds has resolved the four gasolines into two groups of two derived from different crude oils.

## Other Forensic Applications of Biomarkers

In the preceding sections all the emphasis has been placed on the hydrocarbon biomarkers in fossil fuel-derived products and that is where most of the applications have been in the past few years. However, it would be negligent not to mention another very important environmental application of biomarkers to environmental problems: contamination of groundwater, rivers, lakes, oceans, or soil by runoff or discharge from sewage treatment plants. In the introduction, a definition for biomarkers was provided that focused on the hydrocarbon molecule. The precursor of that hydrocarbon is also a biomarker; indeed some argue that this is the major biomarker since this is the actual compound directly derived from the source, whereas the hydrocarbon product could be thought of as a geomarker.

Utilisation of sterols as biomarkers indicative of faecal input into the environment was the original work of Nichols and Leeming (1991), who looked at discharge from the sewage outfalls in Sydney, Australia. Coprostanol (5β-cholestan-3β-ol) is a faecal steroid that has been used to monitor and study the fate of sewage contamination in the environment (Readman, Preston, and Mantoura 1986; Nichols and Leeming 1991; Leeming et al. 1997; Mudge and Seguel 1997; Mudge and Seguel 1999; O'Leary et al. 1999). Its utility as a tracer for sewage-derived pollution in the environment is based on the fact that coprostanol is formed in the digestive tract of humans and other animals through the biohydrogenation of cholesterol (cholest-5-en-3β-ol) and released into the environment in human faeces (Readman et al. 1986; Nichols and Leeming 1991; Mudge and Seguel 1999). Studies by Ferezou et al. (1978), Leeming et al. (1996a and b), and O'Leary et al. (1999) have reported that human faeces contains the highest concentration of coprostanol compared to that of other mammals; of the sterols present in human faeces, coprostanol comprises as much as 60% of the total sterol content (Sinton, Finlay, and Hannah 1998).

Two common isomers of coprostanol include epi-coprostanol (5β-cholestan-3α-ol) and cholestanol (5α-cholestan-3β-ol). Epi-coprostanol has only been observed in treated sewage sludge (McCalley, Cooke, and Nickless 1981; Mudge et al. 1999) and appears to be the preferential product in the anaerobic digestion of sewage sludge in sewage treatment facilities (Sherblom and Kelly 1993). McCalley et al. (1981) and Mudge and Seguel (1999) have utilised the ratio of epi-coprostanol to coprostanol to estimate the relative amount of treated versus untreated sewage. Under anaerobic conditions in the natural environment, cholesterol is preferentially hydrogenated to cholestanol. Faecal steroid profiles can also be used to distinguish sources of faeces (Leeming et al. 1996a, 1997b; Sinton et al. 1998). Herbivores (e.g., sheep and cows) typically consume significant amounts of plants enriched with $C_{29}$ sterols. Faecal matter from sheep and cows is usually composed of significant amounts of 24-ethylcoprostanol and 24-ethylepicoprostanol (Leeming et al. 1997). Faecal matter from dogs and birds can also be distinguished from human faeces, based on sterol profiles. Dogs have abundant amounts of cholesterol, 24-ethylcholesterol, and 24-methylcholesterol; birds typically have higher concentrations of 24-ethylcholesterol and cholesterol (Leeming et al. 1997).

## Stable Isotopes in Environmental Forensics

Relatively recent commercial introduction of the gas chromatography–continuous flow isotope monitoring mass spectrometry technique (GC-IRMS) opened new perspectives for stable isotope work by permitting compound-specific isotope ratio analysis (CSIA)—that is, analysis of isotope ratios in individual organic chemical species. A growing number of environmental and forensic applications in the past 15 years reflects the growing popularity of the CSIA technique.

In typical carbon CSIA, chromatographic separation is followed by online high-temperature conversion of individual compounds to $CO_2$, which is passed into a continuous-flow IRMS for isotope ratio determination (Merritt et al. 1995). Similar principles are followed for hydrogen (Burgoyne and Hayes 1998), oxygen (Brand, Tegtmeyer, and Hilkert 1994), nitrogen (Merritt and Hayes 1994), and sulphur (elemental analyser-IRMS applications were published, but sulphur GC-IRMS was apparently never attempted) isotope analysis, except that different types of online conversion processes are necessary to obtain $H_2$, CO, $N_2$, and $SO_2$, respectively. Compound-specific analysis of Cl (Holt et al. 1997b) and Br (Shouakar-Stash et al. 2005b) is possible after offline chromatographic separation of the target analytes and offline chemical processing to IR-MS-amenable gaseous products ($CH_3Cl$ and $CH_3Br$, respectively).

Most existing work in the environmental field utilises the most robust and most widely available carbon CSIA, followed by hydrogen CSIA. A considerable body of research on chlorine isotope results for various contaminants has been produced as well. Slater (2003) gives an overview of CSIA application to environmental forensics. Several review papers discuss stable isotope technique applications to environmental studies (Schmidt et al. 2004; Meckenstock, Morasch, et al. 2004) or provide discussion of isotope work in the wider context of contaminant studies (e.g., Scow and Hicks 2005; Meckenstock, Safinowski, and Griebler 2004). Little published work concerns

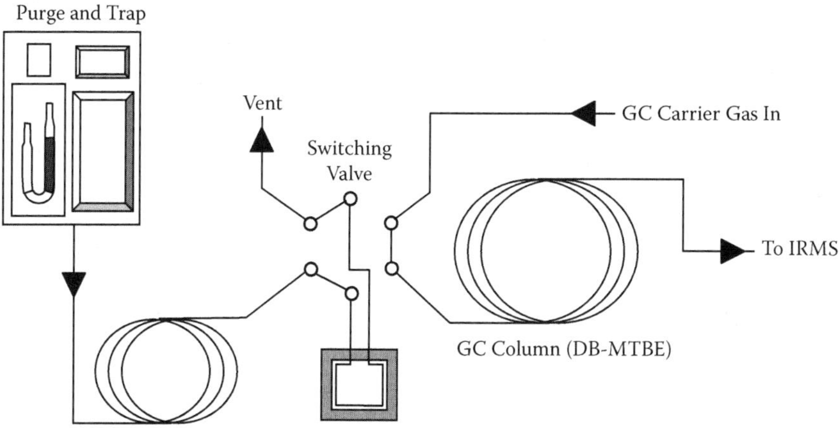


**Figure 4.12** Schematic diagram of purge-and-trap interfaced to GC-IRMS in use at the isotope laboratory of the University of Oklahoma.

other isotopes in environmental contaminants. An example of nitrogen CSIA application is provided by a study of the environmental fate of explosives (Coffin et al. 2001).

Significant research effort has been directed towards improving CSIA method detection limits and allowing work with low, environmentally feasible concentrations of the target analytes present in water, soil, or air. Various extraction, reconcentration, fraction separation, etc. procedures are necessary to make these samples amenable to CSIA. For aqueous, volatile range organic contaminants, the best results are obtained with purge and trap interfaced to the GC-IRMS instrument (Zwank et al. 2003; Kuder et al. 2005). Very good method-quantitation limits are possible with the technique (in low $\mu g.L^{-1}$ range or even $ng.L^{-1}$ for certain species). For example, a purge-and-trap interface in use at a University of Oklahoma laboratory (Figure 4.12), $\delta^{13}C$ of benzene, toluene, etc., can be determined at approximately 700 $ng.L^{-1}$ concentration and MTBE at 1.5 $\mu g.L^{-1}$. Slightly higher limits of detection are practical for polar contaminants (e.g., 200 $\mu g.L^{-1}$ for ethanol and 20 $\mu g.L^{-1}$ for tert-butyl alcohol). Analysis of hydrogen isotope ratios is possible at approximately 10 times as high concentration. For comparison, headspace carbon isotope analysis of volatile species requires concentrations in the range of 500 $\mu g.L^{-1}$ or more (detection limits from Hunkeler and Aravena 2000). Solid-phase microextraction (SPME) improves over the headspace analysis and permits better detection limits (e.g., Hunkeler and Aravena 2000).

Most of the currently available chlorine isotope work follows the procedure of $CH_3Cl$ preparation for inlet IRMS similar to that described by Holt et al. (1997). Shouakar-Stash et al. (2005a) presented a method for offline $CH_3Cl$ preparation for subsequent analysis by continuous flow-IR-MS. Ader et al. (2001) describe chlorine isotope and Bao and Gu (2004) oxygen isotope analysis of perchlorate. A recent review of environmental forensics of perchlorate is given by Boehlke et al. (2005). Alternative approaches to chlorine IRMS have been proposed; for example, isotope ratios of chlorine can be determined with less complicated sample workup by thermal ionisation mass spectrometry (Holmstrand, Andersson, and Gustafsson 2004) or FAB-MS (Westaway et al. 1998). Chlorine isotope ratios cannot be currently determined at as low concentrations as carbon isotope ratios. The best detection limits to date were reported by Wassenaar and Koehler (2004) (0.2 μmol of chlorine per analysis).

For nonvolatile and semivolatile analytes, a standard solvent extraction is usually adequate, optionally followed by column chromatography, molecular sieving, etc. to reduce the complexity of the sample prior to GC separation (e.g., Kim, Kennicutt, and Qian 2005). Wang et al. (2004) described a protocol for semipermeable membrane processing of low-concentration aqueous hydrophobic compounds. Two different concepts of interfacing LC and IRMS have been proposed (Caimi and Brenna 1993; Krummen et al. 2004). No applications of LC-IRMS to contaminant studies are available at present. In their review of atmospheric chemistry of volatile organic compounds, Goldstein and Shaw (2003) described analytical developments relevant to vapour-phase contaminant work.

## Isotope Geochemistry—Principles

For an exhaustive introduction to stable isotope chemistry, the reader is advised to consult textbooks in the field. Hoefs (2004) and Faure and Mensing (2005) are two textbooks of stable isotope geochemistry, while Melander and Saunders (1980), Galimov (1985), and Cook (1991) focus on chemical and biochemical aspects of stable isotopes. The elements comprising molecules of common organic contaminants (such as C, H, O, N, S, and Cl) have each at least two stable isotope species. Stable isotope ratios may potentially provide valuable information characteristic of contaminant source or spill history. In recent decades, by universally accepted consensus, isotope data published in earth and environmental science are reported relative to the same international standard using a so-called delta notation. Delta units are referred to as ‰, permil, or per mill; they represent the deviation of the measured ratio versus the international standard. In chemical literature, some authors prefer to use absolute values of isotope ratios. In the case of carbon, the ratio of $^{13}C$ to $^{12}C$ is represented in the delta notation as follows:

$$\delta^{13}C = (R_{sample}/R_{standard} - 1) \times 1000$$

where R is $^{13}C/^{12}C$, and the standard is VPDB (Vienna Pee Dee Belemnite).

### *Isotope Fractionation*

Changes of isotope ratios caused by various chemical, biochemical, or physical processes are referred to as isotope fractionation or isotope effects. Isotope effects upon degradation result from the different reaction rates for molecules substituted with different isotope species (e.g., $^{13}C$ vs. $^{12}C$). The presence of a 'heavy' species (e.g., $^{13}C$) in the bond targeted by the degradation increases the bond cleavage activation energy and, in consequence, reduces the rate of degradation. During the process of degradation, the residual substrate becomes progressively enriched in the 'heavier' isotope species (i.e., $^{13}C/^{12}C$ increases). Degradation-related isotope effects are strong for the atoms included at the chemical bond being broken (primary isotope effects), while weak secondary isotope effects affect the remaining atoms. At the normal precision of CSIA, only secondary isotope effects of hydrogen are significant. In molecules where only a selected atomic position or positions are targeted by a biodegradation process (e.g., oxidation of a methyl group of toluene), the overall expression of isotope fractionation, as measured by CSIA, is 'diluted'.

The intrinsic isotope effect (the isotope effect with an impact on the atoms directly participating in the reaction) is not affected by the size of the molecule and can be calculated from CSIA data (Elsner et al. 2005). Isotope effects upon biodegradation reflect the first rate-limiting step of the reaction sequence. In certain cases, the slow step may not be a bond cleavage, but rather a phenomenon that does not cause isotope fractionation (e.g., formation of substrate–enzyme complex). If the latter is completely rate limiting, no isotope fractionation will be observed. The rule of thumb is that biodegradation isotope effects are limited to small molecules readily permeating cell membranes of degraders. Chlorinated ethenes and ethanes, MTBE, and mono-aromatic compounds have been most widely studied, and isotope fractionation has been typically observed. Biodegradation of semivolatiles, such as long chain *n*-alkanes, multi-ring PAHs, and PCBs typically does not cause isotope fractionation.

Physical processes such as phase transitions, sorption, and diffusion can also cause isotope fractionation. The mass difference between isotope species results in different kinetic energies of gaseous phase molecules, leading to differential rates of vapour migration, and different bond energies of light isotope-substituted versus heavy isotope-substituted molecules will affect phase partitioning equilibria and evaporation–condensation. While isotope fractionation due to physical processes is generally negligible in environmental science applications, it cannot be completely neglected, especially in the case of systems dominated by gas/vapour phase migration. Various laboratory studies of isotope effects upon volatile organic compound (VOC) volatilisation show, in general, small carbon isotope fractionation, more pronounced chlorine isotope fractionation, and, in select cases, strong hydrogen fractionation (e.g., Poulson and Drever 1999; Wang and Huang 2003 and references therein). Enrichment of $^{13}C$ and D and depletion of $^{37}Cl$ in the vapour phase in equilibrium with NAPL have been reported for various chlorinated compounds, mono-aromatics, and hydrocarbon species.

Open-system progressive evaporation, more relevant to environmental conditions, has not been extensively studied. It appears that in experimental conditions where the NAPL–vapour equilibrium element is significant, the residual NAPL tends to become enriched in the lighter $^{12}C$ and H, which is the reverse of the direction of fractionation upon degradation (Wang and Huang 2003). If the kinetic fractionation element (diffusion of vapour) is more significant, the residual NAPL may become enriched in the heavier carbon isotope $^{13}C$ (unpublished results from the University of Oklahoma). In a study of a controlled kerosene release, Kjeldsen et al. (2003) reported that molecules of VOC with a lighter isotope ($^{12}C$) migrated faster through the vadose zone, resulting in $\delta^{13}C$ reduction by several units over the distance of 3 m of sand (eventually, after the onset of biodegradation, the transport-related isotope effects were obliterated). Isotope effects have been identified in laboratory experiments or proposed to result from sorption phenomena (e.g., Schuth et al. 2003; Kopinke et al. 2005), in particular for VOC compounds. However, at the scale of a contaminant plume sorption should not cause a detectable extent of isotope fractionation (Kopinke et al. 2005).

The so-called Rayleigh fractionation model provides a mathematical framework for interpretation of isotope fractionation. The model is valid for fractionation occurring in unidirectional, irreversible reactions (kinetic isotope fractionation). The model was originally developed for distillation of mixed liquids (Rayleigh 1896). The principle of the Rayleigh-type process is a constant relationship between reaction rates of the elements of the reacting mixture. Where $^{13}k$ is the rate constant of degradation of the $^{13}C$-substituted bond, and $^{12}k$ is the rate constant of degradation of the $^{12}C$-substituted bond, the constant ratio of $^{12}k/^{13}k$ is referred to as the fractionation factor ($\alpha$). The same can be expressed if instantaneous isotope ratios ($R = {}^{13}C/^{12}C$) are substituted for reaction rates:

$$\alpha = R_{product}/R_{substrate} \tag{4.1}$$

Modification of equation 4.1 to substitute the instantaneous isotope ratios with readily measurable isotope ratios and concentrations of remaining reaction substrate yields equation 4.2. $R_t$ is R of substrate at time t; $R_{t=0}$ is R of substrate at time t = 0 (at the beginning of the reaction); F represents the ratio of substrate remaining at time = t ($concentration_t/concentration_{t=0}$); and $\varepsilon$ is the isotopic enrichment factor, $\varepsilon = 1000 \times (\alpha - 1)$.

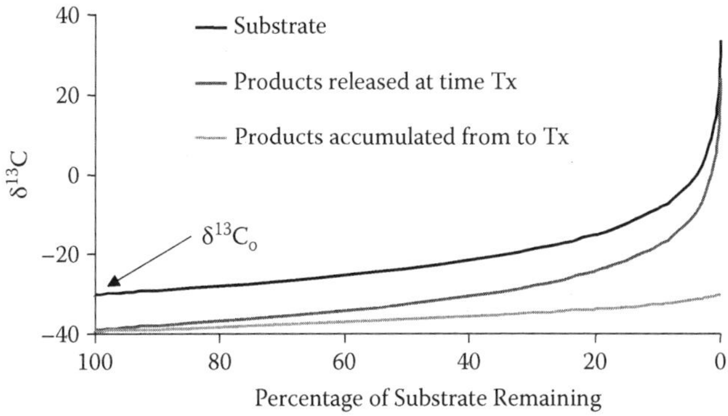


**Figure 4.13** Isotope effects in a Rayleigh-type, closed system kinetic fractionation.

$$1000 \times \ln(R_t / R_{t=0}) = \varepsilon \times \ln F \tag{4.2}$$

Writing the same using delta notation for isotope ratios yields equation 4.2a, where $\delta^{13}C_t$ is $\delta^{13}C$ of the substrate at time t; $\delta^{13}C_{t=0}$ is $\delta^{13}C$ of substrate at time t = 0.

$$1000 \times \ln\{(10^{-3}\,\delta^{13}C_t + 1)/(10^{-3}\,\delta^{13}C_{t=0} + 1)\} = \varepsilon \times \ln F \tag{4.2a}$$

A simplified variant of equation 4.2, reasonably accurate for ε between 0 and −20, was proposed by Mariotti et al. (1981):

$$\delta^{13}C_t = \delta^{13}C_{t=0} + \varepsilon \times \ln F \tag{4.3}$$

In a homogeneous system without a steady supply of substrate, equation 4.2 can be converted to equation 4.4, to calculate the reduction of the remaining fraction of the original mass of the substrate (e.g., the progress of biodegradation). This is the principle used in evaluating in situ degradation that will be further discussed in the section on using isotope data for contaminant attenuation studies.

$$F = \exp\,[1000 \times \ln\{(10^{-3}\,\delta^{13}C_t + 1)/(10^{-3}\,\delta^{13}C_{t-0} + 1)\}/\varepsilon\,] \tag{4.4}$$

Figure 4.13 shows changes of $\delta^{13}C$ of reaction substrate and product as a function of decreasing substrate concentration.

## Stable Isotopes in Tracking Contaminant Sources

Isotope fingerprinting (bulk or compound specific) was originally used in petroleum geochemistry to supplement biomarker techniques (e.g., Sofer 1984). Application to environmental contaminant source apportionment is based on a similar principle that, for example, various manufacturers will use different supplies of crude oil or methane for a given contaminant and possibly different synthetic processes will be involved. As a result, the same chemical compounds provided by different companies, or produced at different periods may have distinct isotope ratios. A combined two-dimensional (carbon–hydrogen or carbon–chlorine) isotope analysis should provide the most robust fingerprinting tool.

An example of two-dimensional bulk isotope fingerprinting was recently shown by Davis et al. (2005), who applied the technique as one of the lines of evidence to delineate hydrocarbon plumes at a former refinery site (Yukon Territory, Canada). Source tracking based on isotope ratios is virtually the only scientifically valid option for single-compound spills or for simple mixtures where the traditional fingerprinting techniques cannot be applied. The isotope approach is also extremely beneficial in the situations where GC-MS fingerprints are altered by sample weathering. As will be discussed in the following sections, weathering and biodegradation of several classes of important contaminants do not affect their stable isotope ratios. The stable isotope approach is not a silver bullet; for certain classes of contaminants, postrelease diagenetic changes (in particular, biodegradation) will alter the original source signatures and prevent reliable source–spill correlations. The following sections will address the method limitations and define the area for successful application.

Stable carbon and hydrogen isotope ratios of the organic contaminants are in most cases relatively similar to those of their manufacturing precursors; carbon and hydrogen isotope ratios reflect those of crude oil or crude oil distillate fractions, methane, C3 biomass or C4 biomass used in manufacturing of organochemicals that eventually end up as environmental pollutants. Similarly, the expected variability of chlorine isotope ratios will reflect the range of $\delta^{35}$Cl in the chlorine sources. An additional degree of variability will be introduced by changes in isotope ratios caused by manufacturing processes and, finally, after the contaminant spill, changes in isotope ratios may occur as the result of various attenuation processes as discussed below in the section on contaminant attenuation.

Few systematic studies of stable isotope composition have been published for industrial chemicals. A study of gasolines from different retailers from the East Coast and southwestern United States has shown different $\delta^{13}$C fingerprints, permitting discrimination between the samples (Smallwood, Philp, and Allen 2002). Similar types of differences between different gasolines are apparent in $\delta$D fingerprints (unpublished results from the University of Oklahoma). Two data sets on MTBE are available, from the United States (Smallwood et al. 2001) and worldwide (O'Sullivan et al. 2004). Carbon and/or chlorine isotope composition of chlorinated ethenes (Van Warmerdam et al. 1995; Beneteau, Aravena, and Frape 1999; Stout et al. 1998), polychlorinated biphenyls (Jarman et al. 1998; Reddy et al. 2000; Drenzek et al. 2002; Horii et al. 2005), polychlorinated naphthalenes (Horii et al. 2005), and pesticides (Drenzek et al. 2002; Reddy et al. 2002) is also available. Additional data on the isotope composition of various chemicals are scattered among the papers cited in this chapter.

## *Data Interpretation—Different Contaminant Sources or Diagenetic Changes?*

As discussed in the introductory section, apart from the predominance of gas/vapour phase attenuation, the only significant process to alter contaminants' isotope ratios is degradation—biodegradation or abiotic chemical degradation. Stable isotope ratios of organic contaminants may potentially yield two types of information: a fingerprint for source tracking or a measure of biodegradation or, in specific cases, other processes of contaminant attenuation. The two categories tend to be mutually exclusive; contaminants susceptible to biodegradation and exhibiting isotope fractionation upon biodegradation have to be considered with special care if source fingerprinting is required.

The behaviour of chlorinated ethene plumes presents an interesting illustration of the dilemma of source signature preservation. Biodegradation of these compounds is known to result in significant isotope fractionation and, if biodegradation is active, the original isotope ratios of spilled contaminant become quickly altered. At a number of sites, the change of isotope compositions in PCE (perchloroethylene), TCE (trichloroethylene), 1,1,1-TCA (1,1,1-trichloroethane), etc. was successfully applied to confirm in situ biodegradation of the compounds as discussed below in the section on contaminant attenuation. On the other hand, particularly in environments not conducive to reductive dechlorination, in situ degradation is negligible and the isotope ratios of these contaminants remain unaffected. If it can be ascertained that the contaminants at a given site were not undergoing degradation, isotope fingerprinting can be considered as a reliable source tracking tool.

## *When CSIA Works for Identification of Source Signatures*

Long-time preservation of stable isotope contaminant source signatures is expected in environmental systems that are not conducive to in situ degradation and for the contaminants that do not fractionate upon degradation. As will be discussed later, even for compounds known to fractionate upon biodegradation, it may be possible to ensure data quality by careful evaluation of the study site context.

The former case is represented by accumulation of free phase product, NAPL (nonaqueous phase liquid), or solid precipitates (such as tar balls). Studies of tar balls demonstrate that source correlations of such samples can be successfully undertaken using a combination of stable isotopes and biomarkers. NAPL samples are a good candidate for retention of source signal: It has been verified experimentally that dissolution into the aquatic phase does not significantly change the isotope ratios of the NAPL contaminant pool (Hunkeler et al. 2005). For DNAPL (dense nonaqueous phase liquids), where evaporative weathering does not apply, the isotope ratios of a contaminant source will remain stable. For LNAPL (light nonaqueous phase liquids), the situation is complicated by the potential of evaporation-related phenomena. Based on a study by Smallwood et al. (2002) and on follow-up work at the University of Oklahoma (unpublished), carbon isotope effects resulting from evaporation of gasoline-range compounds can be neglected for the $C_{8+}$ aromatic fraction of gasoline, including xylenes, ethylbenzene, tri- and tetra-methylbenzenes (TMB), and naphthalenes.

Other systems may be not conducive to biodegradation. In many toxicological studies, limited bioavailability of contaminants resulted in low attenuation rates or no detectable in situ degradation. In particular, semivolatile and nonvolatile compounds, such as PAH, were likely to be protected by various sorption interactions. Often, pools of contaminant with slow and fast turnover times could be identified, corresponding to protected and unprotected fractions. Mazeas and Budzinski (2002b) describe a simulated oil spill experiment where apparent availability difference between background and spill PAH fractions was directly observed. Isotope composition and individual compound distribution demonstrated that the spill-related fraction was completely removed, while the ones originally present in the sediment were not affected. This mechanism is also feasible for other compounds with low water solubility and strong soil sorption potential.

The lack, or minimal amount, of isotope fractionation observed in laboratory biodegradation experiments was reported for a number of compounds and/or biodegradation pathways. Clearly, these compounds appear the most straightforward candidates for stable isotope source fingerprinting in environmental studies. Aerobic degradation of crude oil long chain *n*-alkanes and PAH fraction results in no isotope fractionation* (Mansuy, Philp, and Allen 1997; Huang et al. 1997; Trust et al. 1995; Mazeas, Budzinski, and Raymond 2002). Similarly, aerobic degradation of various aromatic compounds (e.g., toluene, naphthalene, and trichlorobenzene have been published) by organisms utilising the dioxygenase pathway resulted in no or minimal isotope fractionation (Meckenstock, Morasch, et al. 2004). The lack of isotope fractionation is explained in the case of the dioxygenase systems by the fact that the rate-limiting step of the reaction is not a bond cleavage, but rather an interaction between the $\pi$-electron system of the substrate and the oxygen

* Yanik et al. (2003a) observed large, up to eight permil fluctuations of $\delta^{13}C$ of individual PAH compounds and interpreted them as biodegradation effects. This interpretation is problematic due to apparent low precision and accuracy of $\delta^{13}C$ data.

species of dioxygenase. This type of reaction does not result in significant isotope effects in general.

Finally, no isotope effects, either carbon or chlorine, occur during reductive dechlorination of PCB (Drenzek et al. 2001, 2004). As for biodegradation of long chain *n*-alkanes of crude oil and heavy fuels, coal tar PAH fractions, etc., it can be speculated that the lack of isotope fractionation results from 'dilution' of the overall isotope effect by the molecule size (Pond et al. 2002 reported that hydrogen isotope fractionation becomes progressively smaller with the increase of *n*-alkane length, becoming undetectable at $C_{19}$) or from a rate-limiting physiological rather than enzymatic step, inherently not inducing isotope effects. For example, no isotope fractionation would be expected for bacteria feeding on crude oils that employ surfactants and a specific cellular transport mechanism for the oil globules.

Assurance of data quality for the degradable and fractionating compounds is of prime importance. In a recent paper, Hunkeler et al. (2005) showed two examples of plumes where source variations could be observed in high-resolution vertical and horizontal profiles. If reductive dechlorination of chlorinated ethenes is hampered, dissolved phase PCE and/or TCE would not degrade and source fingerprinting is appropriate. Direct and compelling evidence of the lack of in situ biodegradation is in this case provided by basic geochemical characterisation of the plume, in particular its redox status and the history of concentrations of potential dechlorination products. The absence of dechlorination products at the time of and prior to sampling permits safe interpretation of any parent compound data in terms of source fingerprinting. In a specific situation of single-contaminant spill (e.g., PCE only) and dechlorination not advancing beyond the intermediate product, such as *cis*-DCE, it is possible to reconstruct $\delta^{13}C$ of the parent compound even if it is significantly degraded. Concentrations of individual chlorinated ethenes are recalculated into a weighted average $\delta^{13}C$, which, in ideal conditions, corresponds to time zero $\delta^{13}C$ of the parent compound. This method can be used to test a hypothesis of single versus composite source, in that constant values obtained throughout the plume would confirm a single source, while different values for individual samples would imply that the parent compound was spilled several times with different $\delta^{13}C$. In theory, the same approach can be extended to the systems with more advanced dechlorination, but with the limitation of typically low precision of concentration data on the most volatile end members (e.g., vinyl chloride [VC] and ethene) and a possibility of undetected ethene degradation.

Another common group of contaminants is gasoline- or coal tar-related aromatics (benzene through methylnaphthalene). As will be discussed in more detail in the section concerning CSIA application to monitor in situ biodegradation, all of these compounds can be potentially affected by degradation and their isotope ratios accordingly altered. Based on published parameters of isotope fractionation, it can be expected that for benzene, the most isotopically labile compound in this category, at least 20% of the original amount of contaminant has to be removed by biodegradation to alter its $\delta^{13}C$ beyond the typical precision of the GC-IRMS method. In the case of naphthalenes, it may be as much as 50%. As will be discussed in more detail in the monitoring of biodegradation section, these values represent 'worst case scenarios' (from the perspective of source signature tracking) valid for homogeneous aquifer and uniform distribution of degrading organisms. Typically, the observed isotope fractionation will be significantly lower (i.e., even significantly more biodegradation will not alter the source signatures).

Figure 4.14 illustrates the impact of biodegradation on the aromatic compounds' fingerprint appearance between samples 5 and 7 (no significant degradation) and samples 3 and 8 (several compounds have been preferentially depleted). A strong corresponding increase of $\delta^{13}C$ in ethylbenzene and xylenes was observed (toluene could no longer be measured in the degraded samples). It can be concluded that while biodegradation can limit the utility of CSIA in providing a source signal of aromatic fractions of gasoline and coal tars, careful evaluation of the combined isotope and traditional fingerprints can help to focus the data interpretation.

Finally, biodegradation of a number of compounds results in very strong isotope fractionation. MTBE in anaerobic environments, VC, *cis*-DCE, and 1,2-DCA fall into this category. The strong isotope fractionation of these compounds is of great utility in natural attenuation studies, but it makes it difficult to use CSIA of these compounds for source tracking. We will limit the discussion to MTBE. While it may be tempting to use isotope composition of MTBE as a diagnostic tool for plume correlation, the validity of such data would be problematic. Data interpretation would have to account for very strong isotope fractionation characteristics of MTBE biodegradation in anaerobic conditions. In situ biodegradation as low as 5% of the original contaminant would significantly alter its original isotope signature. There are few actual sites where a dense monitoring network and frequent monitoring permit independent verification of the absence of degradation at such a low level to be obtained, making it difficult to defend in a courtroom the validity of MTBE $\delta^{13}C$ for source tracking.

### Field Applications

Existing work in the field can be roughly divided into application of specific matching of contaminants (correlation of spill to a specific source or plume delineation) and to more general apportionment of contaminant sources (e.g., by assessing the relative importance of natural and anthropogenic sources). Examples of the former category are the two papers of Kvenvolden et al. (1995) and Davis et al. (2005) discussed earlier. Mansuy et al. (1997) showed an example of $\delta^{13}C$ fingerprint matching of fuel residue collected from con-

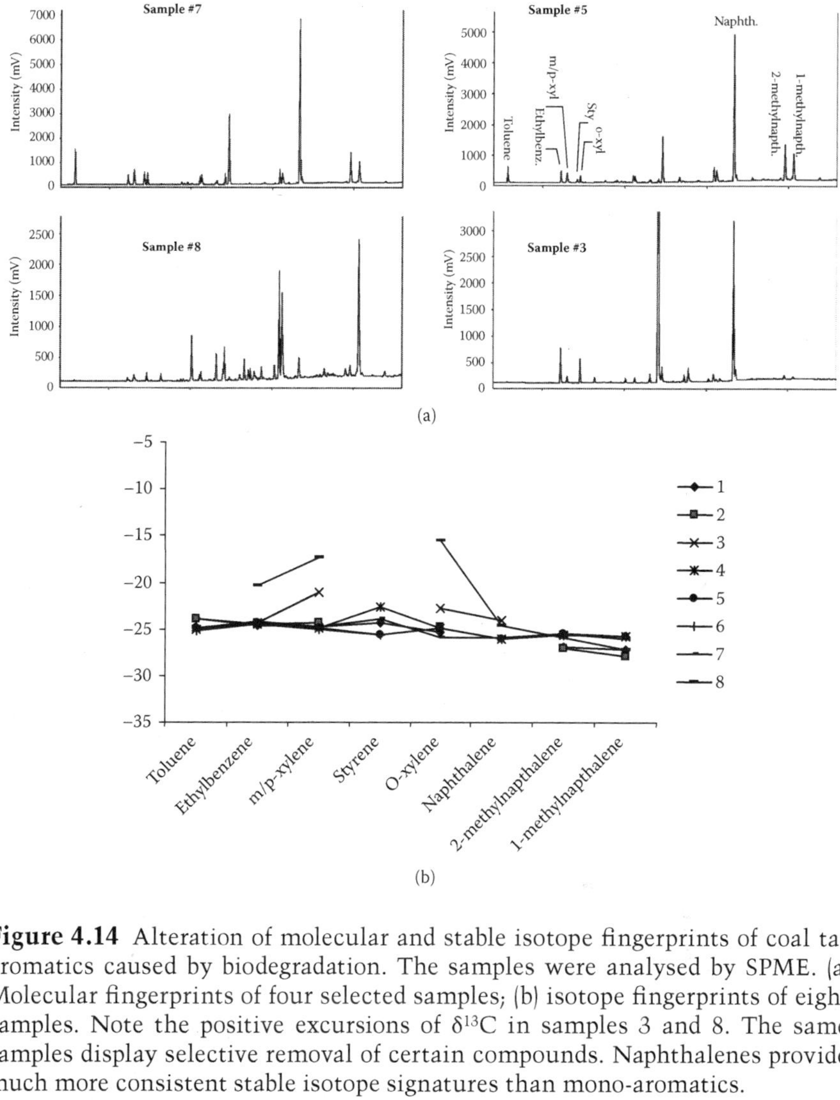


**Figure 4.14** Alteration of molecular and stable isotope fingerprints of coal tar aromatics caused by biodegradation. The samples were analysed by SPME. (a) Molecular fingerprints of four selected samples; (b) isotope fingerprints of eight samples. Note the positive excursions of $\delta^{13}C$ in samples 3 and 8. The same samples display selective removal of certain compounds. Naphthalenes provide much more consistent stable isotope signatures than mono-aromatics.

taminated birds with a suspected fuel spill source (the two fingerprints were identical; Figure 4.15). Mazeas and Budzinski (2001, 2002a) presented results of correlation of environmental oil residues after the *Erika* oil spill (Atlantic coast of France). Unambiguous correlation of pollution with *Erika* oil was obtained for some of the samples, while other samples were apparently related to tar ball incidents following the *Erika* oil spill. In a recent paper, Boyd et al. (2006) applied principal component analysis combined with multivariate analysis of variance (ANOVA) to assess interrelatedness of $\delta^{13}C$ fingerprints of individual *n*-alkanes. Examples of correlation of a fuel spill (Norfolk, Virginia) and samples from potential sources are shown.

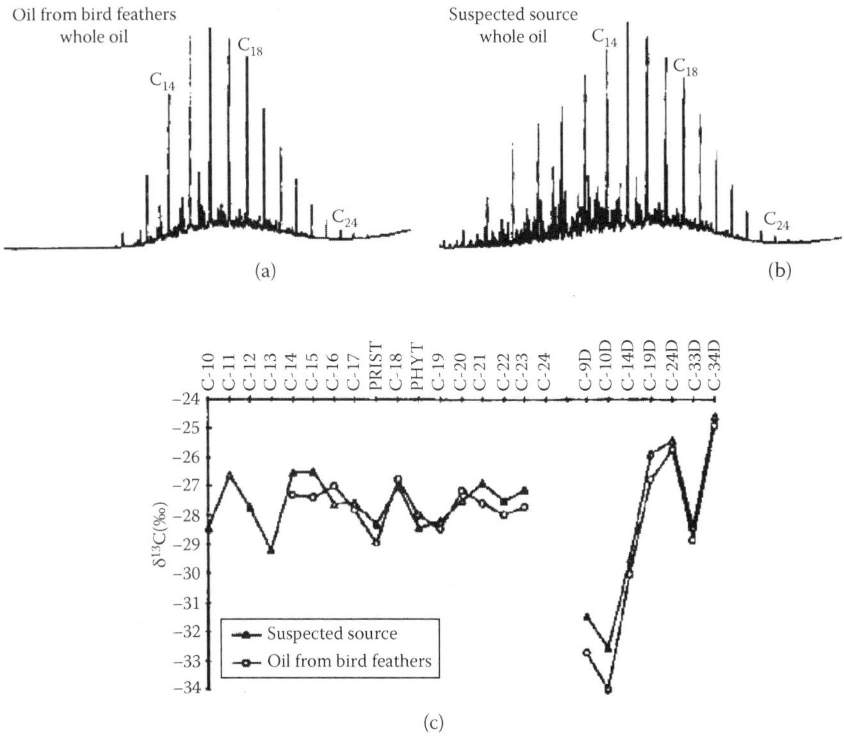


**Figure 4.15** Molecular and isotopic fingerprints of oil from contaminated bird feathers and from the suspected oil source. (From Mansuy, L. et al., *Environmental Science and Technology*, 31: 3417–3425, 1997.)

Examples of the latter type of application are studies of sources of environmental PAH. Differences of stable carbon isotope composition of PAH result from different processes of combustion or pyrolysis of coal, petroleum, or methane (McRae et al. 1999, 2000). Sun, Cooper, and Snape (2003) show a similar data set of carbon and hydrogen CSIA, with distinct isotope fingerprints of PAH derived by four different combustion processes. O'Malley, Abrajano, and Hellou (1996); Fabbri et al. (2003); Stark et al. (2003); Sun, Snape, et al. (2003); Glaser et al. (2005); and Walker et al. (2005) combined molecular fingerprinting and CSIA of PAH compounds to resolve environmental contributions from different pollution sources. For example, Fabbri et al. (2003) studied lagoon sediment from the Adriatic Sea (Ravenna, Italy) to identify the origin of PAH contamination. Molecular PAH fingerprints were consistent with pyrolysis of methane and strongly depleted stable isotope values confirmed that the local biogenic methane supply used by industry was indeed implicated. Yanik et al. (2003b) presented a study comparing compound-specific $\delta^{13}C$ of PCB extracted from aquatic fish and bird tissues with that of commercial Aroclors.

## Stable Isotopes in Contaminant Attenuation Studies

Unless successful remedial action at the contaminant release site is undertaken immediately, resulting contamination of water, soil, air, or biological species has to be dealt with. From the forensic perspective, attenuation processes are important in two ways: First, physical and chemical attenuation results in alteration of samples, imposing limitations on the types of results that can be obtained, and second, liability of the involved parties depends on the environmental impact of the offence. The following section will focus on applications of forensic geochemistry to the study of attenuation of contaminants, primarily through demonstration of in situ degradation of groundwater organic contaminants and measuring its progress.

In particular, protection of groundwater resources requires that specific action is required to (1) contain the spread of contaminant in aquifer, and (2) eventually ensure that the dangerous contaminant concentrations are eliminated. The sensitive cases involving very toxic species and immediate threat to important water supply sites often require engineered solutions, such as installation of pump-and-treat facilities or reactive barriers to stop the spread of the contaminants. On the other hand, at sites with low or moderate priority, natural attenuation is a reasonable and cost-effective, albeit slower, option to deal with the problem. It is the prerogative of the environmental protection authorities to define the criteria for monitored natural attenuation (MNA) application and differences exist within the United States and internationally.

In general, the guidelines for MNA require empirical evidence confirming reduction of plume size and scientifically valid prediction of the future trend for reducing the residual dissolved mass of the plume (e.g., EPA OSWER Directive 9200.4-17P 1999). The latter involves demonstration of in situ degradation of the contaminants (biodegradation or abiotic degradation). A microcosm biodegradation study is the classic means for confirming biodegradation potential of the site. While the microcosm technique is a well-established and accepted approach for this purpose, the technique has its limitations; the study may be time consuming (incubations taking 1 year or more) and inconclusive results are not infrequent. Alternative options for biodegradation testing are provided by compound-specific stable isotope analysis (CSIA) of the target contaminants to detect and/or quantify isotope fractionation attributable to biodegradation. More recently, applications involving in situ introduction of stable isotope-labelled surrogates and monitoring for isotope label transfer into microbial biomass or into biodegradation products have been developed. The benefit of the stable isotope approach is in the virtually immediate delivery of the results and, in some cases, the ability to quantitatively assess the extent of elapsed biodegradation. Biodegradation of organic contaminants can be also studied by bulk stable isotope analysis. The rationale of bulk isotope analysis application is different, relying on isotope composition of the final degradation product—typically, soil $CO_2$, methane, or DIC (e.g., Aggarwal and Hinchee 1991; Landmeyer, Vroblesky, and Chapelle 1996; Kirtland et al. 2003).

Another novel approach to in situ biodegradation studies is in situ application of target contaminant labelled with heavy isotope species. Incorporation of the label into biomass (Geyer et al. 2005) or detection of the isotopic label in degradation metabolites (Reusser et al. 2002) is a direct evidence of active contaminant metabolism. In principle, the difference between this and the CSIA approach is that the respective results refer to real-time contaminant metabolism as opposed to the degradation effect accumulated over time since the contaminant was spilled. The following discussion, for brevity, will refer to biodegradation, although the same approach can be used to verify the performance of abiotic chemical degradation (e.g., in engineered in situ degradation solutions).

Contaminants well suited for CSIA study are those that undergo isotope fractionation upon biodegradation. In laboratory biodegradation studies of VOC-class contaminants, virtually all the species studied exhibited isotope fractionation with at least some types of degrading organisms. Most work to date has concerned biodegradation of chlorinated solvents, BTEX and MTBE. A recent review by Meckenstock, Morasch, et al. (2004) provides an up-to-date summary of isotope effects of diverse biodegrading organisms.

### *Qualitative Evidence of Biodegradation*

On the most basic level, CSIA results can be used simply to confirm the onset of in situ biodegradation. An increase of stable isotope ratios over the value of undegraded contaminant (the difference should exceed the precision limit of the used GC-IRMS method) is typically sufficient to propose that some degradation has occurred. Extensive evaporation can also fractionate isotope ratios of the most volatile species (e.g., $C_5$ and $C_6$ hydrocarbons in gasoline), resulting in a $\delta^{13}C$ change of 1–2‰ (unpublished results, University of Oklahoma), which can be mistaken for a biodegradation effect.* Small changes in the isotope ratios of this type of compound should be evaluated with care to avoid false detection. The change of isotope ratios is measured versus the isotope composition of the undegraded contaminant (a time zero benchmark); in some cases, this reference value can be obtained from NAPL present in the source area.

If the NAPL sample is not available or it is suspected that multiple spills have occurred, possibly each one with a different isotope signature, the

* Strong hydrogen fractionation is possible for certain compounds, as reported by Wang and Huang (2003). Note that this type of effect does not mimic biodegradation but masks it instead.

benchmark value can be provided by using the most positive member value from the reference data range on the isotope ratios in the commercial product of interest. Such information is available for MTBE and various chlorinated compounds, as discussed in the preceding section. An alternative approach valid for sites with good monitoring-well coverage and several years of available geochemical data is to use isotope ratios in upgradient, close to the source part of the plume. This approach requires that the contaminant plume has had enough time to reach a steady state in respect to the source area. It is also necessary to establish the hydrologic relationship between the source zone, for which the benchmark isotope signature is measured, and the postulated biodegradation zone (cf. Richnow et al. 2003). If these two criteria are met, a value of $\delta^{13}C$ more positive in the tested samples than in the source area is strong evidence that in situ degradation has occurred since the spill.

- Qualitative verification of in situ biodegradation requires only that isotope fractionation is confirmed relative to the initial source isotope ratio (enrichment of the heavier isotope, e.g., increase of $\delta^{13}C$ value).
- Is the steady-state source isotope signature of the target contaminant known (measured directly)? A non-steady-state source may involve multiple releases of contaminant over time, each with possibly different isotope composition. The source variability can mimic biodegradation. In this case a conservative estimate (most positive $\delta^{13}C$ of the relevant range) of source benchmark is necessary.

### *Quantitative Interpretation of CSIA*

Quantitative interpretation of CSIA yields not only confirmation of biodegradation but also an estimate of the amount of the contaminant degraded since the spill. The calculated extent of biodegradation accounts for the percentage of degraded contaminant, rather than for the overall reduction of contaminant concentration, caused by biodegradation, dispersion, volatilisation, etc. The basis for the calculation is provided by the Rayleigh fractionation model, where the unknown parameter—the ratio of remaining contaminant—can be mathematically obtained if the initial isotope ratio of the contaminant, the enrichment factor characteristic of the specific reaction (enrichment factor describes the magnitude of isotope fractionation), and the present isotope ratio of the contaminant are known. The latter parameter is measured on field samples by CSIA (e.g., Richnow et al. 2003; Sherwood Lollar et al. 2001; Kuder et al. 2005).

Stringent application of the Rayleigh model to an in situ biodegradation system requires a good hydrogeological site characterisation to confirm that the sampling wells follow a plume flow line, that the plume is at steady state in respect to the source area, and that the plume is at steady state in respect to contaminant sorption. If these criteria are met and the correct value of enrichment factor is used (in this case, ε from various experiments on sulphate reductive biodegradation of toluene appears to be consistent), the results obtained from equation 4.4 should closely match the contaminant concentration profile. This approach was utilised in a study of toluene and o-xylene biodegradation (Richnow et al. 2003).

Laboratory experiments show that for most contaminants, the enrichment factors differ between individual microbial species or cultures. Moreover, for most contaminated sites, the monitoring network is not adequate to accurately determine plume flow line, and the steady-state status of the contaminant sorption and source may also be problematic. CSIA-based evaluation of biodegradation in such cases cannot provide an accurate quantitative assessment, so a conservative approach for data evaluation is necessary. For the two unknown parameters in equation 4.4, the predegradation source isotope composition and the enrichment factor, conservative values are selected to avoid overestimating the biodegradation. The selected ε value should be based on the mechanism of degradation (if known) and, if a range of enrichment factor values is feasible for a given contaminant, the ε value with the largest isotope effect should be selected (i.e., most negative ε), so as not to overestimate the extent of biodegradation. Table 4.3 lists a selection of enrichment factors adopted from Meckenstock, Morasch, et al. (2004), Schmidt et al. (2004), and Elsner et al. (2005).

The conservative value of the predegradation $\delta^{13}C$ of a contaminant is the most positive number for the range of various commercial brands (e.g., based on Smallwood et al. 2001 and O'Sullivan et al. 2004; note that the conservative value of MTBE $\delta^{13}C$ is –27.5 ± analytical precision of the method). This conservative approach is most appropriate for contaminants with a relatively narrow range of potential source isotope ratios and those that fractionate strongly upon degradation; otherwise, most of the studies would yield inconclusive results. Contaminants such as MTBE (Figure 4.16), 1,2-DCA, and vinyl chloride fit this category. The same conservative approach for data evaluation is necessary for other contaminants, including the ubiquitous chlorinated solvents TCE and PCE, where a diverse range of initial isotope ratios is feasible and enrichment factors measured for various degrading organisms are significantly different from each other. A large ε implies that biodegradation can be quantified accurately and precisely and that the early stages of degradation can be studied, because much larger isotope fractionation will result from reactions with strongly negative ε value.

An enrichment factor for anaerobic degradation of MTBE was recently obtained at the University of Oklahoma of approximately –17 (similar values were obtained in three different experiments with homogenous soil microcosms). Assuming that the contaminant source $\delta^{13}C$ is accurately known, the extent of anaerobic biodegradation of MTBE can reasonably accurately be

**Table 4.3 Summary of Isotope Effects in Various Contaminant Degradation Experiments**

| Compound | Range of $^{13}C/^{12}C$ fractionation (ε) | Other elements' fractionation | Conditions |
|---|---|---|---|
| Benzene | −1.5, −3.5 | εH = −11, −13 | Oxic |
| Benzene | −1.9 to −3.6 | εH = −29 to −79 | Various anaerobic |
| Ethylbenzene | −2.2, −3.7 | | Various anaerobic |
| Toluene | No effect to −3.3 | | Oxic |
| Toluene | −0.5 to −2.2 | εH = −12 to −956 | Various anaerobic |
| m-Xylene | −1.7 | | Oxic |
| m-Xylene | −1.8 | | Sulphate red. |
| p-Xylene | −2.3 | | Oxic |
| p-Xylene | −1.1, −1.5, −3.2 | | Sulphate red. |
| m-Cresol | −3.9 | | Sulphate red. |
| p-Cresol | −1.6 | | Sulphate red. |
| Naphthalene | −0.1 (no effect) | | Oxic |
| Naphthalene | −1.1 | | Sulphate red. |
| 2-Methylnaphthalene | −0.9 | | Sulphate red. |
| PCE | −1.8 to −5.5 | εCl = −10 | Red. dechlorination, various electron acceptors |
| PCE | −15.7, −17.7 | | Permanganate oxidation |
| TCE | −2.2 to −13.8 | εCl = −5.5, −5.6, −5.7, −30 | Red. dechlorination, various electron acceptors |
| TCE | −18.2 | | Oxic |
| TCE | −1.1 | | Oxic, cometabolic |
| TCE | −18.5, −26.8 | | Permanganate oxidation |
| *cis-DCE* | −0.4 | | Oxic, cometabolic |
| *cis-DCE* | −14.1 to −20.4 | | Red. dechlorination, various electron acceptors |
| *cis-DCE* | −21.1 | | Permanganate oxidation |
| *trans-DCE* | −3.5, −6.7 | | Oxic, cometabolic |
| *trans-DCE* | −30.3 | | Red. dechlorination |
| 1,1-DCE | −7.3 | | Red. dechlorination |
| VC | −3.2 to −8.2 | | Oxic, cometabolic |
| VC | −21.5 to −31.1 | | Red. dechlorination |
| 1,2-DCA | −3 and −27, −32 | | Oxic |
| 1,2-DCA | −32.1 | | Red. dechlorination |
| 1,1,2-TCA | −2 | | Red. dechlorination |
| 1,2,4-Trichlorobenzene | Not significant | | Oxic |
| 1,2,4-Trichlorobenzene | −3.2, −3.5 | | Red. dechlorination |

Continued

**Table 4.3 Summary of Isotope Effects in Various Contaminant Degradation Experiments (Continued)**

| Compound | Range of $^{13}C/^{12}C$ fractionation (ε) | Other elements' fractionation | Conditions |
|---|---|---|---|
| MTBE | -1.5 to -2.4 | εH = -29 to -66 | Oxic |
| MTBE | -9 to -17[a] | εH = -16α | Various anoxic |
| TBA | -4.2 | | Oxic |

[a] The maximum values for isotope effect in MTBE are based on Kuder, T. et al., *Environmental Science and Technology,* 39: 213–220, 2005, and unpublished University of Oklahoma results.

*Sources:* Meckenstock, Morash et al., 2004; Schmidt, T. C. et al., *Analytical and Bioanalytical Chemistry,* 378: 283–300, 2004; Elsner, M. et al., *Environmental Science and Technology,* 39: 6896–6916, 2005 and references therein.

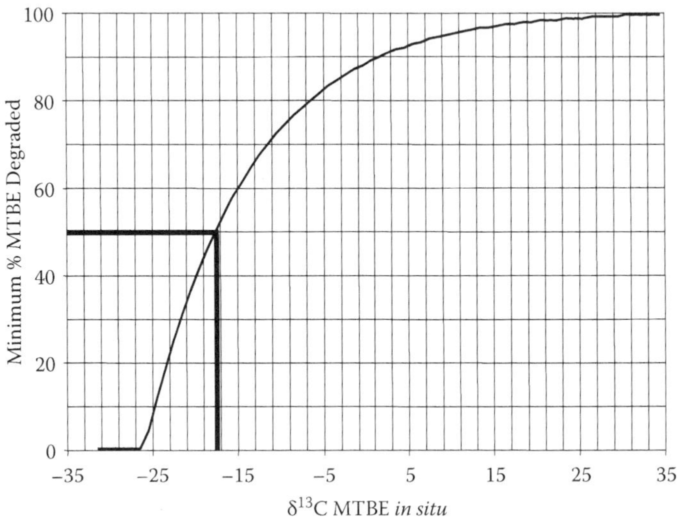


**Figure 4.16** Nomogram showing the projected conservative extent of MTBE biodegradation based on CSIA analysis. See text for additional explanations.

assessed when only 10–20% of the MTBE has been biodegraded. In the case of the anaerobic degradation of toluene (ε between –0.5 and –2.2), reasonably accurate assessment can be attempted after 90% or more is degraded. An example of MTBE biodegradation assessment at a contaminated gasoline station site in Orange County, California, is shown in Figure 4.17. This is a typical monitoring situation at a contaminated site where samples are obtained from a relatively small number of monitoring wells and carbon isotope values and concentrations obtained for the MTBE. At sites such as this, the highest levels of natural attenuation are relatively close to the source of the MTBE and not at the leading edge of the plume, as may have been thought

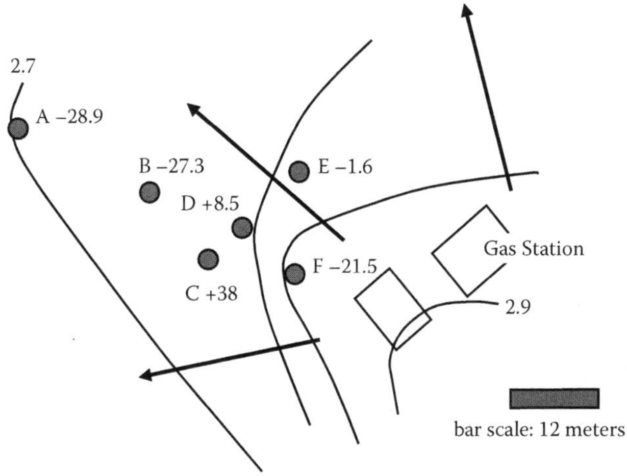


| MW# | $\delta^{13}C$ | Est. Biodegradation |
|---|---|---|
| A | −28.9 | no evidence |
| B | −27.3 | no evidence |
| C | +38 | 98% |
| D | +8.5 | 88% |
| E | −1.6 | 79% |
| F | −21.5 | 30% |

**Figure 4.17** Calculated conservative extent for MTBE biodegradation at a gas station site in California. MTBE $\delta^{13}C$ shown for the six monitoring wells was used to calculate the conservative estimate of biodegradation based on equation 4.4. Contour lines show water table elevation above sea level. See text for additional explanations.

using a conventional interpretation. The carbon isotope values of the MTBE at the leading edge of the plume are similar to the source values and suggest that these are relatively small quantities of the MTBE, which eluded the areas where high levels of natural attenuation were occurring. This figure shows the result of using the isotope data, concentration estimates, and enrichment factors to provide an indication of the amount of MTBE that has been removed from the site.

Finally, an example of a situation when quantitative CSIA should be avoided is a study of aerobic biodegradation of MTBE. The largest ε observed for this reaction is −2.4, so $\delta^{13}C$ will change upon extensive biodegradation by a few permil. Similar magnitudes of $\delta^{13}C$ change would result from a minor, difficult to disprove contribution of anaerobic biodegradation. An attempt to evaluate the small change of $\delta^{13}C$ of MTBE in terms of aerobic biodegradation would imply very advanced biodegradation, while the same change of $\delta^{13}C$ would imply much less degradation in terms of the anaerobic process. A conservative approach, using the generic maximum ε value, would be counterproductive for aerobic MTBE degradation, as virtually all results would be inconclusive.

In summary, for application of CSIA in quantitative evaluation of in situ biodegradation, one should bear in mind the following points:

- The amount of degraded contaminant is calculated from the Rayleigh isotope fractionation model (equation 4.4). The result is not affected by abiotic, nondegradation attenuation processes and is exclusively based on changes of isotope ratios of contaminant.
- Is the steady-state source isotope signature of the target contaminant known (measured directly)? A non-steady-state source may involve multiple releases of contaminant over time, each with a possibly different isotope composition. The source variability can mimic biodegradation. In this case a conservative estimate (most positive $\delta^{13}C$ of the relevant range) of source benchmark is necessary.
- What is the enrichment factor valid for a study site? For example, for sulphate-reducing toluene or anaerobic MTBE degradation, enrichment factors obtained for different microbial cultures are similar to each other. However, for other contaminants such as PCE, this is not the case. For problematic situations, a conservative (most negative $\varepsilon$ of the relevant range) value has to be assumed.
- If accurate values of source isotope composition and $\varepsilon$ are known, accurate determination of the extent of elapsed biodegradation is possible. Otherwise, the conservative estimate yields a minimum (underestimated) amount of the biodegraded contaminant. If the sampled plume is heterogeneous in respect to degradation activity, samples mixed with more or less degraded material will cause underestimation of the calculated value.
- The conservative approach does not overestimate the mass of degraded contaminant. On the other hand, underestimation of the biodegraded amount is very probable, due to real values of source $\delta^{13}C$ and/or $\varepsilon$ being not the same as the conservative estimate values, or due to values of in situ $\delta^{13}C$ being affected by plume heterogeneity.

### *Interpreting the Biodegradation of a Sequence of Intermediates*

The interpretation of results involving biodegradation of a sequence of intermediates, such as in the case of reductive dechlorination of PCE or a more simple case of TBA produced at the expense of MTBE, is more complicated; straightforward application of Rayleigh fractionation model is valid only for the parent compound. This is because the transient pool of the intermediate is replenished by newly generated product, so the isotope signature of

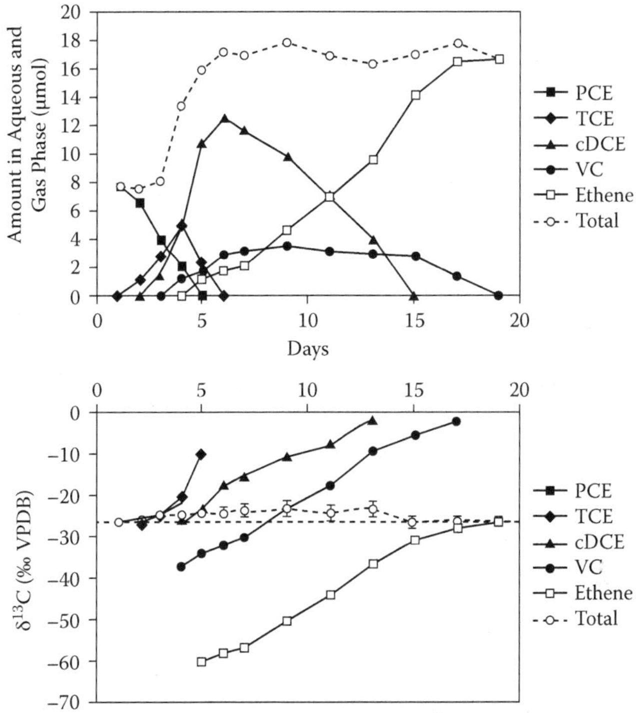


**Figure 4.18** Changes of concentration and $\delta^{13}C$ of PCE and its reductive dechlorination products, measured in a microcosm experiment. (From Hunkeler, D. et al., *Environmental Science and Technology*, 33: 2733–2738, 1999.)

the intermediate reflects the balance between the rates of degradation and production. (Once the parent compound is completely gone, the first intermediate becomes a parent compound in respect to the Rayleigh model.) The isotopically depleted intermediate is indicative of the prevalence of accumulation over degradation, while isotopically enriched values suggest a net degradation of the intermediate pool.

Data on the dechlorination of PCE presented by Hunkeler, Aravena, and Butler (1999) clearly show this type of relationship (Figure 4.18). In the example, the VC produced in the initial stage of the reaction has a strongly negative $\delta^{13}C$, as expected for a biodegradation product. The interaction of production and dechlorination of VC pool results in a transient peak, followed by a decrease of VC concentration as its precursor, *cis*-DCE, disappears. An increase of $\delta^{13}C$ corresponds to the progressive dechlorination of the parent compound. In qualitative terms, the certainty of the onset of biodegradation of an intermediate (e.g., *cis*-DCE) can be safely advocated once its $\delta^{13}C$ exceeds that of the original source (e.g., PCE). This relationship is valid for any parent–daughter system, since a degradation product is always $^{13}C$ depleted relative to the parent compound* (Figure 4.13).

In a recent paper, Van Breukelen, Hunkeler, and Volkering (2005) successfully demonstrated the use of a kinetic model (adopted one-dimensional reactive transport model) to quantify degradation of PCE-TCE-DCE-VC-ethene dechlorination sequence. The model was able to accurately match a set of experimental microcosm data, accounting for degradation and soil sorption effects. A variant of the model accounting for one-dimensional contaminant transport was applied to field site data and, while mixing of more and less degraded material was found to be somewhat of a problem, a reasonable level of agreement between model and field data was achieved. This approach can in principle be extended to three-dimensional simulation and has a good future potential to be useful in evaluation of chlorinated solvent field sites.

### *Identification of the Mechanism of Biodegradation*

Identification of the mechanism of biodegradation may be of value for quantitative site evaluation (to choose the right enrichment factor value for data evaluation) and for design of in situ treatment. A recent paper by Elsner et al. (2005) provides a discussion of isotope effects in various types of biochemical reactions. Identification of the mechanism of biodegradation in environmental samples requires two-dimensional CSIA (Kuder et al. 2005). MTBE appears to be particularly well suited for two-dimensional CSIA because aerobic and anaerobic pathways of biodegradation have very different isotope fractionation patterns (Figure 4.19). Two-dimensional CSIA is a relatively new approach and few field applications have been published.

### *Limitations of CSIA*

Apart from instrumental limitations (detection limits, etc.) discussed earlier and certain compounds not fractionating upon biodegradation (discussed in the tracking of contaminant sources previously), CSIA may yield inconclusive results due to missing the active parts of a plume or 'dilution' of biodegradation signal by sampling of heterogeneous plumes. Monitoring wells installed at most sites sample a depth interval, which may intersect with more or less degraded (or undegraded) portions of a plume. A sample withdrawn from a well represents an averaged value for the plume within the hydraulic radius of the well. A contribution of undegraded material, more likely at higher

* In molecules where intramolecular isotopic differences exist, the position-specific $\delta^{13}C$ has to be considered. In MTBE, t-butyl carbon atoms are systematically enriched in $^{13}C$ by several permil (cf. Kuder et al., 2005, and Zwank et al., 2005) so that initially $\delta^{13}C$ of TBA resulting from biodegradation is in effect more positive than that of the precursor MTBE.

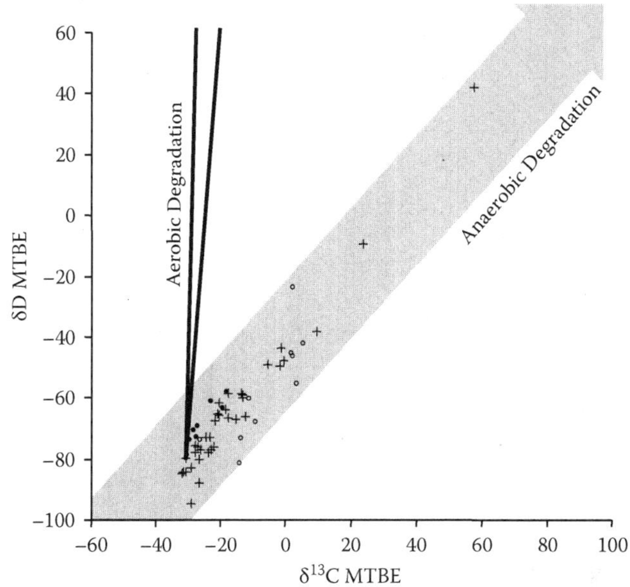


**Figure 4.19** Combined carbon and hydrogen CSIA results (2-D CSIA) of anaerobic (data points) and aerobic (trends calculated based on published laboratory work) degradation of MTBE. (From Kuder, T. et al., *Environmental Science and Technology*, 39: 213–220, 2005.)

concentration than in the degraded zone of a plume, will cause underestimation of the net isotope effect or, in extreme cases, will nullify the isotope effect resulting from degradation. An interference of the same type is caused by proximity of free phase product replenishing the degraded contaminant, resulting in a mixed, underestimated isotope signal.

At multiple sites of MTBE contamination studied by the authors, it was found that isotope ratios at distal plume fringes were not showing any evidence of biodegradation, even if a strong signal was observed closer to the source. It appears that while the bulk of MTBE was removed by biodegradation, part of the contaminant would break through with no or minimum extent of biodegradation, retaining the original isotope ratio. In consequence, samples collected down gradient from the biodegradation zones would be dominated by the undegraded contaminant. It appears that a similar phenomenon was observed at the Port Hueneme site, where an aerobic biobarrier was installed to degrade MTBE (Lesser et al. 2005). (Port Hueneme is a naval base in Ventura County, California, with a significant groundwater plume containing elevated levels of MTBE. This plume has been extensively investigated and characterised.) Low concentrations of MTBE detected down gradient from the barrier did not exhibit isotope ratios indicative of extensive degradation.

### *Applications of CSIA*

In recent years, a growing number of published studies has demonstrated the utility of stable isotope techniques for demonstrating and quantifying in situ degradation of chlorinated compounds, MTBE and BTEX. Selected contributions in the field will be briefly summarised. The framework for data isotope interpretation was provided by laboratory biodegradation studies of chloro-ethenes, aromatic compounds, MTBE, and others (Meckenstock, Morasch, et al. 2004 and references therein). Table 4.3 summarises isotope effects in various experiments published in the literature. At multiple sites, enrichments of $^{13}C$, D, and/or $^{37}Cl$ in the aforementioned contaminants were indicative of in situ attenuation.

Richnow et al. (2003) applied CSIA to in situ biodegradation monitoring of aromatic compounds at an industrial site in Germany. Carbon isotope fractionation was observed in toluene, o-xylene, and naphthalene; for the two former compounds, the quantitative estimation of the extent of biodegradation matched the observed contaminant concentration profiles.On the other hand, no isotope effects were observed for indane and indene degradation. Griebler et al. (2004) and Steinbach et al. (2004) studied in situ biodegradation of aromatics at a site of a former coal gas plant (Germany). The latter group presented carbon and hydrogen isotope data indicating the loss of toluene, xylenes, and 2-methylnaphthalene, while apparent attenuation of 1-methylnaphthalene and methylbenzofuran was not accompanied by detectable isotope fractionation. Griebler et al. presented carbon isotope compositions of benzene, toluene, xylenes, naphthalene, and 1- and 2-methylnapthalenes, confirming in situ biodegradation. The same study showed the presence of specific carboxylated metabolites of ethylbenzene, xylenes, naphthalenes, and benzothiophene. Peter et al. (2004) used concentration time series and $\delta^{13}C$ of o-xylene from a pumping test to obtain a quantitative estimate of in situ biodegradation.

Hunkeler et al. (1999) presented $\delta^{13}C$ values of PCE and its dechlorination products from microcosms and field samples. The pattern of isotope fractionation matched complete dechlorination of PCE in the microcosm; a similar pattern was present in the field data set. Sherwood Lollar et al. (2001) studied stable carbon isotope fractionation of TCE and PCE at Dover Air Force Base (AFB) in the United States. Isotope fractionation of TCE, in particular, was indicative of extensive dechlorination. Song et al. (2002) used CSIA to monitor TCE dechlorination in a pilot study of in situ lactate amendment at a test site of Idaho National Engineering and Environmental Laboratory in the United States. CSIA results confirmed quantitative conversion of TCE to ethene. Hunkeler et al. (2005) presented data on a chlorinated hydrocarbon plume study (an undisclosed former solvent disposal site). CSIA data on 14 chemical species—either the original spilled compounds or their degradation products—permitted identification of the degradation pathways. Instead, typical reductive dechlorination of chlorinated ethenes was found to be of minor importance, and the predominant reaction pathways were dehydrochlorination of 1,1,2,2-PCE to TCE, reductive dechlorination of chloroform to dichloromethane, dichloroelimination of 1,1,2-TCA to vinyl chloride, and dichloroelimination of 1,2-TCA to ethene. Morrill et al. (2005) showed that enrichment of $^{13}C$ of *cis*-DCE validated in situ dechlorination of the compound in a bioaugmentation project (Kelly AFB, United States). Stable isotope quantification of *cis*-DCE degradation suggested a rate constant lower by a factor of two to four than in a parallel study using chlorinated ethene concentration data.

In the last discussed chlorinated solvent study (Chartrand et al. 2005), CSIA was used to monitor reductive dechlorination processes in a fractured bedrock plume during a pilot-scale bioaugmentation project. Chlorinated ethene concentration profiles were not convincing evidence for the progress of dechlorination due to variable hydraulic gradients in the fractured bedrock and ongoing flux of contaminants from the NAPL source. On the other hand, enrichments of $^{13}C$ in *cis*-DCE and VC were indicative of dechlorination and a quantitative estimate of the CSIA data permitted evaluation of the efficiency of the bioaugmentation treatment. Examples of CSIA application to monitor dechlorination in abiotic systems are shown for chlorinated ethene reactions with iron (Vanstone et al. 2004) and permanganate (Hunkeler et al. 2003).

The last group of applications is in the study of attenuation of gasoline oxygenates. Kolhatkar et al. (2002) showed enrichment of $^{13}C$ in anaerobic MTBE plume (gasoline retail station in New Jersey). Kuder et al. (2005) presented combined carbon and hydrogen CSIA data for nine MTBE sites (gasoline retail stations in California and New Jersey) and from attendant microcosm experiments. The isotopic enrichments were indicative of biodegradation and the two-dimensional isotope fractionation pattern was indicative of an anaerobic biodegradation pathway for all of the studied sites. The pattern of isotope fractionation was indicative of demethylation of MTBE to TBA. Assessment of the progress of in situ degradation based on a Rayleigh model (equation 4.4) demonstrated significant removal of MTBE, exceeding 90% for some of the samples. Zwank et al. (2005) presented carbon and hydrogen data sets from a chemical waste storage site in Brasilia, reaching similar conclusions on the MTBE degradation pathway. Day et al. (2002) showed an increase of $\delta^{13}C$ in TBA corresponding with plume downgradient concentration reduction at an undisclosed chemical plant in the United States. This is apparently the only available example of CSIA confirmation of in situ TBA degradation.

## Dating of Contaminant Spills

A favourite question from anyone involved in the litigation process is: 'How long has the product been in the environment?' The reasons for such a question are obvious: Clearly, a specific answer could readily absolve a defendant of guilt if it was shown the client did not own the site at the time of the proposed spill. In order to address this question, numerous methods have been developed, some of which appear to be valid and have some justification for their use whereas others are totally irrelevant and can be extremely misleading. It should be noted at the outset that stable isotopes cannot be used to age date contaminant pools, primarily due to uncertain rates of biodegradation. As discussed in the section on contaminant attenuation, the isotope ratios measured in a contaminant tend to be biased towards values implying lower than actual degrees of degradation; even if an accurate degradation rate were known for a given set of environmental conditions, stable isotopes dating would tend to variably underestimate the real age of the onset of biodegradation.

Most age-dating methods commonly used in environmental forensics fall into two categories. For refined products, particularly gasoline, the most commonly used methods revolve around the presence or absence of certain compounds or additives. Many papers have appeared where attempts have been made to age date gasoline based on the presence or absence of certain compounds or changes in the composition of the gasoline. For example, the presence of MTBE would suggest the presence of a gasoline spilled after approximately 1990; the presence of lead would be a gasoline spilled prior to approximately 1985; high concentrations of benzene would suggest pre-1985; presence of high abundance of *n*-alkanes would suggest a straight run gasoline of a vintage prior to the more sophisticated refinery reformulation process. With regards to the presence or absence of MTBE, a point that is often overlooked in environmental forensics is the fact that MTBE was often used as an alkylate for many years prior to its use as an oxygenate. This factor should be taken into consideration when using MTBE as an age-dating tool. However, at the time MTBE was being used as an alkylate, the composition of the gasoline would be different from the composition of the gasoline when the MTBE was being used as an oxygenate. Other factors based on various ratios of branched hydrocarbons can be used to discriminate between gasoline samples derived from a refinery using an HF catalytic process versus $H_2SO_4$ catalytic processes. This, in turn, can be used as a possible clue to when a spill occurred based on when a particular refinery was in operation.

Changes in gasoline composition with time have been quantified by Schmidt et al. (2002). Examination of a large number of gasolines covering the time period 1970–2006—in particular the ratio of toluene + 2,2,4-trimethylpentane/$nC_7$ + $nC_8$—produced the diagram shown in Figure 4.20. This ratio was chosen since it was not affected to any large extent by evaporation. It would appear from this diagram that there was a fairly significant change in this ratio around 1994, most likely as a result of regula-

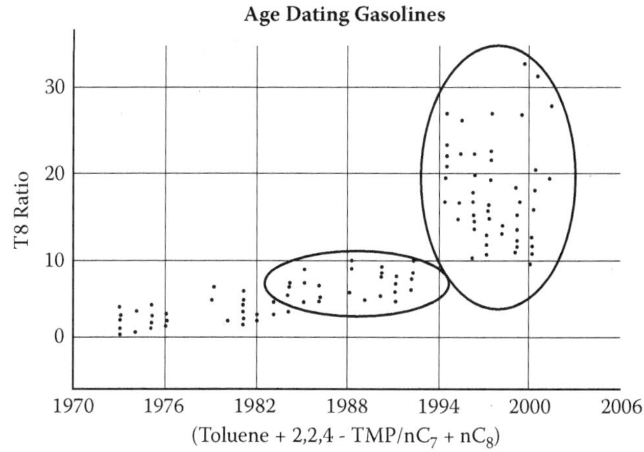


**Figure 4.20** Gasoline age dating diagram. (After Schmidt et al., 2002.)

tory changes. Hence, this ratio could be used to differentiate gasolines before and after 1994 based on this ratio. As with any of these ratios, if one is serious about using them, wherever possible they should be verified with other ratios and, if possible, any historical data that may be available.

Another method for age dating gasolines was published by Kaplan et al. (1997) based on the ratio of benzene + toluene/ethylbenzene + xylenes. From the examination of a large number of gasolines, primarily from California, an exponential decay curve was constructed resulting from solubility variations for these different components and the length of time they had been in contact with the groundwater. The ratio, defined earlier, is determined for the gasoline to be age dated and then the approximate age determined from the calibration curve. While the method has been used for litigation purposes, many of these cases have been in California, where the calibration was developed. The universal applicability of this has to be questioned since the whole concept primarily depends on the contact between the gasoline and the water and factors such as groundwater flow rate, lithology, and porosity of the matrix, among others.

A controversial age dating method for gasolines is based on the use of lead isotopes for the dating of leaded gasolines. This is based on the work of Hurst (2000) and it is claimed that for certain age periods the precision of this method is a few months. However, the method is not without its critics and there are indeed a number of very obvious potential problems. For example, the method is based on the concept that the lead used to make the tetraethyl lead changed with time and the sources used had very distinct isotopic signatures. However, it would appear that no consideration was given to recycling of the lead, mixing from various sources, contamination from refinery sources, and a number of other problems. If this method is going to be used, any resulting age date should certainly be verified by an independent age-dating process.

Unlike the gasoline methods described previously, the primary age dating method that has been used for age dating crude oils and refined products such as diesel is based on the publication of Christensen and Larsen (1993). This approach is based on the changes is the $n$-$C_{17}$/Pr resulting from biodegradation. Once again, it has been well documented in the petroleum geochemistry literature that the *n*-alkanes will be removed by biodegradation much more rapidly than the branched isoprenoids. Several years ago Christensen and Larsen monitored changes in this ratio from refined products spilled in the Netherlands at known times; they measured the changes in this ratio with time and constructed the calibration curve. The calibration curve is fine since it was constructed based on known samples that had been in the environment for a known period of time. However, the problem that has developed with this calibration curve is that it is now being used extensively for spills worldwide. It appears that in many cases little consideration is given to the initial starting values for the $n$-$C_{17}$/Pr, which vary considerably for oils coming from different sources. The fact that the calibration curve was developed for a certain set of environmental conditions is another factor often completely ignored. In brief, while it does seem to work in certain cases, a great deal of caution needs to be exercised when applying this approach.

## Summary

The characterisation of any product spilled in the environment is an extremely important part of any investigation to determine the source of such a product. In other cases the origin of the product may be known but rather than utilise expensive cleanup techniques, the responsible parties may wish to take advantage of natural attenuation for cleanup purposes. The biomarker content of crude oils and refined products combined with the isotopic characterisation of the same provides a powerful tool for solving many of these environmental issues.

Utilisation of biomarkers in petroleum exploration and environmental studies is a tool that has been with us for three or four decades and is well established for the characterisation and correlation of crude oils and most refined products, with the exception of gasoline. The utilisation of stable isotopes, particularly of individual compounds, is a much more recent development. As has been shown in this chapter, stable isotopes can be used to complement the biomarker data for crude oils or the heavier refined products or PAHs. In other cases, where the contaminant might be a single component, the stable isotope composition might be the only tool for correlation or source determination purposes. The role of stable isotopes in evaluating the progress of natural attenuation cannot be overestimated and it is fast becoming the method of choice in groundwater studies for this purpose.

Thus, in brief, this chapter has highlighted the concept and applications of biomarkers and stable isotopes in environmental studies. It was not meant to be an exhaustive review but rather to provide sufficient information to illustrate the value of these techniques both as stand-alone tools and in combination. References to other important review articles have been provided to supplement the information provided in this chapter.

## References

Ader, M., M. L. Coleman, S. P. Doyle, M. Stroud, and D. Wakelin (2001) Methods for the stable isotopic analysis of chlorine in chlorate and perchlorate compounds. *Analytical Chemistry,* 73: 4946–4950.

Aggarwal, P. K., and R. E. Hinchee. (1991) Monitoring in situ biodegradation of hydrocarbons by using stable carbon isotopes. *Environmental Science and Technology,* 25: 1178–1180.

Albaiges, J., and P. Albrecht. (1979) Fingerprinting marine pollutant hydrocarbons by computerized gas chromatography–mass spectrometry. *International Journal of Environmental Analytical Chemistry,* 6: 171.

Alexander, R., R. I. Kagi, R. Noble, and J. K. Volkman. (1984) Identification of some bicyclic alkanes in petroleum. *Organic Geochemistry,* 6: 63–72.

Bao, H., and B. Gu. (2004) Natural perchlorate has a unique oxygen isotope signature. *Environmental Science and Technology,* 38: 5073–5077.

Bence, A. E., K. A. Kvenvolden, and M. C. Kennicutt. (1996) Organic geochemistry applied to environmental assessments of Prince William Sound, Alaska, after the *Exxon Valdez* spill. A review. *Organic Geochemistry,* 24: 7–42.

Beneteau, K. M., R. Aravena, and S. K. Frape. (1999) Isotopic characterization of chlorinated solvents—Laboratory and field results. *Organic Geochemistry,* 30: 739–753.

Blumer, M. (1972). Submarine seeps; are they a major source of open ocean oil pollution? *Science,* 176: 1257–1258.

Blumer, M., G. Souza, and J. Sass. (1970) Hydrocarbon pollution of edible shellfish by an oil spill. *Marine Biology,* 5: 195–202.

Boehlke, J. K., N. C. Sturchio, B. Gu, J. Horita, G. M. Brown, W. A. Jackson, J. Batista, and P. B. Hatzinger. (2005) Perchlorate isotope forensics. *Analytical Chemistry,* 77: 7838–7842.

Bost, F. D., R. Frontera-Suau, T. J. McDonald, K. E. Peters, and P. J. Morris. (2001) Aerobic biodegradation of hopanes and norhopanes in Venezuelan crude oils. *Organic Geochemistry,* 32: 105–115.

Boyd, T. J., C. L. Osburn, K. J. Johnson, K. B. Birgl, and R. B. Coffin. (2006) Compound-specific isotope analysis coupled with multivariate statistics to source-apportion hydrocarbon mixtures. *Environmental Science and Technology,* 40: 1916–1924.

Brand, W. A., A. R. Tegtmeyer, and A. Hilkert. (1994) Compound-specific isotope analysis: Extending toward $^{15}N/^{14}N$ and $^{18}O/^{16}O$. *Organic Geochemistry,* 21: 585–594.

Brooks, P. W., G. Eglinton, S. J. Gaskell, J. R. Maxwell, D. J. McHugh, and R. P. Philp. (1976) Lipids of recent sediments. Part I: Straight-chain hydrocarbons and carboxylic acids of some temperate lacustrine and subtropical lagoonal/tidal flat sediments. *Chemical Geology,* 18: 21–38.

______. (1977) Lipids of recent sediments. Part II: Branched and cyclic alkanes and alkanoic acids of some temperate lacustrine and subtropical lagoonal/tidal flat sediments. *Chemical Geology,* 20: 189–204.

Burgoyne, T. W., and J. M. Hayes. (1998) Quantitative production of $H_2$ by pyrolysis of gas chromatographic effluents. *Analytical Chemistry,* 70: 5136–5141.

Burlingame, A. L., P. Haug, T. Belsky, and M. Calvin. (1965) Occurrence of biogenic steranes and pentacyclic triterpanes in an Eocene shale (52 million years) and in an early Precambrian shale (2.7 million years): A preliminary report. *Proceedings of the National Academy of Science, U.S.A.,* 54: 1406–1412.

Caimi, R. J., and J. T. Brenna. (1993) High-precision liquid chromatography-combustion isotope ratio mass spectrometry. *Analytical Chemistry,* 65: 3497–500.

Calvin, M. (1969) *Chemical evolution.* Oxford: Oxford University Press, 278 pp.

Chartrand, M. M. G., P. L. Morrill, G. Lacrampe-Couloume, and B. Sherwood Lollar. (2005) Stable isotope evidence for biodegradation of chlorinated ethenes at a fractured bedrock site. *Environmental Science and Technology,* 39: 4848–4856.

Christensen, L. B., and T. H. Larsen. (1993). Method for determining the age of diesel oil spills in the soil. *Ground Water Monitoring and Remediation,* 13: 142–149.

Coffin, R. B., P. H. Miyares, C. A. Kelley, L. A. Cifuentes, and C. M. Reynolds. (2001) Stable carbon and nitrogen isotope analysis of TNT: Two-dimensional source identification. *Environmental Toxicology and Chemistry,* 20: 2676–2680.

Cook, P. F. (1991) *Enzyme mechanism from isotope effects.* Boca Raton, FL: CRC Press, 500 pp.

Dahl, J. E., J. M. Moldowan, K. E. Peters, G. E. Claypool, M. A. Rooney, G. E. Michael, M. R. Mello, and M. L. Kohnen. (1999) Diamondoid hydrocarbons as indicators of natural oil cracking. Nature, 399: 54–58.

Davis, A., B. Howe, A. Nicholson, S. McCaffery, and K. A. Hoenke. (2005) Use of geochemical forensics to determine release eras of petrochemicals to groundwater, Whitehorse, Yukon. *Environmental Forensics,* 6: 253–271.

Day, M., R. Aravena, D. Hunkeler, and T. Gulliver. (2002) Application of carbon isotopes to document biodegradation of tert-butyl alcohol under field conditions. *Contaminated Soil Sediment & Water,* August: 88–92.

Drenzek, N. J., T. I. Eglinton, C. O. Wirsen, H. D. May, Q. Wu, K. R. Sowers, and C. M. Reddy. (2001) The absence and application of stable carbon isotopic fractionation during the reductive dechlorination of polychlorinated biphenyls. *Environmental Science and Technology,* 35: 3310–3313.

Drenzek, N. J., T. I. Eglinton, C. O. Wirsen, N. C. Sturchio, L. J. Heraty, K. R. Sowers, Q. Wu, H. D. May, and C. M. Reddy. (2004) Invariant chlorine isotopic signatures during microbial PCB reductive dechlorination. *Environmental Pollution,* 128: 445–448.

Drenzek, N. J., C. H. Tarr, T. I. Eglinton, L. J. Heraty, N. C. Sturchio, V. J. Shiner, and C. M. Reddy. (2002) Stable chlorine and carbon isotopic compositions of selected semivolatile organochlorine compounds. *Organic Geochemistry,* 33: 437–444.

Eglinton, G., and M. Calvin. (1967) Chemical fossils. *Scientific American,* 261: 23–32.
Elsner, M., L. Zwank, D. Hunkeler, and R. P. Schwarzenbach. (2005) A new concept linking observable stable isotope fractionation to transformation pathways of organic pollutants. *Environmental Science and Technology,* 39: 6896–6916.
EPA OSWER Directive 9200.4-17P. Use of monitored natural attenuation at superfund, RCRA corrective action, and underground storage tank sites. April 21, 1999.
Fabbri, D., I. Vassura, C. Sun, C. E. Snape, C. McRae, and A. E. Fallick. (2003) Source apportionment of polycyclic aromatic hydrocarbons in a coastal lagoon by molecular and isotopic characterization. *Marine Chemistry,* 84: 123–135.
Farran, A., J. Grimalt, J. Albaiges, A. V. Botello, and S. Macko. (1987) Assessment of petroleum pollution in a Mexican river by molecular markers and carbon isotope ratios. *Marine Pollution Bulletin,* 18: 284–289.
Faure, G., and T. M. Mensing. (2005) *Isotopes: Principles and applications.* Hoboken, NJ: Wiley, 897 pp.
Ferezou, J., E. Gouffier, T. Coste, and F. Chevallier. (1978) Daily elimination of faecal sterols by humans. *Digestion,* 18: 201–212.
Galimov, E. M. (1985) *The biological fractionation of isotopes.* Orlando, FL: Academic Press, 261 pp.
Geyer, R., A. D. Peacock, A. Miltner, H. H. Richnow, D. C. White, K. L. Sublette, and M. Kaestner. (2005) In situ assessment of biodegradation potential using biotraps amended with $^{13}C$-labeled benzene or toluene. *Environmental Science and Technology,* 39: 4983–4989.
Glaser, B., A. Dreyer, M. Bock, S. Fiedler, M. Mehring, and T. Heitmann. (2005) Source apportionment of organic pollutants of a highway-traffic-influenced urban area in Bayreuth (Germany) using biomarker and stable carbon isotope signatures. *Environmental Science and Technology,* 39: 3911–3917.
Goldstein, A. H., and S. L. Shaw. (2003) Isotopes of volatile organic compounds: An emerging approach for studying atmospheric budgets and chemistry. *Chemical Reviews,* 103: 5025–5048.
Griebler, C., M. Safinowski, A. Vieth, H. H. Richnow, and R. U. Meckenstock. (2004) Combined application of stable carbon isotope analysis and specific metabolites determination for assessing in situ degradation of aromatic hydrocarbons in a tar oil-contaminated aquifer. *Environmental Science and Technology,* 38: 617–631.
Han, J., and M. Calvin. (1969) Hydrocarbon distribution of algae and bacteria and microbiological activity in sediments. *Proceedings of the National Academy of Science, U.S.A.,* 64(2): 436–443.
Hoefs, J. (2004) *Stable isotope geochemistry.* Berlin: Springer, 244 pp.
Holmstrand, H., P. Andersson, and O. Gustafsson. (2004) Chlorine isotope analysis of submicromole organochlorine samples by sealed tube combustion and thermal ionization mass spectrometry. *Analytical Chemistry,* 76: 2336–2342.
Holt, B. D., N. C. Sturchio, T. A. Abrajano, and L. J. Heraty. (1997) Conversion of chlorinated volatile organic compounds to carbon dioxide and methyl chloride for isotopic analysis of carbon and chlorine. *Analytical Chemistry,* 69: 2727–2733.
Horii, Y., K. Kannan, G. Petrick, T. Gamo, J. Falandysz, and N. Yamashita. (2005) Congener-specific carbon isotopic analysis of technical PCB and PCN mixtures using two-dimensional gas chromatography–isotope ratio mass spectrometry. *Environmental Science and Technology,* 39: 4206–4212.

Hostettler, F. D., and K. A. Kvenvolden. (1994) Geochemical changes in crude oil spilled from the *Exxon Valdez* supertanker into Prince William Sound, Alaska. *Organic Geochemistry,* 21: 927–936.

Huang, Y., G. Eglinton, P. Ineson, P. M. Latter, R. Bol, and D. D. Harkness. (1997) Absence of carbon isotope fractionation of individual *n*-alkanes in a 23-year field decomposition experiment with *Calluna vulgaris. Organic Geochemistry,* 26: 497–501.

Hunkeler, D., and R. Aravena. (2000) Determination of compound-specific carbon isotope ratios of chlorinated methanes, ethanes, and ethenes in aqueous samples. *Environmental Science and Technology,* 34: 2839–2844.

Hunkeler, D., R. Aravena, K. Berry-Spark, and E. Cox. (2005) Assessment of degradation pathways in an aquifer with mixed chlorinated hydrocarbon contamination using stable isotope analysis. *Environmental Science and Technology,* 39: 5975–5981.

Hunkeler, D., R. Aravena, and B. J. Butler. (1999) Monitoring microbial dechlorination of tetrachloroethene (PCE) in groundwater using compound-specific stable carbon isotope ratios: Microcosm and field studies. *Environmental Science and Technology,* 33: 2733–2738.

Hunkeler, D., R. Aravena, B. L. Parker, J. A. Cherry, and X. Diao (2003) Monitoring oxidation of chlorinated ethenes by permanganate in groundwater using stable isotopes: Laboratory and field studies. *Environmental Science and Technology,* 37: 798–804.

Hurst, R. W. (2000) Applications of anthropogenic lead archaeostratigraphy (ALAS model) to hydrocarbon remediation. *Environmental Forensics,* 1(1): 11–24.

Jarman, W. M., A. Hilkert, C. E. Bacon, J. W. Collister, K. Ballschmiter, and R. W. Risebrough. (1998) Compound-specific carbon isotopic analysis of Aroclors, Clophens, Kaneclors, and Phenoclors. *Environmental Science and Technology,* 32: 833–836.

Jiang, Z., R. P. Philp, and C. A. Lewis. (1988) Identification of novel bicyclic alkanes from steroid precursors in crude oils from Kelamayi oilfield in China. *Geochimica Cosmochimica Acta,* 52: 491–498.

Kaplan, I. R., Y. Galperin, S.-T. Lu, and R.-P. Lee. (1997) Forensic environmental geochemistry: Differentiation of fuel types, their sources and release time. *Organic Geochemistry,* 27(5–6): 289–317.

Kim, M., M. C. Kennicutt, II, and Y. Qian. (2005) Polycyclic aromatic hydrocarbon purification procedures for compound specific isotope analysis. *Environmental Science and Technology,* 39: 6770–6776.

Kimble, B. J. (1972) Ph.D. thesis, The geochemistry of tritertrenoid hydrocarbons. University of Bristol. 302 pp.

Kirtland, B. C., C. M. Aelion, P. A. Stone, and D. Hunkeler. (2003) Isotopic and geochemical assessment of in situ biodegradation of chlorinated hydrocarbons. *Environmental Science and Technology,* 37: 4205–4212.

Kjeldsen, P., M. Christophersen, M. Broholm, P. Höhener, R. Aravena and D. Hunkeler (2003) Biodegradation of fuel vapours in the vadose zone at Airbase Værløse, Denmark. 2nd International Workshop on Groundwater Risk Assessment at Contaminated Sites and Integrated Soil and Water Protection, ed. D. Halm and P. Grathwohl, Tuebingen, March 20–21, 2003. *Tübinger Geowissenschaftliche Arbeiten,* 69: 31–39.

Kolhatkar, R., T. Kuder, P. Philp, J. Allen, and J. T. Wilson. (2002) Use of compound-specific stable carbon isotope analyses to demonstrate anaerobic biodegradation of MTBE in groundwater at a gasoline release site. *Environmental Science and Technology,* 36: 5139–5146.

Kopinke, F., A. Georgi, M. Voskamp, and H. H. Richnow. (2005) Carbon isotope fractionation of organic contaminants due to retardation on humic substances: Implications for natural attenuation studies in aquifers. *Environmental Science and Technology,* 39: 6052–6062.

Krummen, M., A. W. Hilkert, D. Juchelka, A. Duhr, H. Schlueter, and R. Pesch. (2004) A new concept for isotope ratio monitoring liquid chromatography/mass spectrometry. *Rapid Communications in Mass Spectrometry,* 18: 2260–2266.

Kuder, T., J. T. Wilson, P. Kaiser, R. Kolhatkar, P. Philp, and J. Allen (2005) Enrichment of stable carbon and hydrogen isotopes during anaerobic biodegradation of MTBE: Microcosm and field evidence. *Environmental Science and Technology,* 39: 213–220.

Kvenvolden, K. A., F. D. Hostettler, P. R. Carlson, J. B. Bapp, C. N. Threlkeld, and A. Warden. (1995) Ubiquitous tar balls with a California-source signature on shorelines of Prince William Sound, Alaska. *Environmental Science and Technology,* 29(10): 2684–2694.

Kvenvolden, K. A., F. D. Hostettler, J. B. Rapp, and P. R. Carlson. (1993) Hydrocarbons in oil residues on beaches of islands of Prince William Sound, Alaska. *Marine Pollution Bulletin,* 26: 24–29.

Landmeyer, J. E., D. A. Vroblesky, and F. H. Chapelle. (1996a) Stable carbon isotope evidence of biodegradation zonation in a shallow jet-fuel contaminated aquifer. *Environmental Science and Technology,* 30: 1120–1128.

Leeming, R., A. Ball, N. J. Ashbolt, G. Jones, and P. Nichols. (1996a) Distinguishing between human and animal sources of faecal pollution. *Chemistry in Australia Waters* (supplement), 434–435.

Leeming, R., A. Ball, N. J. Ashbolt, and P. D. Nichols. (1996b) Using fecal sterols from humans and animals to distinguish faecal pollution in receiving waters. *Water Research,* 30: 2893–2900.

Leeming, R., V. Latham, M. Rayner, and P. Nichols. (1997) Detecting and distinguishing sources of sewage pollution in Australian inland and coastal waters and sediments. *ACS Symposium Series,* 671: 306–319.

Lesser, L. E., G. Spinnler, P. C. Johnson, and R. Aravena. (2005) Assessment of stable carbon isotopes as a tool for assessing MTBE biodegradation at a field site. *IAHS Publication,* 297: 290–295.

Mansuy, L., R. P. Philp, and J. Allen. (1997) Source identification of oil spills based on the isotopic composition of individual components in weathered oil samples. *Environmental Science and Technology,* 31: 3417–3425.

Mariotti, A., J. C. Germon, P. Hubert, P. Kaiser, R. Letolle, A. Tardieux, and P. Tardieux. (1981) Experimental determination of nitrogen kinetic isotope fractionation: Some principles, illustration for the denitrification and nitrification processes. *Plant and Soil,* 62: 413–430.

Mazeas, L., and H. Budzinski. (2001) Polycyclic aromatic hydrocarbon $^{13}C/^{12}C$ ratio measurement in petroleum and marine sediments. Application to standard reference materials and a sediment suspected of contamination from the Erika oil spill. *Journal of Chromatography, A,* 923: 165–176.

______. (2002a). Molecular and stable carbon isotopic source identification of oil residues and oiled bird feathers sampled along the Atlantic coast of France after the *Erika* oil spill. *Environmental Science and Technology,* 36(2): 130–137.

______. (2002b) Stable carbon isotopic study ($^{12}C/^{13}C$) of the fate of petrogenic PAHs (methylphenanthrenes) during an in situ oil spill simulation experiment. *Organic Geochemistry,* 33: 1253–1258.

Mazeas, L., H. Budzinski, and N. Raymond. (2002) Absence of stable carbon isotope fractionation of saturated and polycyclic aromatic hydrocarbons during aerobic bacterial biodegradation. *Organic Geochemistry,* 33: 1259–1272.

McCalley, D. V., M. Cooke, and G. Nickless. (1981) The effect of sewage treatment on faecal sterols. *Water Research,* 15: 1019–1025.

McRae, C., C. E. Snape, C. Sun, D. Fabbri, D. Tartari, C. Trombini, and A. E. Fallick. (2000) Use of compound-specific stable isotope analysis to source anthropogenic natural gas-derived polycyclic aromatic hydrocarbons in a lagoon sediment. *Environmental Science and Technology,* 34: 4684–4686.

McRae, C., C. Sun, C. E. Snape, A. E. Fallick, and D. Taylor. (1999) $\delta^{13}C$ values of coal-derived PAHs from different processes and their application to source apportionment. *Organic Geochemistry,* 30: 881–889.

Meckenstock, R. U., B. Morasch, C. Griebler, and H. H. Richnow. (2004) Stable isotope fractionation analysis as a tool to monitor biodegradation in contaminated aquifers. *Journal of Contaminant Hydrology,* 75: 215–255.

Meckenstock, R. U., M. Safinowski, and C. Griebler. (2004) Anaerobic degradation of polycyclic aromatic hydrocarbons. *FEMS Microbiology Ecology,* 49: 27–36.

Melander, L., and W. H. Saunders. (1980) *Reaction rates of isotopic molecules.* New York: Wiley, 331 pp.

Merritt, D. A., K. H. Freeman, M. P. Ricci, S. A. Studley, and J. M. Hayes. (1995) Performance and optimization of a combustion interface for isotope ratio monitoring gas chromatography/mass spectrometry. *Analytical Chemistry,* 67: 2461–2473.

Merritt, D. A., and J. M. Hayes. (1994) Nitrogen isotopic analyses by isotope-ratio-monitoring gas chromatography/mass spectrometry. *Journal of the American Society for Mass Spectrometry,* 5: 387–397.

Moldowan, J. M., J. E. Dahl, B. J. Huizinga, S. R. Jacobson, and D. W. Taylor. (1993) The relationship of angiosperms and oleanane in petroleum through geologic time. *AAPG Bulletin,* 77: 334–335.

Moldowan, J. M., W. K. Seifert, E. Arnold, and J. Clardy. (1984) Structure proof and significance of stereoisomeric 28.30-bisnorhopanes in petroleum and petroleum source rocks. *Geochimica Cosmochimica Acta,* 48: 1651–1661.

Morrill, P. L., G. Lacrampe-Couloume, G. F. Slater, B. E. Sleep, E. A. Edwards, M. L. McMaster, D. W. Major, and B. Sherwood Lollar. (2005) Quantifying chlorinated ethene degradation during reductive dechlorination at Kelly AFB using stable carbon isotopes. *Journal of Contaminant Hydrology,* 76: 279–293.

Mudge, S. M., and C. G. Seguel. (1997) Trace organic contaminants and lipid biomarkers in Concepcion and San Vicente Bays. *Boletín de la Sociedad Chilena de Química,* 42: 5–15.

______. (1999) Organic contamination of San Vicente Bay, Chile. *Marine Pollution Bulletin,* 38: 1011–1021.

Nichols, P. D., and R. Leeming. (1991) Tracing sewage in the marine environment. *Chemistry in Australia,* 58: 274–276.

O'Leary, T., R. Leeming, P. D. Nichols, and J. K. Volkman. (1999) Assessment of the sources, transport, and fate of sewage-derived organic matter in Port Phillip Bay, Australia, using the signature lipid coprostanol. *Marine and Freshwater Research,* 50: 547–556.

O'Malley, V. P., T. A. Abrajano, and J. Hellou. (1996) Stable carbon isotope apportionment of individual polycyclic aromatic hydrocarbons in St. John's Harbor, Newfoundland. *Environmental Science and Technology,* 30: 634–639.

O'Sullivan, G., G. Boshoff, A. Downey, and R. M. Kalin. (2004) Carbon isotope effect during the abiotic oxidation of methyl-tert-butyl ether (MTBE). *Proceedings of the 7th International In Situ and On-Site Bioremediation Symposium,* Orlando, FL, June 2–5, 2003, 1566–1573.

Peter, A., A. Steinbach, R. Liedl, T. Ptak, W. Michaelis, and G. Teutsch. (2004) Assessing microbial degradation of o-xylene at field-scale from the reduction in mass flow rate combined with compound-specific isotope analyses. *Journal of Contaminant Hydrology,* 71: 127–154.

Peters, K. E., C. C. Walters, and J. M. Moldowan. (2005), *The biomarker guide,* 1–1155. Cambridge, U.K.: Cambridge University Press.

Philp, R. P. (1985) *Fossil fuel biomarkers. Applications and spectra,* 294. Amsterdam: Elsevier.

Philp, R. P., T. D. Gilbert, and J. Friedrich. (1981) Bicyclic sesquiterpenoids and diterpenoids in Australian crude oils. *Geochemica Cosmochimica Acta,* 45: 1173–1180.

Pond, K. L., Y. Huang, Y. Wang, and C. F. Kulpa. (2002) Hydrogen isotopic composition of individual *n*-alkanes as an intrinsic tracer for bioremediation and source identification of petroleum contamination. *Environmental Science and Technology,* 36: 724–728.

Poulson, S. R., and J. I. Drever. (1999) Stable isotope (C, Cl, and H) fractionation during vaporization of trichloroethylene. *Environmental Science and Technology,* 33: 3689–3694.

Prince, R. C., D. L. Elmendorf, J. R. Lute, C. S. Hsu, C. E. Halth, J. D. Senlus, G. J. Dechert, G. S. Douglas, and E. L. Butler. (1994) 17α(H)-21β(H)-hopane as a conserved internal marker for estimating the biodegradation of crude oil. *Environmental Science and Technology,* 28: 142–145.

Rayleigh, J. W. S. (1896) Theoretical considerations respecting the separation of gases by diffusion and similar processes. *Philosophical Magazine,* 42: 493–498.

Readman, J. W., M. R. Preston, and R. F. C. Mantoura. (1986) An integrated technique to quantify sewage, oil and PAH pollution in estuarine and coastal environments. *Marine Pollution Bulletin,* 17: 298–308.

Reddy, C. M., N. J. Drenzek, T. I. Eglinton, L. J. Heraty, N. C. Sturchio, and V. J. Shiner. (2002) Stable chlorine intramolecular kinetic isotope effects from the abiotic dehydrochlorination of DDT. *Environmental Science and Pollution Research International,* 9: 183–186.

Reddy, C. M., L. J. Heraty, B. D. Holt, N. C. Sturchio, T. I. Eglinton, N. J. Drenzek, L. Xu, J. L. Lake, and K. A. Maruya. (2000) Stable chlorine isotopic compositions of Aroclors and Aroclor-contaminated sediments. *Environmental Science and Technology,* 34: 2866–2870.

Requejo, A. G., and H. I. Halpern. (1989) An unusual hopane biodegradation sequence in tar sands from the Pt. Arena (Monterey) formation. *Nature,* 342: 670–673.

Reusser, D. E., J. D. Istok, H. R. Beller, and J. A. Field. (2002) In situ transformation of deuterated toluene and xylene to benzylsuccinic acid analogues in BTEX-contaminated aquifers. *Environmental Science and Technology,* 36: 4127–4134.

Richnow, H. H., E. Annweiler, W. Michaelis, and R. U. Meckenstock. (2003) Microbial in situ degradation of aromatic hydrocarbons in a contaminated aquifer monitored by carbon isotope fractionation. *Journal of Contaminant Hydrology,* 65: 101–120.

Schmidt, G. W., D. D. Beckmann, and B. E. Torkelson. (2002) A technique for estimating the age of regular/mid-grade gasolines released to the subsurface since the early 1970s. *Environmental Forensics,* 3: 145–162.

Schmidt, T. C., L. Zwank, M. Elsner, M. Berg, R. U. Meckenstock, and S. B. Haderlein. (2004) Compound-specific stable isotope analysis of organic contaminants in natural environments: A critical review of the state of the art, prospects, and future challenges. *Analytical and Bioanalytical Chemistry,* 378: 283–300.

Schuth, C., H. Taubald, N. Bolano, and K. Maciejczyk. (2003) Carbon and hydrogen isotope effects during sorption of organic contaminants on carbonaceous materials. *Journal of Contaminant Hydrology,* 64: 269–281.

Scow, K. M., and K. A. Hicks. (2005) Natural attenuation and enhanced bioremediation of organic contaminants in groundwater. *Current Opinion in Biotechnology,* 16: 246–253.

Seifert, W. K., and J. M. Moldowan. (1978) Applications of steranes, terpanes and monoaromatics to the maturation, migration and source of crude oils. *Geochimica Cosmochimica Acta,* 42: 77–95.

______. (1981). Paleoreconstruction by biological markers. *Geochimica Cosmochimica Acta,* 45: 783–794.

Seifert, W. K., J. M. Moldowan, G. W. Smith, and E. V. Whitehead. (1978) First proof of a $C_{28}$ pentacyclic triterpane in petroleum. *Nature,* 271: 436–437.

Sherblom, P., and D. Kelly. (1993) Chemical tracer analysis of sediments around the Key West, Florida, ocean outfall. U.S. Environmental Protection Agency. Mote Marine Laboratory, Technical Report No. 338. 15 pp.

Sherwood Lollar, B., G. F. Slater, B. Sleep, M. Witt, G. M. Klecka, M. Harkness, and J. Spivack. (2001) Stable carbon isotope evidence for intrinsic bioremediation of tetrachloroethene and trichloroethene at Area 6, Dover Air Force Base. *Environmental Science and Technology,* 35: 261–269.

Shouakar-Stash, O., R. J. Drimmie, and S. K. Frape. (2005a) Determination of inorganic chlorine stable isotopes by continuous flow isotope ratio mass spectrometry. *Rapid Communications in Mass Spectrometry,* 19: 121–127.

Shouakar-Stash, O., S. K. Frape, and R. J. Drimmie. (2005b) Determination of bromine stable isotopes using continuous-flow isotope ratio mass spectrometry. *Analytical Chemistry,* 77: 4027–4033.

Sinton, L. W., R. K. Finlay, and D. J. Hannah. (1998) Distinguishing human from animal faecal contamination in water: A review. *New Zealand Journal of Marine and Freshwater Research,* 32: 323–348.

Slater, G. F. (2003) Stable isotope forensics—When isotopes work. *Environmental Forensics,* 4: 13–23.

Smallwood, B. J., R. P. Philp, and J. D. Allen. (2002) Stable carbon isotopic composition of gasolines determined by isotope ratio monitoring gas chromatography mass spectrometry. *Organic Geochemistry,* 33: 149–159.

Smallwood, B. J., R. P. Philp, T. W. Burgoyne, and J. D. Allen. (2001) The use of stable isotopes to differentiate specific source markers for MTBE. *Environmental Forensics,* 2: 215–221.

Sofer, Z. (1984) Stable carbon isotope compositions of crude oils: Application to source depositional environments and petroleum alteration. *AAPG Bulletin,* 68: 31–49.

Song, D. L., M. E. Conrad, K. S. Sorenson, and L. Alvarez-Cohen. (2002) Stable carbon isotope fractionation during enhanced in situ bioremediation of trichloroethene. *Environmental Science and Technology,* 36: 2262–2268.

Stark, A., T. Abrajano, J. Hellou, and J. L. Metcalf-Smith. (2003) Molecular and isotopic characterization of polycyclic aromatic hydrocarbon distribution and sources at the international segment of the St. Lawrence River. *Organic Geochemistry,* 34: 225–237.

Steinbach, A., R. Seifert, E. Annweiler, and W. Michaelis. (2004) Hydrogen and carbon isotope fractionation during anaerobic biodegradation of aromatic hydrocarbons—A field study. *Environmental Science and Technology,* 38: 609–616.

Stout, S. A., R. M. Uhler, R. P. Philp, J. Allen, and A. D. Uhler. (1998) Source differentiation of individual chlorinated solvents dissolved in groundwater using compound specific carbon isotopic analysis. Preprints of *Extended Abstracts presented at the ACS National Meeting, American Chemical Society, Division of Environmental Chemistry,* 38: 2–5.

Sun, C., M. Cooper, and C. E. Snape. (2003) Use of compound-specific $\delta^{13}C$ and $\delta D$ stable isotope measurements as an aid in the source apportionment of polyaromatic hydrocarbons. *Rapid Communications in Mass Spectrometry,* 17: 2611–2613.

Sun, C., C. E. Snape, C. McRae, and A. E. Fallick. (2003) Resolving coal and petroleum-derived polycyclic aromatic hydrocarbons (PAHs) in some contaminated land samples using compound-specific stable carbon isotope ratio measurements in conjunction with molecular fingerprints. *Fuel,* 82: 2017–2023.

Trust, B. A., J. G. Mueller, R. B. Coffin, and L. A. Cifuentes. (1995) The biodegradation of fluoranthene as monitored using stable carbon isotopes. In *Monitoring and verification of bioremediation* [papers from the 3rd International In Situ and On-Site Bioreclamation Symposium], San Diego, April 1995, 233–239.

Van Breukelen, B. M., D. Hunkeler, and F. Volkering. (2005) Quantification of sequential chlorinated ethene degradation by use of a reactive transport model incorporating isotope fractionation. *Environmental Science and Technology,* 39: 4189–4197.

Vanstone, N. A., R. M. Focht, S. A. Mabury, and B. Sherwood Lollar. (2004) Effect of iron type on kinetics and carbon isotopic enrichment of chlorinated ethylenes during abiotic reduction on Fe(0). *Ground Water,* 42: 268–276.

Van Warmerdam, E. M., S. K. Frape, R. Aravena, R. J. Drimmie, H. Flatt, and J. A. Cherry. (1995) Stable chlorine and carbon isotope measurements of selected chlorinated organic solvents. *Applied Geochemistry,* 10: 547–552.

Volkman, J. K., A. T. Revill, and A. P. Murray. (1997) Applications of biomarkers for identifying sources of natural and pollutant hydrocarbons in aquatic environments. In *Molecular markers in environmental geochemistry,* ed. R. P. Eganhouse, 83–99. Washington, D.C.: American Chemical Society.

Walker, S. E., R. M. Dickhut, C. Chisholm-Brause, S. Sylva, and C. M. Reddy. (2005) Molecular and isotopic identification of PAH sources in a highly industrialized urban estuary. *Organic Geochemistry,* 36: 619–632.

Wang, Z., M. Fingas, and G. Sergy. (1994) Study of 22-year-old *Arrow* oil samples using biomarker compounds by GC/MS. *Environmental Science and Technology,* 28: 1733–1746.

Wang, Y., and Y. Huang. (2003) Hydrogen isotopic fractionation of petroleum hydrocarbons during vaporization: Implications for assessing artificial and natural remediation of petroleum contamination. *Applied Geochemistry,* 18: 1641–1651.

Wang, Y., Y. Huang, J. N. Huckins, and J. D. Petty. (2004) Compound-specific carbon and hydrogen isotope analysis of subparts per billion level waterborne petroleum hydrocarbons. *Environmental Science and Technology,* 38: 3689–3697.

Wassenaar, L. I., and G. Koehler. (2004) Online technique for the determination of the $\delta^{37}Cl$ of inorganic and total organic Cl in environmental samples. *Analytical Chemistry,* 76: 6384–6388.

Westaway, K. C., T. Koerner, Y.-R. Fang, J. Rudzinski, and P. Paneth. (1998) A new method of determining chlorine kinetic isotope effects. *Analytical Chemistry,* 70: 3548–3552.

Yanik, P. J., T. H. O'Donnell, S. A. Macko, Y. Qian, and M. C. Kennicutt. (2003a) The isotopic compositions of selected crude oil PAHs during biodegradation. *Organic Geochemistry,* 34: 291–304.

______. (2003b) Source apportionment of polychlorinated biphenyls using compound specific isotope analysis. *Organic Geochemistry,* 34: 239–251.

Zakaria, M. P., A. Horinouchi, S. Tsutsumi, H. Takada, S. Tanabe, and A. Ismail. (2000) Oil pollution in the Straits of Malacca, Malaysia: Application of molecular markers for source identification. *Environmental Science and Technology,* 34(7): 1189–1196.

Zwank, L., M. Berg, M. Elsner, T. C. Schmidt, R. P. Schwarzenbach, and S. B. Haderlein. (2005) New evaluation scheme for two-dimensional isotope analysis to decipher biodegradation processes: Application to groundwater contamination by MTBE. *Environmental Science and Technology,* 39: 1018–1029.

Zwank, L., M. Berg, T. C. Schmidt, and S. B. Haderlein. (2003) Compound-specific carbon isotope analysis of volatile organic compounds in the low-microgram per litre range. *Analytical Chemistry,* 75: 5575–5583.

# 5

# Volatile Organic Compound (VOC) Analysis in Water, Sediments, and Soils and Its Application in Environmental Forensics

CLAUDIO BRAVO-LINARES
STEPHEN M. MUDGE

## Contents

Introduction .......... 172
Lifetimes .......... 173
Henry's Law Constant .......... 173
Anthropogenic Contributions .......... 174
Marine Natural Sources of Halocarbons .......... 176
Degradation and Fates .......... 176
Analysis of VOCs .......... 178
Solvent Extraction .......... 179
Head Space .......... 179
Dynamic Head Space (Purge and Trap) .......... 179
Solid-Phase Microextraction (SPME) .......... 180
Analysis of VOCs in Different Matrices Using SPME .......... 181
Seawater Analyses .......... 181
Source Identification in Seawater Samples .......... 183
Classed Posting for Source Identification .......... 183
Sediments .......... 184
Signatures .......... 185
Statistical Approaches for Data Interpretation .......... 186
PCA in the Analysis of VOCs .......... 188
Summary .......... 188
References .......... 189

## Introduction

The atmosphere contains a range of volatile compounds that are man-made (anthropogenic) or produced from natural sources (biogenics). These compounds may have significant effects on atmospheric chemistry as some are ozone depleting chemicals, especially the halocarbons, and some may have global warming potential. Legislation and agreements have focused on the reduction of the man-made compounds that have the greatest effects (e.g., CFCs, $CCl_4$) and there is some evidence that the deleterious processes in the atmosphere have reduced (WMO 2006). Chlorinated solvents in the soils have been a problem for a number of years (Rivett, Feenstra, and Clark 2006) and their presence in or on groundwaters has been the focus of much work, especially in the United States (Lohman 2002). As these compounds (e.g., trichloroethene and tetrachloroethene) do not mix with water, they form a dense nonaqueous phase layer (DNAPL) and may be found at locations remote from their initial source (Chartrand et al. 2007).

The natural production of biological volatile organic compounds (BVOCs), especially in the marine environment, constitutes an important source of gases in the atmosphere (Chuck et al. 2005). Marine organisms can produce different types of trace gases, such as halocarbons, dimethylsulphide (DMS), nonmethane hydrocarbons (NMHCs), and other gases that may be exchanged across the ocean–atmosphere boundary. Their contribution, particularly for halocarbons, has a direct influence on the composition and reactivity of the atmosphere.

Many iodine- and bromine-containing halocarbons can be broken down by sunlight in the troposphere (the lower layer of the atmosphere) to form very reactive halogen radicals (Saiz-Lopez et al. 2007). In this way they differ from chlorofluorocarbons (CFCs), which are man-made halogen-containing chemical compounds. CFCs can only be broken down to halogen radicals by ultraviolet radiation in the stratosphere. Biogenic halocarbons participate in atmospheric photochemical reactions and may be partially responsible for the depletion of the stratospheric ozone layer (Abrahamsson et al. 2003; Scarratt and Moore 1998; Thunis and Cuvelier 2000).

These chemicals are typically referred to as ozone depleting substances (ODSs). However, they are also potent greenhouse gases, potentially exceeding the impact of the greenhouse effect of some gases such as carbon dioxide, methane, and nitrogen-oxide compounds on a per-molecule basis (WMO 2006). In recognition of the harmful effects of these compounds on the ozone layer, many governments signed the Montreal Protocol in 1987 restricting uses and production on substances that deplete the ozone layer (Buchmann, Stemmler, and Reimann 2003).

**Table 5.1 Atmospheric, Global, and Seawater Lifetimes of Selected Halocarbons**

| Compound | Lifetime | Ref. |
|---|---|---|
| $CH_2Cl_2$ | G: 0.38 year; A: 0.46 year | WMO 2006; Keene et al. 1999 |
| $CHCl_3$ | G: 0.41 year; A: 0.5 year | WMO 2006; Keene et al. 1999 |
| $CCl_4$ | G: 26 years; A: 500 years; S: 30 years | WMO 2006; Bullister and Wisegarver 1998; Huhn et al. 2001 |
| $CHBr_3$ | G: 0.6–1.4 years; S: 2.4–6.5 years | Yvon and Butler 1996; Yvon and Butler 1996 |
| $CH_3I$ | G: 4 years; A: 5 days | Keene et al. 1999; Bey et al. 2001 |
| $CH_2I_2$ | G: Few minutes | WMO 2006 |
| $CH_2ClI$ | G: Few hours; A: 100 minutes | WMO 2006; Keene et al. 1999 |
| $CH_2BrCl$ | G: 0.37 year; A: 0.49 year; S: 0.42 year | WMO 2006; Louis et al. 2001; Mellouki et al. 1992 |
| $CHBr_2Cl$ | G: 0.19 year; A: 0.16 year | WMO 2006; Louis et al. 2001 |

*Notes:* G = global, A = atmospheric, and S = seawater lifetimes.

## Lifetimes

The lifetime of a gas is the approximate amount of time a chemical will spend in the atmosphere before either being converted into another chemical compound or being taken out of the atmosphere via a sink. It is dependent on photo-chemical reactions and transformation, along with reactions with the OH radical or other chemical species, temperature, radiation, and oxidant concentration. In seawater, lifetimes are dependent on temperature and biodegradation. The detection of certain compounds may indicate the proximity to a source, as some compounds have very short lifetimes in both the water and atmosphere. Table 5.1 shows the lifetimes of a range of halocarbons.

## Henry's Law Constant

Henry's law constant (the concentration of a compound in air divided by that in water at equilibrium) is a basic parameter used to determine the partitioning of VOCs between various environmental compartments (Sander 1999). This constant is used in environmental applications such as an air-stripping process to remediate VOCs in contaminated waters (Bobadilla et al. 2003), analytical applications (head-space gas chromatography) (Bakierowska and Trzeszczynski 2003), determination of surface seawater gas saturation, and the calculation of exchange velocity or fluxes (Liss et al. 1993, 1994). This constant determines the extent to which resistance to transfer occurs across the gas–liquid interface. High values mean that the gas has a low solubility in the liquid phase and low values mean that the gas has a high solubility in the

**Table 5.2 Henry's Law Constant (Dimensionless) in Seawater**

| Compound | Temperature (°C) | Henry's law constant | Ref. |
|---|---|---|---|
| $CH_2Cl_2$ | 25 | 0.25 | Dewulf et al. 1995 |
| $CCl_4$ | 25 | 1.36 | Dewulf et al. 1995 |
| $CH_2ClCH_2Cl$ | 25 | 0.05 | Dewulf et al. 1995 |
| $CCl_2$=CHCl | 25 | 0.43 | Dewulf et al. 1995 |
| $CCl_3CH_3$ | 25 | 0.81 | Dewulf et al. 1995 |
| $CCl_2$=$CCl_2$ | 25 | 0.77 | Dewulf et al. 1995 |
| $CH_2Br_2$ | 20 | 0.03 | WMO 2006 |
| $CHBr_3$ | 20 | 0.02 | WMO 2006 |
| $CH_3I$ | 22 | 0.22 | WMO 2006 |
| $CH_2I_2$ | 20 | 0.01 | WMO 2006 |
| $CHBrCl_2$ | 20 | 0.08 | WMO 2006 |
| CHClI | 20 | 0.03 | WMO 2006 |
| $CHBr_2Cl$ | 20 | 0.04 | WMO 2006 |

liquid phase, such as halocarbons in seawater. Henry's law constants (dimensionless) for some halocarbons are presented in Table 5.2.

## Anthropogenic Contributions

The sources of VOCs can be divided according to their origins into two basic groups: man-made sources and naturally occurring sources (Kuran and Sojak 1996). The identified emissions of the halogenated compounds come predominantly from their use and production in industrial and commercial processes. The main halocarbon compounds reported in the literature from an anthropogenic origin are the CFCs and chlorinated and brominated compounds. Conversely, iodinated compounds are rarely reported as man-made contaminants.

The anthropogenic sources of some halo-compounds are, for example, the industrial production of flame retardants, the use of bromomethane and bromoethane as fumigants, the utilisation of 1,2-dibromoethane as a gas-oil additive, the use of chlorine for the chlorination of water and industrial cleaning or bleaching processes (Class, Kohnle, and Ballschmiter 1986), and the use and production of trichloroethene and tetrachloroethane as by-products of gasoline, coal combustion, and industrial production (McCulloch et al. 1999). Further source information is summarised in Table 5.3. Their uses, at present, have been restricted and reduced after the Montreal and Kyoto protocols.

**Table 5.3 Production and Historical Uses of Man-Made Halocarbons (USEPA)**

| Compound | Anthropogenic production and uses |
|---|---|
| $CH_3Cl$ | Medication, anaesthetic, aerosol propellant, foaming agent in the plastics industry, methylating agent, manufacture of silicone resins and rubbers, as a refrigerant |
| $CH_2Cl_2$ | Solvent for cellulose acetate, medication, cleaning and industrial solvent, fumigant, used in aerosol formulations, in solid phase peptide synthesis |
| $CHCl_3$ | Solvent, fire extinguishers, insecticidal, fumigant extraction, and solvent purification |
| $CCl_4$ | Additives in refrigerants, grain fumigant, solvent, cleaning agent, in synthesis of nylon-7 and other organic chlorination processes, used in polymer technology as a reaction medium, catalyst in organic synthesis for chlorination of organic compounds, in soap perfumery and insecticides |
| $CH_3CH_2Cl$ | Refrigerant, solvent, alkylating agent, synthesis, insecticides, used in manufacture of dyes and drugs, used as a propellant in aerosols, used in manufacture of perfumes |
| $CH_3CHCl_2$ | Solvent of plastics, oil, and fats, as a fumigant and insecticide spray, fire extinguishing, medical, extractant for heat-sensitive substances, manufacture of high vacuum rubber, coupling agent in antiknock gasoline; in paint, varnish, and paint removers, in ore flotation |
| $CH_2ClCH_2Cl$ | Manufacture of acetyl cellulose, tobacco extract, in paint, varnish, and finish removers, soaps and scouring compounds, wetting and penetrating agents, ore flotation, lead scavenger in antiknock gasoline, fumigant, industrial solvent and cleaner, catalyst, in photography, xerography, and water softening, and in the production of cosmetics |
| $CHCl{=}CCl_2$ | Degreasing, dry cleaning, pharmaceuticals, industrial solvent, wool-fabric scouring, extractant for spice oleoresins, intermediate in the production of pentachloroethane, carrier solvent for the active ingredients of insecticides and fungicides, medication, former use |
| $CCl_2{=}CCl_2$ | Used in the textile industry for dry-cleaning, processing and finishing, used in both cold cleaning and vapour degreasing of metals, chemical intermediate in the synthesis of fluorocarbon 113, 114, 115, and 116, textile manufacturer, medication |
| $CH_3Br$ | Soil fumigant, refrigerant and in fire extinguishers, solvent in aniline dyes, methylating agent, medicinal agent to destroy malignant tissue and as an anesthetic in dentistry, fungicide, nematicide, herbicide, insecticide, and miticide, organic synthesis, extraction solvent for vegetable oils |
| $CH_2Br_2$ | Organic synthesis, solvent, ingredient of fire extinguishing, gauge fluid |
| $CHBr_3$ | Pharmaceutical uses, ingredient in fire-resistant chemicals, industrial solvent, in medicine, mineral flotation, catalyst, initiator or sensitizer in polymer reactions, and in vulcanization of rubber |
| $CH_2BrCH_2Br$ | Catalyst, solvent for resins, gums, and waxes, chemical intermediate in the synthesis of dyes and pharmaceuticals, fumigant, insecticide, nematicide, former uses scavenger for lead in gasoline, general solvent, water-proofing preparations, organic synthesis, in antiknock gasoline |

Continued

**Table 5.3 Production and Historical Uses of Man-Made Halocarbons (USEPA) (Continued)**

| Compound | Anthropogenic production and uses |
|---|---|
| $CHBrCl_2$ | Fire retardant, solvent, intermediate in synthesis of other compounds, fire-extinguisher fluid ingredient, heavy liquid for mineral and salt separations and laboratory use |
| $CHBr_2Cl$ | Organic synthesis, as a chemical intermediate in the manufacture of fire extinguishing agents, aerosol propellants, refrigerants, and pesticides |
| $CH_2BrCl$ | Chemical intermediate in industrial processes and fire extinguishing agent |

## Marine Natural Sources of Halocarbons

Biological contributions to the atmosphere, such as the formation of halocarbons by marine and terrestrial bacteria, are also important; the findings of Amachi et al. (2001) suggest that the bacteria contribute iodine from terrestrial and marine ecosystems into the atmosphere. Marine biota produce a great variety of VHOCs and these compounds are of diverse biosynthetic origins. Bromine rather than chlorine is the most prevalent halogen found in marine VOCs and has greater ozone depletion potential (Fenical 1981). Halogenation in marine biota is produced by diverse organisms such as marine bacteria; green, blue-green, brown, and red algae; and several classes of marine invertebrates, such as sponges, molluscs, coelenterates, several marine worms, corals, sea slugs, tunicates, sea fans, bacteria, microbes, and some marine mammals (Gribble 2000).

Marine organisms excrete many organic compounds directly into the seawater as a mechanism for removal of metabolic wastes (Gagosian and Lee 1981); as chemical communicants (Gagosian and Lee 1981); for chemical protection (Laturnus 1996); as secondary waste compounds formed with the help of peroxidases in order to lower the high concentration of hydrogen peroxide formed in algal cells (Collen and Pedersen 1996); maybe as a form of oxidative stress relief (Abrahamsson et al. 2003; Mtolera et al. 1996); for antiherbivory activity (Young, McConnell, and Fenical 1981); for antimicrobial properties (Fenical, 1981, 1982; Neidleman and Geigert 1987) to facilitate food gathering; or as hormones (Gribble 2000) and other reasons not yet clearly understood.

Terrestrial sources are also an important contribution of natural halocarbons (Keppler et al. 2003). This production can be abiotic, mediated by photochemical reactions (Prilepsky et al. 1998), and biotic mediated by degradation of organic matter (Hoekstra et al. 2001).

## Degradation and Fates

The degradation of halocarbons by biological and chemical mechanisms has been reported by some authors. Tokarczyk, Goodwin, and Saltzman

(2003) point out that methyl chloride is rapidly degraded in coastal water by microbial activity. Other evidence of halocarbon degradation by biological pathways has been found in the Pacific Ocean, the Atlantic Ocean, and the Caribbean Sea using isotopic techniques (Tokarczyk et al. 2001, 2003; Tokarczyk and Saltzman 2001).

Chemical degradation may be through different pathways (see the following equations). The work of Tanhua, Fogelqvist, and Basturk (1996) with some halocarbons in the Black Sea revealed evidence that the concentration of halocarbons decrease rapidly in anoxic areas.

These reactions represent the transformation mechanism of methyl halides in seawater:

$$CH_3X \rightarrow CH_3X_{(g)} \tag{5.1}$$

$$CH_3X + Y^- \rightarrow CH_3Y + X^- \tag{5.2}$$

$$CH_3X + H_2O \rightarrow CH_3OH + X^- + H^+ \tag{5.3}$$

$$RX + H^+ + 2e^- \rightarrow RH + X^- \tag{5.4}$$

$$RX - RX + 2e^- \rightarrow R = R + 2X^- \tag{5.5}$$

$$2RX + 2e^- \rightarrow R - R + 2X^- \tag{5.6}$$

$$X - CH_2 - X + HS^-(H_2S) \rightarrow X - CH_2\text{-}SH + X^- (+H^+) \tag{5.7}$$

$$CH_2Cl_2 + GSH \xrightarrow{-HCl} \left[GSCH_2Cl \xrightarrow{+H_2O/-HCl} GSCH_2OH\right] \rightarrow CH_2O \tag{5.8}$$

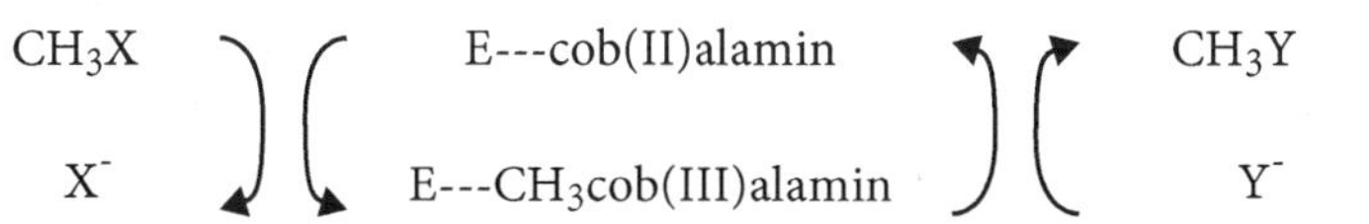

(5.9)

$$C_2Cl_4 \xrightarrow{\text{Corrinoid-dependent PCE reductive dehalogenase}} C_2HCl_3 \tag{5.10}$$

where X = $Cl^-$, $Br^-$, or $I^-$, and Y = $Cl^-$, $Br^-$, $I^-$, or another arbitrary ion.

Equation 5.1 represents vapourisation, 5.2 represents the nucleophilic substitution reaction with another halide, and 5.3 shows the possible reaction between halocarbons with water and the subsequent hydrolysis to methanol. Equations 5.4, 5.5, and 5.6 symbolise different pathways of the reduction process: hydrogenolysis, dihalo-elimination, and coupling,

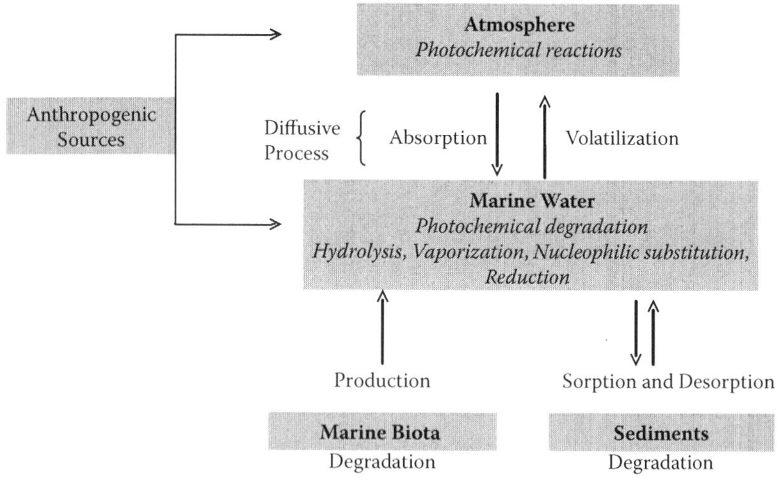


**Figure 5.1** Schematic representation of the sources and fates of volatile halocarbons. (Modified from Dewulf, J. et al., *Atmospheric Environment,* 29(3): 323–331, 1995; Dewulf, J., and H. VanLangenhove, *Atmospheric Environment,* 31(20): 3291–3307, 1997.)

respectively. Equation 5.7, proposed by Roberts, Sanborn, and Gschwend (1992), represents the nucleophilic attack of dihalomethanes by sulphide species in solution. The last three mechanisms (equations 5.8, 5.9, and 5.10) denote the thiolytic dehalogenation of dichloromethane catalysed by a glutathione transferase, dehalogenation of halomethanes catalysed by cobalamin-containing methyltransferases, and corrinoid-dependent reductive dehalogenation of tetrachloroethane under anaerobic conditions coupled to energy metabolism (halorespiration), respectively (van Pee and Unversucht 2003). Some possible mechanisms and fates of halocarbons in seawater are shown in Figure 5.1.

## Analysis of VOCs

Analytical techniques for VOCs have been changing over the last few years, with a lowering of the detection limits to trace levels (e.g., pg $L^{-1}$ in seawater, PPTv in air). One of the key aspects of VOC analysis is the sampling method. Sampling must be carefully designed in order to avoid losses and the photochemical degradation or production of analytes. The main instrumental techniques to analyse VOCs are gas chromatography (GC) coupled to flame ionisation (FID), electron capture (ECD), or mass spectrometric (MS) detection. When used with single ion monitoring (SIM), the latter detector can significantly increase the sensitivity of the technique and considerably lower the limit of detection compared to scan mode. This chapter concentrates on the potential application of the newer methods of solid phase microextraction (SPME) in environmental analysis of VOCs with in situ and ex situ sampling in marine seawater and sediments and outlines the older methods as well.

## Solvent Extraction

Solvent extraction has been widely used in the past to analyse VOCs in the marine environment (Amaral et al. 1994). The method is based on the equilibrium between two phases (polar and nonpolar), where the compounds with affinity for the organic nonpolar solvent employed are moved from an aqueous phase to an organic phase. The solvents mostly used are pentane and hexane. However, this technique does not have good detection limits, is not environmentally friendly, and is highly dependent upon the concentration and the partition coefficients between the solvent employed and the analytes in the sample.

## Head Space

Head-space analysis is one of the most commonly employed techniques and is still widely used to analyse VOCs because the instrumental methods can be automated. This enables control of several factors, such as temperature, sample agitation, head-space volume, volume to be injected, sampling and injection speed, etc. Here the volatiles are sampled from the head space, which is in equilibrium with the liquid phase in a closed system. The relative low sensitivity of this technique makes it only suitable for analysis when samples are very concentrated (e.g., sewage samples and sediments) (Ebrahimzadeh et al. 2007; Golfinopoulos, Lekkas, and Nikolaou 2001).

## Dynamic Head Space (Purge and Trap)

In this methodology, VOCs are stripped from a liquid by a continuous stream of an inert gas (commonly, nitrogen or helium). The purged volatiles are trapped on a sorbent cartridge (e.g., Tenax) or a cryotrap. The analytes are desorbed by thermal desorption and transferred to a capillary GC column. This methodology provides good and reliable data, but it is time demanding and the procedure is complicated when various sample matrices are involved (Huybrechts, Dewulf, and Van Langenhove 2003). This methodology has been applied with different matrices, such as sediments (Roose et al. 2001), waters (Zoccolillo et al. 2005), seawater (Connan, LeCorre, and Morin 1996; Hashimoto et al. 2001), and drinking water (Antoniou, Koukouraki, and Diamadopoulos 2006), among others.

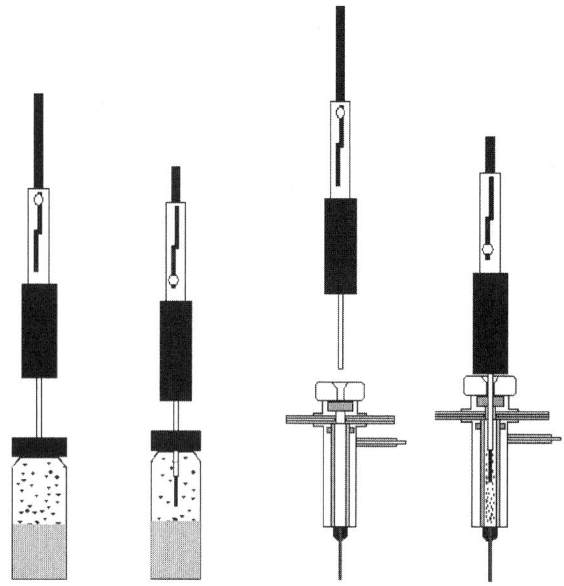


**Figure 5.2** Absorption and desorption process using SPME for VOCs analysis from the head space. (a) SPME device is placed in the sample's head space. (b) The fibre is released and the VOCs absorbed on it. (c) The fibre is retrieved and ready to be injected into a GC injector. (d) The fibre is injected and the compounds desorbed into the GC column for further analysis.

## Solid-Phase Microextraction (SPME)

This technique is one of the newest approaches to VOC sampling and analysis. Solid-phase microextraction (SPME) was developed by Pawliszyn and co-workers (Chai et al. 1993). It is a technique of extraction that combines sampling and concentration in a single step. It requires no solvent and provides good results for a wide range of analyte concentrations. Recent work (Bravo-Linares and Mudge 2007; Bravo-Linares et al. 2007) has shown that over 50 VOCs can be quantified in a single analysis. This technique has been used to analyze different compounds in gaseous or liquid samples using direct immersion, head space, or gas streams (dynamic head space). This absorption is based on the partition between a coated fibre and the analytes (Eisert and Levsen 1996). Figure 5.2 shows schematically the absorption and desorption process in the analysis of VOCs from the head space. The type of fibre to be employed depends on the target compounds. Table 5.4 provides an overview of which fibre should be used according to the molecular weight of the target compounds, as well as polarities.

**Table 5.4 Fibre Selection Guide according to VOCs Molecular Weight and Polarities Provided by Supelco®**

| Fibre type | Common uses |
|---|---|
| 75 μm/85 μm Carboxen/ polydimethylsiloxane | Gases and low molecular weight compounds (MW 30–225) |
| 100 μm Polydimethylsiloxane | Volatiles (MW 60–275) |
| 65 μm Polydimethylsiloxane/divinylbenzene | Volatiles, amines, and nitro-aromatic compounds (MW 50–300) |
| 85 μm Polyacrylate | Polar semivolatiles (MW 80–300) |
| 7 μm Polydimethylsiloxane | Nonpolar high molecular weight compounds (MW 125–600) |
| 30 μm Polydimethylsiloxane | Nonpolar semivolatiles (MW 80–500) |
| 65 μm/70 μm Carbowax/divinylbenzene polydimethylsiloxane | Alcohols and polar compounds (MW 40–275) |
| 50 μm/30 μm Divinylbenzene/carboxen on polydimethylsiloxane on a stableflex fibre | Flavour compounds: volatiles and semivolatiles ($C_3$–$C_{20}$) (MW 40–275) |
| 50 μm/30 μm Divinylbenzene/carboxen on polydimethylsiloxane on a 2-cm stableflex fibre | Trace compound analysis ($C_2$–$C_{20}$) (MW 40–275) |

## Analysis of VOCs in Different Matrices Using SPME

### *Seawater Analyses*

Sampling for VOCs in seawater must be done carefully and, if possible, the compounds must be analysed immediately. The volume of water to be sampled depends on the concentration of the analytes. The sample is normally taken by fully filling a glass container; it is then kept at 4°C and, if possible, inverted to avoid losses of volatile components. Some authors use preservatives such as sodium azide and suggest analysis within 58 days (Kristiansen et al. 1993). Even so, is not recommended to leave the sample for a long time. Results have shown that for samples taken simultaneously, the VOC concentrations in those stored for 2–4 hours decreased at 5–30% per hour (Bravo-Linares 2007).

There are not many references published using SPME to analyse VOCs in seawater. This application is relatively new, and purge and trap is the most frequently used technique for this purpose. One advantage of SPME in VOC analysis of seawater is that it can identify and quantify a wide variety of VOCs, such as sulphur-containing compounds; halogenated, nonmethane hydrocarbons (NMHC); BTEXs and other mono-aromatic compounds; linear and branched hydrocarbons; aldehydes; and terpenes in a single analysis (Bravo-Linares et al. 2007). A system of purging and further concentration on an SPME fibre is shown in Figure 5.3. This methodology allowed detection and quantification of concentrations of a wide variety of 50+ VOCs at the levels of picograms per litre to nanograms per litre. Single ion monitoring improves the limit of detection on mass spectrometers, as typical diagnostic fragments can be used most sensitively (Table 5.5).

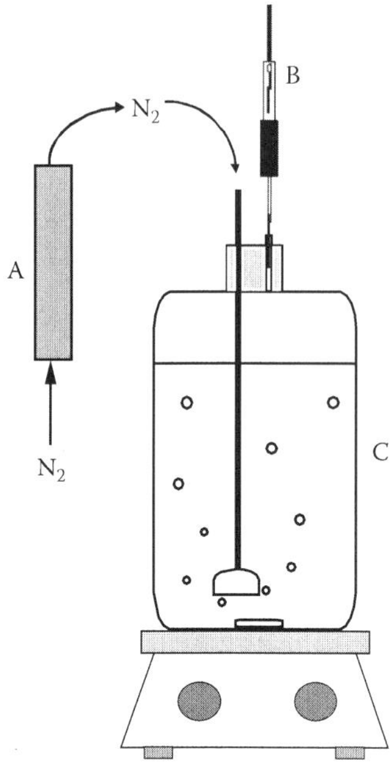


**Figure 5.3** Diagram showing the purging and concentration system employed for VOCs analysis in seawater. (A) Organic filter; (B) SPME fibre; and (C) amber bottle (variable volume) with continuous stirring.

**Table 5.5 Diagnostic Fragments (*m/z*) for Identification and Quantification of Selected VOCs Detected in Seawater**

| Compounds | *m/z* | Compounds | *m/z* |
|---|---|---|---|
| Terpenes (e.g.) | | BTEXs and mono-aromatic | |
| Limonene, α-pinene | 93 | compounds (e.g.) | 78–91–105–106 |
| | | Benzene, toluene, xylenes, ethylbenzene | |
| Sulphur-containing | | Halogenated compounds (e.g.) | |
| compounds (e.g.) | | Dichloromethane | 49 |
| Hydrogen sulphide | 32 | Chloroform | 83 |
| Dimethyl sulphide | 62 | Bromoform | 173 |
| Methyl mercaptan | 47 | Iodomethane | 142 |
| Carbon disulphide | 76 | Diiodomethane | 268 |
| Dimethyl disulphide | 94 | Carbon tetrachloride | 117 |
| Dimethyl sulphoxide | 63 | 1,1,1-Trichloroethane | 97 |
| Dimethyl sulphone | 79 | | |
| Aldehydes (e.g.) | 44–43 | NMHC (e.g.) | 43–57 |
| Decanal, hexanal | | C5–C11 | |

## Source Identification in Seawater Samples

### *Classed Posting for Source Identification*

Classed posting is a useful tool to visually investigate the spatial relationship between potential sources of specific chemicals such as chlorinated halocarbons, although it does not demonstrate cause and effect. As an example, Figure 5.4 shows how the concentrations of some chlorinated compounds are higher near industrial areas of the River Mersey, United Kingdom, and dispersed to the sea by tide and currents.

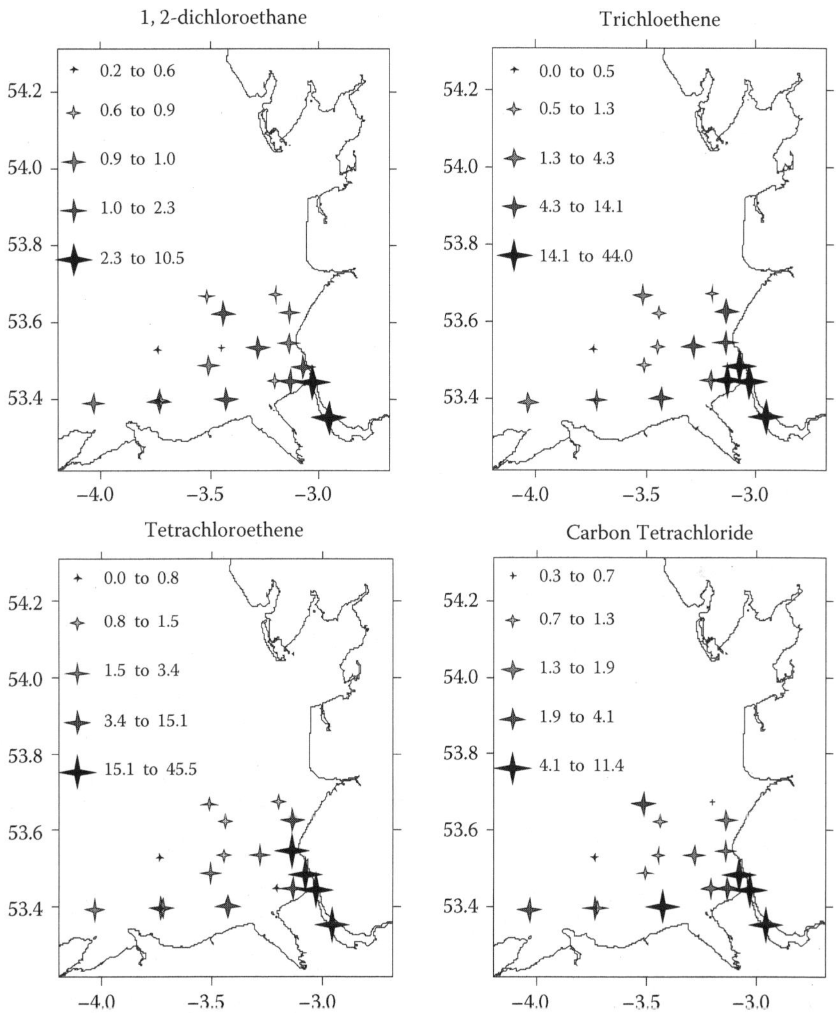


**Figure 5.4** Halogenated solvents' dispersion in Liverpool Bay waters (concentrations in nanograms per liter). (Bravo-Linares, C. M. et al., *Marine Pollution Bulletin*, 54 (11): 1742–1753, 2007.)

**Table 5.6 Proposed Source Allocation for VOCs in Coastal Seawater**

| Sources | Type of compound | Examples |
|---|---|---|
| Sediments | BTEXs | Toluene, xylenes |
| | NMHC | Linear $C_9$–$C_{11}$ |
| | Sulphur containing | DMS, $DMS_2$, and others |
| | Chlorinated | Dichloromethane, chloroform |
| | Aldehydes | Benzaldehyde, propanal, hexanal, decanal |
| Macroalgae | Brominated | Bromoform, dibromomethane |
| | Chlorinated | Tetrachloroethene, chloroform, dichloromethane |
| | Iodinated | Diiodomethane, iodoethane |
| | Mixed halogen | Bromodichloromethane |
| | Sulphur containing | DMS, $DMS_2$ |
| Phytoplankton | Iodinated | Iodomethane, 1-iodobutane, diiodomethane |
| | Halogenated | 1,1,2-Trichloroethene, 2-chloropropane |
| | Mixed halogen | Chloroiodomethane, dibromochloromethane |
| | Brominated | 1-Bromopentane, 1-bromopropane |
| | NHMC | $C_5$–$C_8$ |
| | Sulphur containing | DMS, $DMS_2$ |
| | Aldehydes | Hexanal |
| | Others | 2,4-Dimethylfuran |
| Anthropogenic | Chlorinated | 1,2-Dichloroethane, 1,1,1-trichloroethane, trichloroethene, tetrachloroethene, and carbon tetrachloride |
| | BTEXs | Xylenes, ethyl-benzene, and other monoaromatics |
| | NMHC | Linear $C_9$–$C_{11}$ |

## Sediments

Sediments are not typically referred to as a source of VOCs (Dewulf et al. 1996). However, they play an important role in the absorption of many chemicals, as well as the biogeochemical processes of organic matter decomposition, which may convert complex organic molecules into VOCs, especially in coastal areas. Different methodologies have been applied to assess the types and amounts of VOCs present in sediments (Bianchi, Varney, and Phillips 1991), most of which involve taking the sample into the laboratory and using head-space (Bianchi et al. 1991) or purge-and-trap (Fu et al. 2005) analysis (Campillo et al. 2004). Some methods involve the use of solvents such as methanol to force the VOCs to migrate to the methanolic phase to make the extraction process easier (Amaral et al. 1994). The versatile application of SPME methodology, however, makes it an ideal methodology for field sampling (Table 5.6).

Sampling design is very important in order to avoid losses of analytes. Handling and storing samples prior to analysis for VOCs tend to lead to losses. This can be improved, however, if sample collection and concentration are performed in situ (Kuran and Sojak 1996). Results comparing these

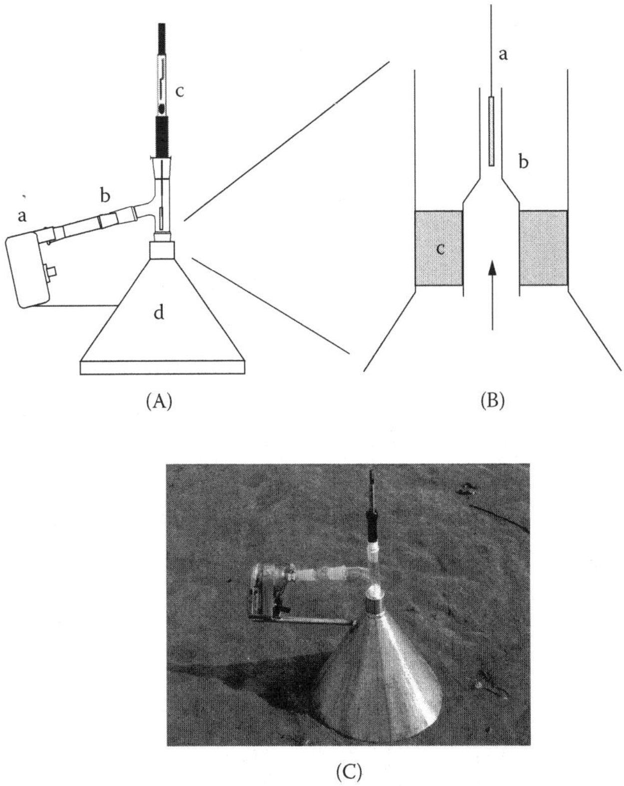


**Figure 5.5** (A) Diagram of the system employed to collect VOCs in coastal sediments. (a) Vacuum pump with a flow controller set at ~100 ml $min^{-1}$. (b) Glass connectors. (c) SPME fibre and manual holder. (d) Stainless-steel funnel; internal volume: 3.8 L, height: 23 cm, internal diameter: 24.5 cm. (B) Detail of the connection between the still head and the steel funnel. (a) SPME fibre. (b) Cut down Pasteur pipette. (c) PTFE tape coated rubber bung. (C) Picture of the system employed for sampling VOCs in sediments in situ.

methods (Bravo-Linares and Mudge 2007) have shown that when ex situ sampling was used, the sensitivity of the technique was lower compared to in situ SPME sampling. The amount of compounds detected and the limit of detections can be significantly improved by using in situ methods.

Sampler devices used in field sampling are diverse; an example of a device especially designed to sample VOCs coming from coastal sediments is shown in Figure 5.5. Figure 5.6 shows a sample trace obtained using the SPME in situ sampling method over anoxic intertidal sediments.

## *Signatures*

According to the intrinsic properties of coastal sediments such as organic matter content, particle size, pH, temperature, and redox potential can

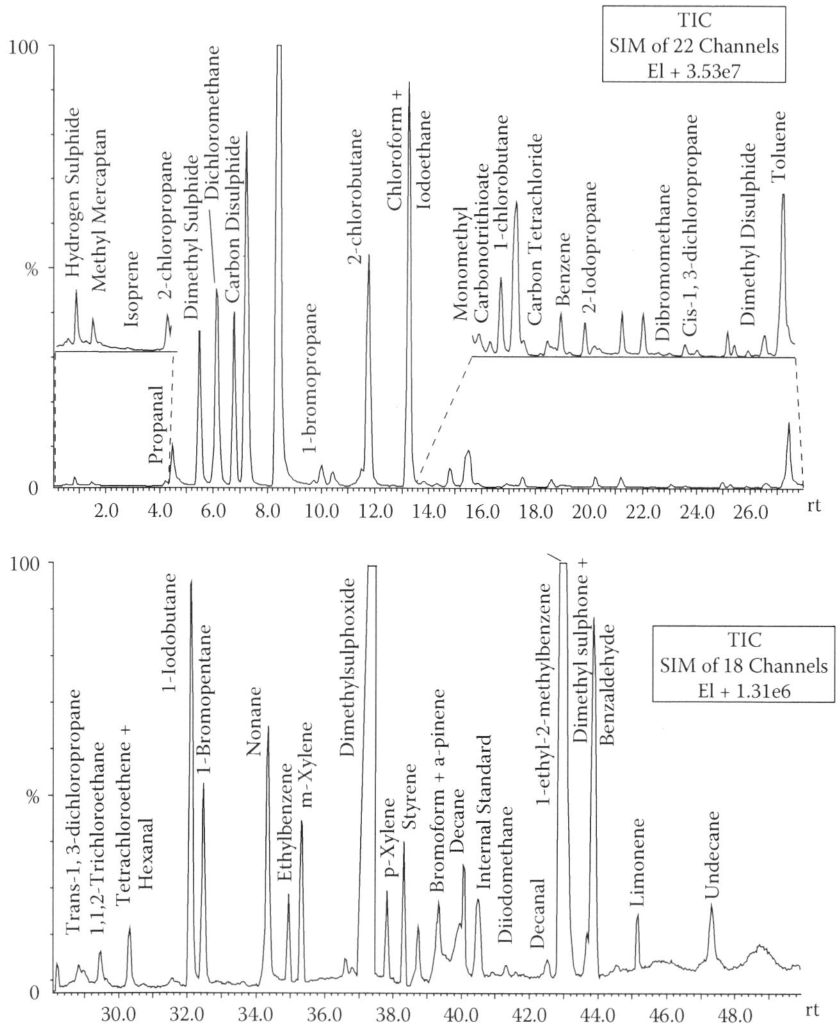


**Figure 5.6** An example trace for VOCs extracted in one hour in situ from anoxic sediments.

influence the VOC signatures (Bravo-Linares and Mudge 2007). Figure 5.7 shows how sampling at different sites with sediments having similar textural properties can lead to a range of signatures related to the sediment type.

## Statistical Approaches for Data Interpretation

Environmental data collected in this manner may be large and complex with many variables and observations. To analyse such data sets, it may be necessary to use a tool able to summarise and find the underlying relationships among the data. Multivariate statistical analysis has become a widely

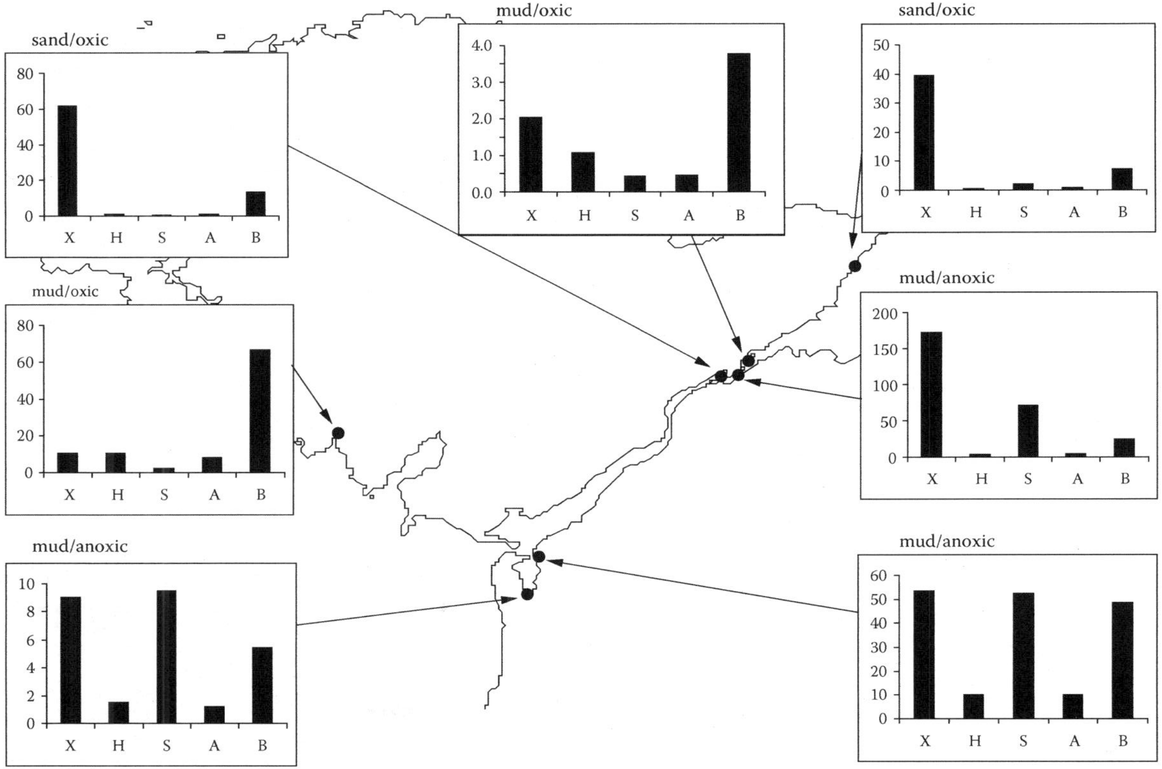


**Figure 5.7** Signatures found in different samples according to sediment physicochemical properties. Total VOCs are expressed in picograms per gram, where X is total halogenated compounds, H is total hydrocarbons, S is total sulphur-containing compounds, A is total aldehydes, and B is total BTEXs and mono-aromatic compounds.

used technique within the recent past. One of the most used analyses in environmental chemometrics and forensics is principal component analysis (PCA). However, partial least squares (PLS) is also an important tool for predicting and modelling environmental data (see chapter 9). These methods are useful tools to investigate the spatial and temporal variability in the occurrence and production of VOCs and also to find a causal relationship between the variables and observations (Glenn 2002).

### *PCA in the Analysis of VOCs*

PCA can be used to identify the relationships between VOCs collected in different areas. The concentrations should be standardised using proportions to remove concentration effects (Mudge 2007). The loadings and scores from the PCA model (Figure 5.8) constructed with the sediment samples shown in Figure 5.7 indicate that VOCs are grouped according to the source sediment properties. For instance, all the muddy samples are in the same group; the same is true for the mud–sand mixtures and sandy sediments. Therefore, the substrate texture has the major effect on the types as well the amount of the VOCs produced or adsorbed and then released in the sediments. Muddy sediments are, in the majority of cases, anoxic, and concentrations of the reduced sulphur-containing compounds are in greater proportions in such sediments.

Other factors that may influence the production of VOCs are redox potential, sediment temperature, and organic matter input, among others. Mixed mud/sand sites overlap the two end members with VOCs from both environments (Figure 5.8). Sandy samples may be characterised by having a major presence of BTEXs, hydrocarbons, aldehydes, and terpenes. The presence of these compounds may be indicative of the subsurface biochemical processes. For instance, the aldehydes may result from the aerobic degradation of fatty acids (Goni et al. 2000).

## Summary

VOC analysis may be of great interest in environmental forensics in the identification of natural (biotic and abiotic) emissions from sediments, soils, and waters and compared to suspected anthropogenic sources. The ability to quantify a wide range of compounds in a single analysis is of particular use, especially at picogram-per-liter levels. The in situ nature of the sampling certainly improves the sensitivity of the method, but would ideally require GC-MS analysis in the field to improve sample throughput. To date, these methods have been used in determining the VOC present in seawater throughout the year at one location and at several locations where different inputs contribute to the overall signature. Sewage sludges, soils, and sedi-

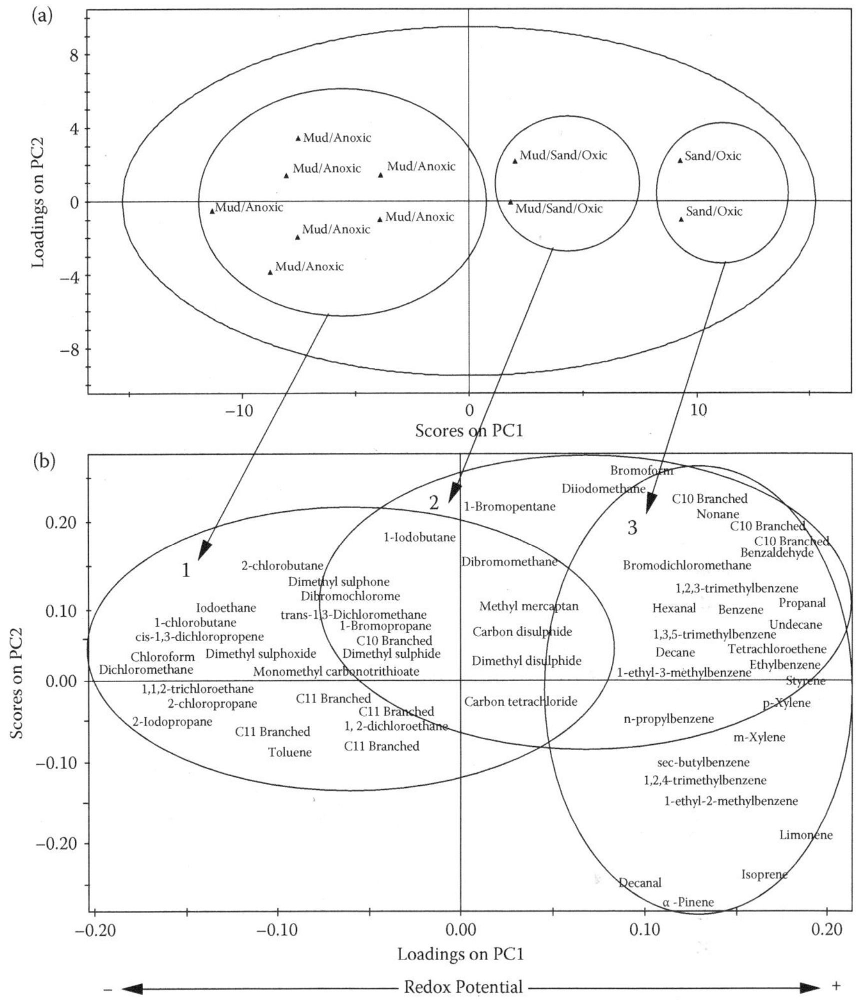


**Figure 5.8** PCA analysis for the samples collected in different sites. (a) Score plot for sediment samples showing the grouping of sites according to sediment properties. (b) Loading plot for sediment samples showing the groups of VOCs formed according to the substrate characteristics. Groupings in circles are related to sediments: (1) muddy/anoxic, (2) muddy/sandy oxic, and (3) sandy/oxic.

ments and the corresponding atmospheric concentrations have been investigated; sources have been identified and remedial action taken.

## References

Abrahamsson, K., K. S. Choo, M. Pedersen, G. Johansson, and P. Snoeijs. (2003) Effects of temperature on the production of hydrogen peroxide and volatile halocarbons by brackish-water algae. *Phytochemistry,* 64(3): 725–734.

Amaral, O. C., L. Olivella, J. O. Grimalt, and J. Albaiges. (1994) Combined solvent-extraction purge-and-trap method for the determination of volatile organic-compounds in sediments. *Journal of Chromatography A,* 675(1–2): 177–187.

Amachi, S., Y. Kamagata, T. Kanagawa, and Y. Muramatsu. (2001) Bacteria mediate methylation of iodine in marine and terrestrial environments. *Applied and Environmental Microbiology* 67(6): 2718–2722.

Antoniou, C. V., E. E. Koukouraki, and E. Diamadopoulos. (2006) Determination of chlorinated volatile organic compounds in water and municipal wastewater using headspace-solid phase microextraction-gas chromatography. *Journal of Chromatography A,* 1132(1–2): 310–314.

Bakierowska, A. M., and J. Trzeszczynski. (2003) Graphical method for the determination of water/gas partition coefficients of volatile organic compounds by a headspace gas chromatography technique. *Fluid Phase Equilibria,* 213(1–2): 139–146.

Bey, I., D. J. Jacob, R. M. Yantosca, J. A. Logan, B. D. Field, A. M. Fiore, Q. B. Li, H. G. Y. Liu, L. J. Mickley, and M. G. Schultz. (2001) Global modeling of tropospheric chemistry with assimilated meteorology: Model description and evaluation. *Journal of Geophysical Research—Atmospheres,* 106(D19): 23073–23095.

Bianchi, A. P., M. S. Varney, and J. Phillips. (1991) Analysis of volatile organic-compounds in estuarine sediments using dynamic headspace and gas-chromatography mass-spectrometry. *Journal of Chromatography,* 542(2): 413–450.

Bobadilla, R., T. Huybrechts, J. Dewulf, and H. Van Langenhove. (2003) Determination of the Henry's constant of volatile and semivolatile organic compounds of environmental concern by the BAS (batch air stripping) technique: A new mathematical approach. *Journal of the Chilean Chemical Society,* 48(3): 7–11.

Bravo-Linares, C. M. (2007) Temporal variation, occurrence and factors controlling the production and concentrations of VOCs in the marine environment. *Ocean Sciences,* Bangor University: 234.

Bravo-Linares, C. M., and S. M. Mudge. (2007) Analysis of volatile organic compounds (VOCs) in sediments using in situ SPME sampling. *Journal of Environmental Monitoring,* 9(5): 411–418.

Bravo-Linares, C. M., S. M. Mudge, and R. H. Loyola. (2007) Occurrence of volatile organic compounds (VOCs) in Liverpool Bay, Irish Sea. *Marine Pollution Bulletin,* 54(11): 1742–1753.

Buchmann, B., K. Stemmler, and S. Reimann. (2003) Regional emissions of anthropogenic halocarbons derived from continuous measurements of ambient air in Switzerland. *Chimia,* 57(9): 522–528.

Bullister, J. L., and D. P. Wisegarver. (1998) The solubility of carbon tetrachloride in water and seawater. *Deep-Sea Research Part I—Oceanographic Research Papers,* 45(8): 1285–1302.

Campillo, N., N. Aguinaga, P. Vinas, I. Lopez-Garcia, and M. Hernandez-Cordoba. (2004) Speciation of organotin compounds in waters and marine sediments using purge-and-trap capillary gas chromatography with atomic emission detection. *Analytica Chimica Acta,* 525(2): 273–280.

Chai, M., C. L. Arthur, J. Pawliszyn, R. P. Belardi, and K. F. Pratt. (1993) Determination of volatile chlorinated hydrocarbons in air and water with solid-phase microextraction. *Analyst,* 118(12): 1501–1505.

Chartrand, M. M. G., S. K. Hirschorn, G. Lacrampe-Couloume, and B. S. Lollar. (2007) Compound specific hydrogen isotope analysis of 1,2-dichloroethane: Potential for delineating source and fate of chlorinated hydrocarbon contaminants in groundwater. *Rapid Communications in Mass Spectrometry,* 21(12): 1841–1847.

Chuck, A. L., S. M. Turner, and P. S. Liss. (2005) Oceanic distributions and air–sea fluxes of biogenic halocarbons in the open ocean. *Journal of Geophysical Research—Oceans,* 110(C10).

Class, T., R. Kohnle, and K. Ballschmiter. (1986) Chemistry of organic traces in air.7. Bromochloromethanes and bromochloromethanes in air over the Atlantic Ocean. *Chemosphere,* 15(4): 429–436.

Collen, J., and M. Pedersen. (1996) Production, scavenging and toxicity of hydrogen peroxide in the green seaweed *Ulva rigida. European Journal of Phycology,* 31(3): 265–271.

Connan, O., P. LeCorre, and P. Morin. (1996) Purge and trap GC determination of volatile halocarbon compounds in seawater. *Analysis,* 24(2): 55–59.

Dewulf, J., T. Dewettinck, A. DeVisscher, and H. VanLangenhove. (1996) Sorption of chlorinated C-1- and C-2-hydrocarbons and monocyclic aromatic hydrocarbons on sea sediment. *Water Research,* 30(12): 3130–3138.

Dewulf, J., D. Drijvers, and H. Vanlangenhove. (1995) Measurement of Henry's law constant as function of temperature and salinity for the low-temperature range. *Atmospheric Environment,* 29(3): 323–331.

Dewulf, J., and H. VanLangenhove. (1997) Analytical techniques for the determination and measurement data of 7 chlorinated C-1- and C-2-hydrocarbons and 6 monocyclic aromatic hydrocarbons in remote air masses: An overview. *Atmospheric Environment,* 31(20): 3291–3307.

Ebrahimzadeh, H., Y. Yamini, F. Kamarei, and M. Khalili-Zanjani. (2007) Application of headspace solvent microextraction to the analysis of mononitrotoluenes in waste water samples. *Talanta,* 72(1): 193–198.

Eisert, R., and K. Levsen. (1996) Solid-phase microextraction coupled to gas chromatography: A new method for the analysis of organics in water. *Journal of Chromatography A,* 733(1–2): 143–157.

Fenical, W. (1981). Natural halogenated organics. In *Marine organic chemistry, evolution, composition, interactions and chemistry of organic mater in seawater,* ed. E. K. Duursma and R. Dawson, 375–393. New York: Elsevier.

______. (1982) Natural-products chemistry in the marine-environment. *Science,* 215(4535): 923–928.

Fu, J. H., X. T. Ai, H. S. Liu, D. Q. Han, D. Y. Chen, and W. Y. Ma. (2005) Determination of trace benzene, toluene, ethylbenzene and xylenes in seabed sediments and seawater by purge and trap gas chromatography. *Chinese Journal of Analytical Chemistry,* 33(12): 1753–1756.

Gagosian, R. B., and C. Lee. (1981) Processes controlling the distribution of biogenic organic compounds in seawater. In *Marine organic chemistry, evolution, composition, interactions and chemistry of organic mater in seawater,* ed. E. K. Duursma and R. Dawson, 91–107. New York: Elsevier.

Glenn, W. J. (2002) State of the art on multivariate chemometric methods in environmental forensics. *Environmental Forensics,* 3: 59–79.

Golfinopoulos, S. K., T. D. Lekkas, and A. D. Nikolaou. (2001) Comparison of methods for determination of volatile organic compounds in drinking water. *Chemosphere,* 45(3): 275–284.

Goni, M. A., M. B. Yunker, R. W. Macdonald, and T. I. Eglinton. (2000) Distribution and sources of organic biomarkers in arctic sediments from the Mackenzie River and Beaufort Shelf. *Marine Chemistry,* 71(1–2): 23–51.

Gribble, G. J. (2000) The natural production of organobromine compounds. *Environmental Science and Pollution Research,* 7(1): 37–49.

Hashimoto, S., T. Tanaka, N. Yamashita, and T. Maeda. (2001) An automated purge and trap gas chromatography-mass spectrometry system for the sensitive shipboard analysis of volatile organic compounds in seawater. *Journal of Separation Science,* 24(2): 97–103.

Hoekstra, E. J., J. H. Duyzer, E. W. B. de Leer, and U. A. T. Brinkman. (2001) Chloroform-concentration gradients in soil air and atmospheric air, and emission fluxes from soil. *Atmospheric Environment,* 35(1): 61–70.

Huhn, O., W. Roether, P. Beining, and H. Rose. (2001) Validity limits of carbon tetrachloride as an ocean tracer. *Deep-Sea Research Part I—Oceanographic Research Papers,* 48(9): 2025–2049.

Huybrechts, T., J. Dewulf, and H. Van Langenhove. (2003) State-of-the-art of gas chromatography-based methods for analysis of anthropogenic volatile organic compounds in estuarine waters, illustrated with the river Scheldt as an example. *Journal of Chromatography A,* 1000(1–2): 283–297.

Keene, W. C., M. A. K. Khalil, D. J. Erickson, A. McCulloch, T. E. Graedel, J. M. Lobert, M. L. Aucott, et al. (1999). Composite global emissions of reactive chlorine from anthropogenic and natural sources: Reactive chlorine emissions inventory. *Journal of Geophysical Research—Atmospheres,* 104(D7): 8429–8440.

Keppler, F., R. Borchers, P. Elsner, I. Fahimi, J. Pracht, and H. F. Scholer. (2003) Formation of volatile iodinated alkanes in soil: Results from laboratory studies. *Chemosphere,* 52(2): 477–483.

Kristiansen, N. K., E. Lundanes, M. Froshaug, and G. Becher. (1993) Evaluation of the open-loop stripping technique used for the determination of volatile organic-compounds in water. *Analytica Chimica Acta,* 280(1): 111–117.

Kuran, P., and L. Sojak. (1996) Environmental analysis of volatile organic compounds in water and sediment by gas chromatography. *Journal of Chromatography A,* 733(1–2): 119–141.

Laturnus, F. (1996) Volatile halocarbons released from Arctic macroalgae. *Marine Chemistry,* 55(3–4): 359–366.

Liss, P. S., A. J. Watson, M. I. Liddicoat, G. Malin, P. D. Nightingale, S. M. Turner, and R. C. Upstill-Goddard. (1993) Trace gases and air–sea exchanges. *Philosophical Transactions of the Royal Society of London Series A—Mathematical Physical and Engineering Sciences,* 343(1669): 531–541.

______. (1994) Trace gases and air-sea exchanges. In *Understanding the North Sea system,* ed. H. Charnock, K. R. Dyer, J. M. Huthnance, P. S. Liss, J. H. Simpson, and P. B. Tett, 153–163. London: Chapman & Hall for the Royal Society.

Lohman, J. H. (2002). A history of dry cleaners and sources of solvent releases from dry cleaning equipment. *Environmental Forensics,* 3(1): 35–58.

Louis, F., C. A. Gonzalez, R. E. Huie, and M. J. Kurylo. (2001) An ab initio study of the kinetics of the reactions of halomethanes with the hydroxyl radical. 3. Kinetic parameters predictions for the potential halon replacements $CH_2FBr$, $CHFBr_2$, CHFClBr, $CHCl_2Br$, and $CHClBr_2$. *Journal of Physical Chemistry A,* 105(9): 1599–1604.

McCulloch, A., M. L. Aucott, T. E. Graedel, G. Kleiman, P. M. Midgley, and Y. F. Li. (1999) Industrial emissions of trichloroethene, tetrachloroethene, and dichloromethane: Reactive chlorine emissions inventory. *Journal of Geophysical Research—Atmospheres,* 104(D7): 8417–8427.

Mellouki, A., R. K. Talukdar, A. M. Schmoltner, T. Gierczak, M. J. Mills, S. Solomon, and A. R. Ravishankara. (1992) Atmospheric lifetimes and ozone depletion potentials of methyl-bromide ($CH_3Br$) and dibromomethane ($CH_2Br_2$). *Geophysical Research Letters,* 19(20): 2059–2062.

Mtolera, M. S. P., J. Collen, M. Pedersen, A. Ekdahl, K. Abrahamsson, and A. K. Semesi. (1996) Stress-induced production of volatile halogenated organic compounds in *Eucheuma denticulatum* (Rhodophyta) caused by elevated pH and high light intensities. *European Journal of Phycology,* 31(1): 89–95.

Mudge, S. M. (2007) Multivariate statistical methods in environmental forensics. *Environmental Forensics,* 8(1–2): 155–163.

Neidleman, S. L., and J. Geigert. (1987) Biological halogenation—Roles in nature, potential in industry. *Endeavour,* 11(1): 5–15.

Prilepsky, E. B., V. G. Povarov, N. V. Bredelev, and V. A. Isidorov. (1998) Formation of halocarbons in the methane-alkaline halide crystal system under UV radiation. *Russian Chemical Bulletin,* 47(10): 1910–1913.

Rivett, M. O., S. Feenstra, and L. Clark. (2006) Lyne and McLachlan (1949): Influence of the first publication on groundwater contamination by trichloroethene. *Environmental Forensics,* 7(4): 313–323.

Roberts, A. L., P. N. Sanborn, and P. M. Gschwend. (1992) Nucleophilic-substitution reactions of dihalomethanes with hydrogen-sulfide species. *Environmental Science & Technology,* 26(11): 2263–2274.

Roose, P., J. Dewulf, U. A. T. Brinkman, and H. Van Langenhove. (2001) Measurement of volatile organic compounds in sediments of the Scheldt Estuary and the southern North Sea. *Water Research,* 35(6): 1478–1488.

Saiz-Lopez, A., A. S. Mahajan, R. A. Salmon, S. J. B. Bauguitte, A. E. Jones, H. K. Roscoe, and J. M. C. Plane. (2007) Boundary layer halogens in coastal Antarctica. *Science,* 317(5836): 348–351.

Sander, R. (1999) Compilation of Henry's law constants for inorganic and organic species of potential importance in environmental chemistry (version 3). From http://www.henrys-law.org.

Scarratt, M. G., and R. M. Moore. (1998) Production of methyl bromide and methyl chloride in laboratory cultures of marine phytoplankton II. *Marine Chemistry,* 59(3–4): 311–320.

Tanhua, T., E. Fogelqvist, and O. Basturk. (1996) Reduction of volatile halocarbons in anoxic seawater, results from a study in the Black Sea. *Marine Chemistry,* 54(1–2): 159–170.

Thunis, P., and C. Cuvelier. (2000) Impact of biogenic emissions on ozone formation in the Mediterranean area—A BEMA modelling study. *Atmospheric Environment,* 34(3): 467–481.

Tokarczyk, R., K. D. Goodwin, and E. S. Saltzman. (2001) Methyl bromide loss rate constants in the North Pacific Ocean. *Geophysical Research Letters,* 28(23): 4429–4432.

______. (2003) Methyl chloride and methyl bromide degradation in the Southern Ocean. *Geophysical Research Letters,* 30(15): art. no. 1808.

Tokarczyk, R., and E. S. Saltzman. (2001) Methyl bromide loss rates in surface waters of the North Atlantic Ocean, Caribbean Sea, and eastern Pacific Ocean (8 degrees–45 degrees N). *Journal of Geophysical Research—Atmospheres,* 106(D9): 9843–9851.

van Pee, K.-H., and S. Unversucht. (2003) Biological dehalogenation and halogenation reactions. *Chemosphere,* 52(2): 299–312.

WMO (World Meteorological Organization). (2006) Scientific assessment of ozone depletion 2006, Report 50, WMO Global Ozone Research and Monitoring Project.

Young, D. N., O. J. McConnell, and W. Fenical. (1981) In vivo biosynthesis of tribromoheptene oxide in bonnemaisonia-nootkana. *Phytochemistry,* 20(10): 2335–2337.

Yvon, S. A., and J. H. Butler. (1996) An improved estimate of the oceanic lifetime of atmospheric $CH_3Br$. *Geophysical Research Letters,* 23(1): 53–56.

Zoccolillo, L., L. Amendola, C. Cafaro, and S. Insogna. (2005) Improved analysis of volatile halogenated hydrocarbons in water by purge-and-trap with gas chromatography and mass spectrometric detection. *Journal of Chromatography A,* 1077(2): 181–187.

# Application of Molecular Microbiology to Environmental Forensics

# 6

ANDREW S. BALL
JULES N. PRETTY
RAKHI MAHMUD
ERIC ADETUTU

## Contents

Introduction .......... 195
Microbiology and Environmental Forensics .......... 196
Microbial Analyses and Environmental Forensics .......... 197
The Basis of Molecular Microbial Forensic Techniques .......... 197
Ribosomes .......... 199
Ribosomal RNA and Taxonomy .......... 200
Polymerase Chain Reaction (PCR) .......... 201
PCR-Based DNA Fingerprinting Techniques .......... 202
Denaturing Gradient Gel Electrophoresis and Its Derivatives .......... 202
DGGE, T-RFLP, and Forensic Science .......... 208
Single-Stranded Conformation Polymorphism Analysis (SSCP) .......... 209
Terminal-Restriction Fragment Length Polymorphism (T-RFLP) .......... 210
Limitations of PCR-Based Methodologies .......... 211
Forensic Interpretation of Profiles .......... 214
Conclusions .......... 214
References .......... 215

## Introduction

The application of microbiology to environmental forensic investigations includes a range of subdisciplines, including microbial physiology, molecular microbial ecology, and microbial biochemistry. Microbial forensics employs a range of techniques to trace a contaminant through the environment using a microbial marker. Many of these techniques, such as selective isolation plating, are well established and have been successfully employed for many years (Ball 2004).

The purpose of this chapter is to provide an introduction to the application of molecular microbial ecology, an emerging subdiscipline within microbiology that has applications in environmental forensics. This review will focus on the benefits of community fingerprinting to the field of environmental forensics and outline the methods commonly used, as well as indicating the potential developments in this emerging field of forensics. Illustrations of the application of these technologies through examples are presented.

## Microbiology and Environmental Forensics

Molecular environmental forensics can be defined as the application of molecular microbiology to environmental forensics. Molecular environmental forensics provides a means by which a profile of a microbial community is used to trace a contaminant source (Petrisor et al. 2006). Molecular environmental forensics can be applied to both terrestrial and aquatic environments, although most studies and examples have been based in terrestrial systems. The techniques that can be applied are numerous; however, the basic premise on which these techniques are applied to environmental forensics is that micro-organisms are indicators of the contamination event. In a simple example, the presence and fate of faecal contamination in the environment can be followed by determining the number of faecal bacteria (e.g., faecal coliforms) in the environmental sample (Mudge and Ball 2006).

Micro-organisms are generally good indicators of environmental contamination as they are ubiquitous in all environments; that is not to say that all bacteria are everywhere. Particular contaminants have an associated microbial community that consists of micro-organisms capable of surviving in the presence of the contaminant. It is possible that these organisms have also developed the metabolic capacity to utilise these contaminants, offering the opportunity to remediate the contaminant. Two broad classes of micro-organisms associated with the contaminant can be described:

- Those organisms that constituted part of the contaminant. Sewage is an example of an environmental contaminant that has an associated microflora (Mudge and Ball 2006). In this instance, specific genera of micro-organisms are used to both identify and quantify the level of contamination. In this scenario, it is desirable to monitor a microbial population that is capable of long-term survival in the environment, but unable to grow in the environment.
- The monitoring of a microbial population present in the environment but that may have not been associated with the contaminant at the

source. However, when the contaminant is released into the environment, naturally occurring micro-organisms become associated with the contaminant through their utilisation of the contaminant. In this instance, the identification of the micro-organisms allows identification of the contaminant in the environment but does not quantify any changes in concentration of the contaminant as it moved through an environment. An example here would be an oil spill where naturally occurring micro-organisms capable of degrading components of the pollutant can be detected (Fahy et al. 2005, 2006).

The molecular microbiological fingerprinting techniques that can be applied to the monitoring of these two populations have considerable overlap. This chapter presents an overview of molecular microbial fingerprinting techniques together with examples of their application in environmental forensics.

## Microbial Analyses and Environmental Forensics

Traditional microbiological techniques such as selective isolation plating (Ball 1997) have been used either to follow a particular micro-organism or to follow changes in a microbial community after a contaminant event (Budowle et al. 2003) (Figure 6.1).

Bacteria are generally regarded as good indicators of environmental contamination as they are ubiquitous in nature; they are found in all environments, even in extreme conditions such as low pH, high temperature, or high salinity (Ball 2004). Bacterial communities are also able to assimilate a wide range of contaminating chemicals, such as polychlorinated biphenyls (PCBs) (Truper 1992).

Over the past decade culture-independent techniques have been increasingly used to study microbial communities (Girvan et al. 2003; Turpeinen, Kairesalo, and Häggblom 2004). These techniques generally use DNA and ribonucleic acid (RNA) structures to examine the diversity and activity of microbial communities (Girvan et al. 2004). This approach does not require any microbial isolations and the DNA or RNA extracted from the environmental sample represents the sum of the community DNA.

## The Basis of Molecular Microbial Forensic Techniques

The analysis of DNA represents the most widely used technology in molecular environmental forensics. The following section provides an overview

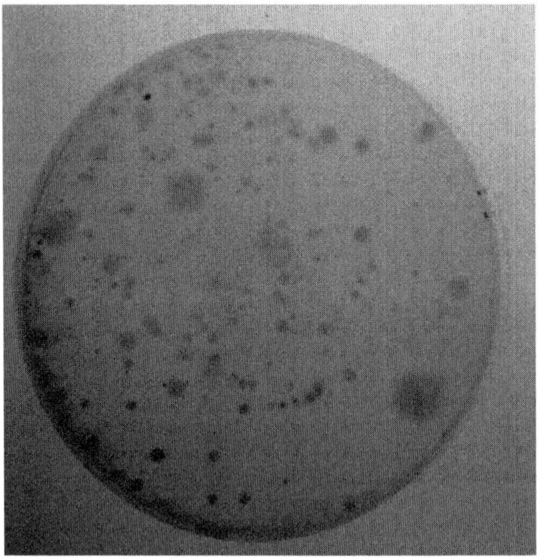

**Figure 6.1 (See colour insert following page 336.)** Selective isolation plating of an environmental sample.

of DNA and its application to environmental forensics. DNA is present in almost all known organisms. DNA stores information in genes, discreet sequences of nucleotides. DNA is a polymer consisting of a large repetition of monomeric sequences, called nucleotides. Each nucleotide consists of a deoxyribose (a five-carbon sugar), a nitrogen-containing base, and a phosphate group. Deoxyribose and phosphate components are common in all nucleotides, while the nitrogen-containing bases may be one of four types. These bases belong to two main classes: purines (adenine [A] and guanine [G]) and pyrimidines (cytosine [C] and thymine [T]) (Figure 6.2). It is the arrangement of these bases that regulates the production of specific proteins inside the cell. The information in genes is transcribed (written) into RNAs. These contain uracil (U) instead of thymine (T) [AGCU]. The 'message' in the RNA is read (translated) and proteins synthesised in ribosomes.

DNA represents the basic identity (genotype) of an organism, which in turn determines the physical features (phenotype) of an organism. The DNA

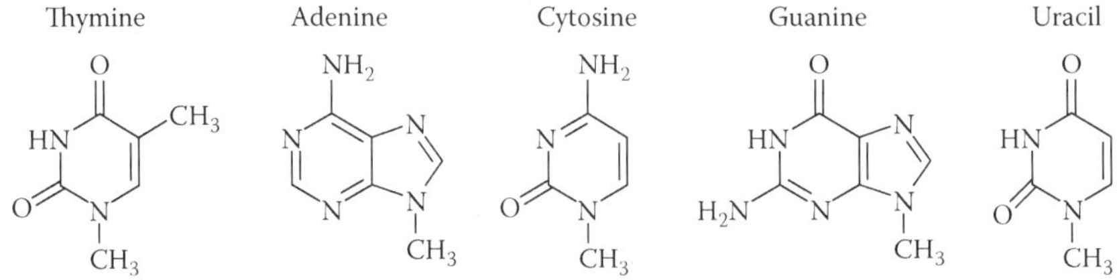


**Figure 6.2** The deoxyribonucleotides present in DNA and RNA.

profile, therefore, represents a unique fingerprint that is specific to each individual. In a perfect world, we would manipulate and compare whole genomes to determine relationships among microbes. However, based on current technology, this is not yet feasible due to time and cost constraints. Instead, comparisons between organisms are based on phylogenetic markers. This is generally a gene whose sequence is used to infer phylogenetic relationships among microbes. One major assumption with this technique is that gene phylogeny more or less reflects the evolutionary history of the microbes possessing the gene of interest (Girvan et al. 2003; Petrisor et al. 2006).

## Ribosomes

Ribosomes are large, abundant ribonucleoprotein complexes upon which protein synthesis occurs. They are found free in the cytoplasm and in eukaryotic cells, associated with the membranes of the rough endoplasmic reticulum. Ribosomes are the site of protein synthesis. Ribosomes therefore perform a vital function in all cells and are critical to cell function. Ribosomes are present in large numbers in active cells; usually, 10,000–20,000 ribosomes are present per cell. Ribosomes may occupy up to 25% of cell volume. Protein synthesis is an energetically demanding process and therefore ribosomes may also utilise up to 90% of the cell's energy.

Ribosomes may be differentiated on the basis of size into large and small subunits. All ribosomes comprise two dissimilarly sized subunits—the large and the small subunits that attach to the mRNA at the beginning of protein synthesis and detach when the polypeptide has been translated (Madigan and Martinko 2006). Each subunit consists of several ribosomal RNAs (rRNAs) and numerous ribosomal proteins (r-proteins). Their relative size is usually expressed in Svedburg units. In *Escherichia coli,* the 70S ribosome is composed of a small 30S subunit and a large 50S subunit. The large subunit comprises 34 proteins and the 23S and 5S rRNAs. The small subunit contains 21 different proteins and the 16S rRNA. In eukaryotic organisms, the ribosomes are larger (80S) and the large subunit (60S) contains 50 proteins and 28S, 5.8S, and 5S rRNAs. The smaller subunit (40S) contains 33 proteins and the 18S rRNA (Madigan and Martinko 2006). In Archaea, a prokaryotic form of life that is distinct from bacteria and form a domain in the tree of life, ribosomes resemble those of bacteria but may contain extra subunits similar to those of eukaryotic cells. While bacteria and Archaea look similar in structure, they have very different metabolic and genetic activities. One defining physiological characteristic of Archaea is their ability to live in extreme environments. They are often called extremophiles and, unlike bacteria and eukarya, depend on high salt, high or low temperature, high pressure, or high or low pH (Madigan and Martinko 2006).

The function of ribosomes is crucial to the cell; it would therefore be expected that their structural RNAs should not evolve rapidly, as any sequence change may disable the ribosome. Consequently, ribosomal gene sequences are highly conserved (i.e., they do not change much over time). It is estimated that the divergence rate for 16S rRNA is 1% per 50 million years, although this estimate may vary by an order of magnitude. The 16S ribosomal RNA therefore represents a universally conserved DNA sequence possessed by all bacteria. Also, importantly, with very few exceptions, the 16S ribosomal RNA is not horizontally transferred; that is, the 16S rRNA is rarely transferred via a process in which an organism imparts genetic material to another cell that is not its offspring. In contrast, vertical transfer occurs when an organism receives genetic material from its ancestor (e.g., its parent or a species from which it evolved).

## Ribosomal RNA and Taxonomy

A closer inspection of the 16S ribosomal RNA gene (Figure 6.3) reveals that the 16S rRNA can fold into a pattern of hairpins and loops that constitute its secondary structure. This folding pattern probably serves as a molecular signpost for allowing recognition of rRNA segments by proteins during assembly of the ribosomal subunits. The 16S rRNA sequence has hypervariable regions, where sequences have diverged over evolutionary time. These are often flanked by strongly conserved regions. The highly conserved sec-

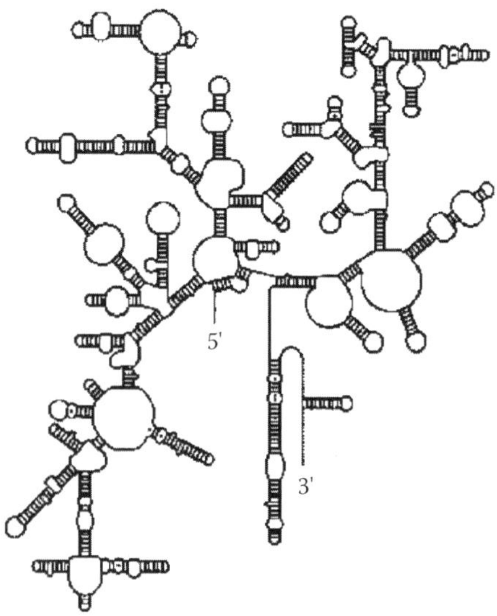


**Figure 6.3** Structure of the 16S ribosomal RNA.

ondary structure is useful for detecting polymerase chain reaction (PCR) and sequencing artefacts and errors. However, the extent of variation in sequence conservation allows the 16S rRNA to have a broad range of utility in phylogenetic analyses. This was first utilised in the late 1970s, when Carl Woese and colleagues studied the evolutionary relationships among prokaryotes through the comparison of rRNA gene sequences (Woese and Fox 1977). One of the most important findings of their work was the discovery that not all prokaryotes are related. One group of bacteria, the Archaea, possess rRNA gene sequences that were as unrelated to the eubacteria as they are to eukaryotes. Some 30 years later, we now routinely investigate phylogenetic relationships between prokaryotes by comparing nucleotide sequences (AGCT) of their 16S (small subunit) rRNA genes (Arias et al. 2005).

## Polymerase Chain Reaction (PCR)

In terms of its application to environmental forensics, PCR represents a way of finding a needle in a haystack and subsequently producing a pile of needles from the hay. For example, we may be looking for a specific 300-base pair (bp) strand of DNA amongst a sequence of 3,000,000,000 bp. The technique, developed in the 1980s, requires only small amounts of sample DNA—in this case, DNA isolated from the environment. This makes PCR highly applicable to environmental forensic investigations. To carry out PCR, primers (strings of nucleotides) are required. This string of nucleotides (usually 15–30) is complementary to the first part of the segment of DNA that is being copied. This primer attaches to the beginning of the template strand by base pairing. For any target gene, two primers are required to amplify the target; these two primers bind to conserved regions of the rRNA by flanking the target, a variable region of the 16S rRNA that is being amplified. To make primers of the correct sequence that will bind to the template DNA, it is necessary to know a little bit of the template sequence on either side of the region of DNA to be amplified. The section of target DNA bounded by the two primers is called the amplicon. The length of the amplicon is variable but ideally should be about 400–600 bp.

The PCR process requires three main steps (Figure 6.1). First, denaturation of the DNA sample at high temperature (94°C) occurs. In this stage the DNA denatures, splitting the double-stranded DNA into single stands. Second, annealing takes place at about 54°C. In the presence of selected primers, primases and polymerase enzymes are used to identify the target DNA sequence and then to produce a copy of them. Finally, extension occurs at about 72°C, where bases complementary to the template are coupled to the primer on the 3′ side by polymerases. The cycle is repeated a number of times, with each cycle resulting in a doubling in the number of amplicons. The result of the PCR of DNA extracted from an environmental sample is the

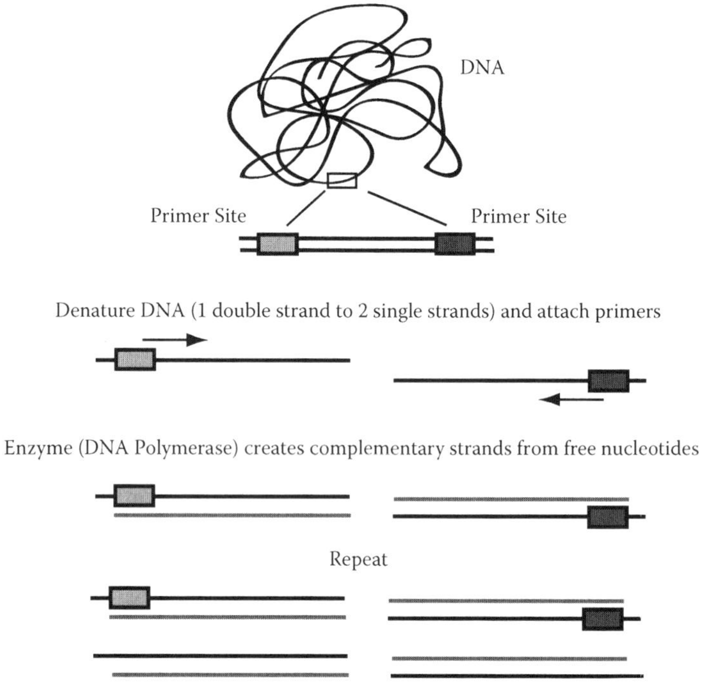


**Figure 6.4** The PCR method.

production of large numbers of copies of the original template DNA present in the sample. Theoretically, any DNA to which the primers can bind will be replicated. Therefore, at the end of the PCR reaction, many copies of the same sequence of DNA will be present, and subtle differences in the base pair composition of the 16S rRNA sequence will reflect evolutionary divergence between organisms. These differences in base pair composition can be exploited to provide a fingerprint of the community. The methods associated with DNA fingerprinting are discussed next.

## PCR-Based DNA Fingerprinting Techniques

The most common PCR-based community DNA analytical techniques with applicability to environmental forensics include:

- denaturing gradient gel electrophoresis (DGGE) and its derivatives;
- single-stranded conformation polymorphisms (SSCPs); and
- terminal restriction fragment length polymorphisms (T-RFLP).

### *Denaturing Gradient Gel Electrophoresis and Its Derivatives*

Molecular-based fingerprinting based on community DNA and RNA has recently been shown to be an effective technique for the examination of microbial communities (Girvan et al. 2003). In particular, this technique has been used to study the impact of various environmental factors, including pollutants, climate change, and agricultural practice, on the microbial community (Girvan et al. 2004). The technique is based upon the PCR amplification of short 16S rRNA gene sequence and involves the separation of amplicons of the same length (200–300 bp) but with a different base pair sequence on polyacrylamide gels containing a linear gradient of a DNA denaturant. The factor that allows for the separation of the same-size DNA fragments is the inherent thermal instability of DNA fragments caused by differences in base pair sequences. These differences, in turn, will result in differences in the melting behaviour of the DNA in a suitable linear gradient. The unzipping of the DNA molecules is achieved when it reaches its critical denaturant concentration, at which point the DNA stops moving. The increasing gradient of denaturant (20–70% in this illustration), which causes the unzipping of double stranded DNA fragments, is obtained through increasing concentrations of formamide and urea. During DGGE, fragments move down the gradient; with the exception of the GC clamp, they denature, halting the mobility of the amplicon (Figure 6.5).

The key steps in DGGE analysis (as shown in Figure 6.6) for community profiling include:

- DNA extraction from an environmental sample;
- amplification of DNA fragments by PCR using primers for the target gene (usually rRNA); amplicons are designed to include a GC clamp—a stretch of GC-rich sequences of 20–36 bp used to introduce a high melting temperature (Tm) domain to each of the target amplicons;
- incomplete denaturation of the amplicons and separation of the fragments in gels containing a linear gradient of DNA denaturant; and
- staining for visualisation of separated fragments.

Following extraction of DNA from the environmental samples, PCR amplification of the extracted DNA is carried out, usually through the targeting of the 16S rRNA gene. A range of primers targeting the conserved regions of the gene is widely available (Girvan et al. 2004). The primers used to amplify the 16S rRNA genes contain a GC clamp that prevents the complete denaturation of DNA fragments. Amplification of the 16S rRNA genes within a community and subsequent analysis by DGGE give rise to a banding pattern in which each band (made visible by the use of appropriate staining reagents) may correspond to a single species. Different stains, such as SYBR gold, SYBR green, silver staining, and ethidium bromide, can be used to stain DGGE gels; however, SYBR gold and silver staining are believed to be the best (Tuma et al. 1999). The staining of the gel also reveals a community fingerprint, which represents the complex band profile of the genetic structure of the community being investigated (Muyzer, Dewaal, and Uitterlinden

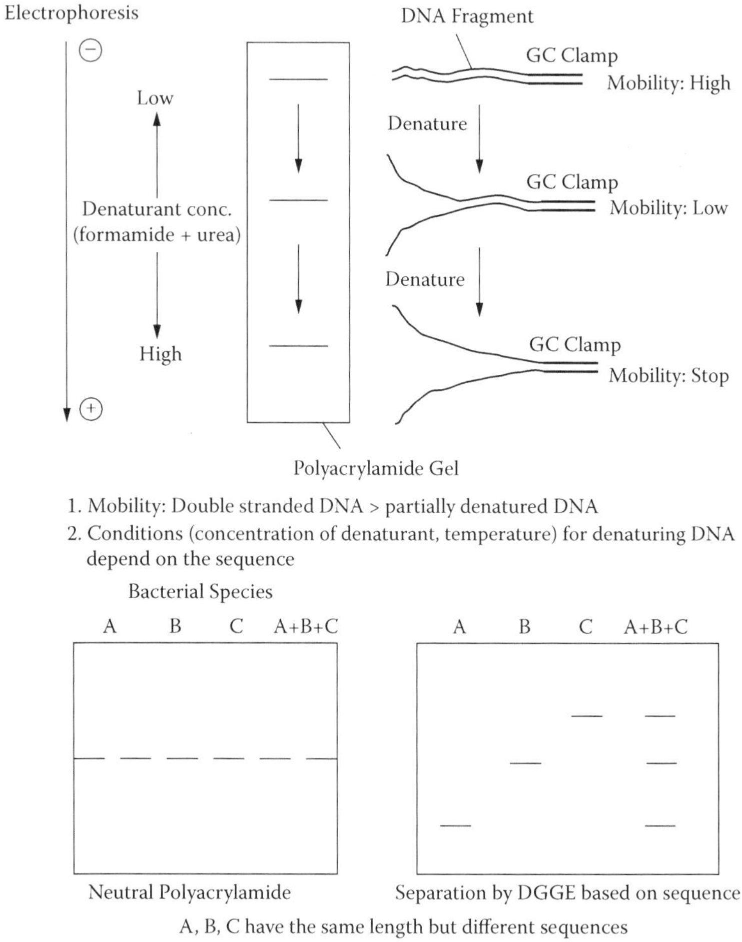


**Figure 6.5** The principle of DNA separation in DGGE. (Iwamoto, T., and M. Nasu, *Journal of Bioscience and Bioengineering*, 92: 1–8, 2001.)

1993). Following gel electrophoresis and staining, bands may be excised from the gel; following a cleanup procedure, they may be sequenced and the identity of members of the community determined.

Temperature gradient gel electrophoresis (TGGE) is a variation of DGGE. While both are based around the same principle—differentiating DNA based on the thermal properties of different sequences (of the same length)—TGGE and DGGE differ in the method used to induce the denaturation of the double-stranded DNA fragment. TGGE uses a linearly increasing temperature gradient in place of an increasing chemical denaturant gradient to achieve the separation of the double-stranded DNA amplicons.

Analysis of PCR amplified 16S rDNA gene fragments from environmental samples has been widely used in molecular microbial ecology (Girvan et

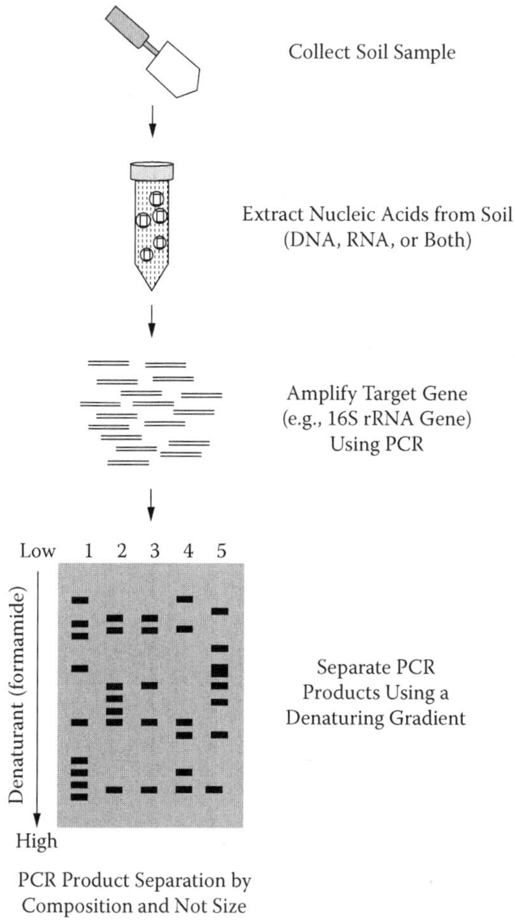


**Figure 6.6** Flow diagram of PCR-DGGE. (Nakatsu, C. H., *Soil Science Society of America Journal,* 71: 562–571, 2007.)

al. 2003), where it has been employed as a tool to investigate mainly bacterial communities. Analysis of gels (bands) is usually carried out using a variety of software such as Phoretix 1D (TL 120 TotalLab), Applied Maths (BioNumerics and GelCompar II), and Fingerprinting II Informatix (Nakatsu 2007). Similarity indices can easily be generated based on the presence or absence of bands across the gel.

Figure 6.7 shows an example of soil DNA profiles illustrating the influence of land management on the temporal changes of a soil microbial community DNA fingerprint. While similarities in the banding patterns of the soil bacterial communities can be seen, differences in banding patterns over time and with treatment can also be seen (Figure 6.7). Figure 6.7 also shows the analysis of the data using unpaired-mean group analysis (UPMGA). Each ribotype (band) was identified and its intensity measured after image capture

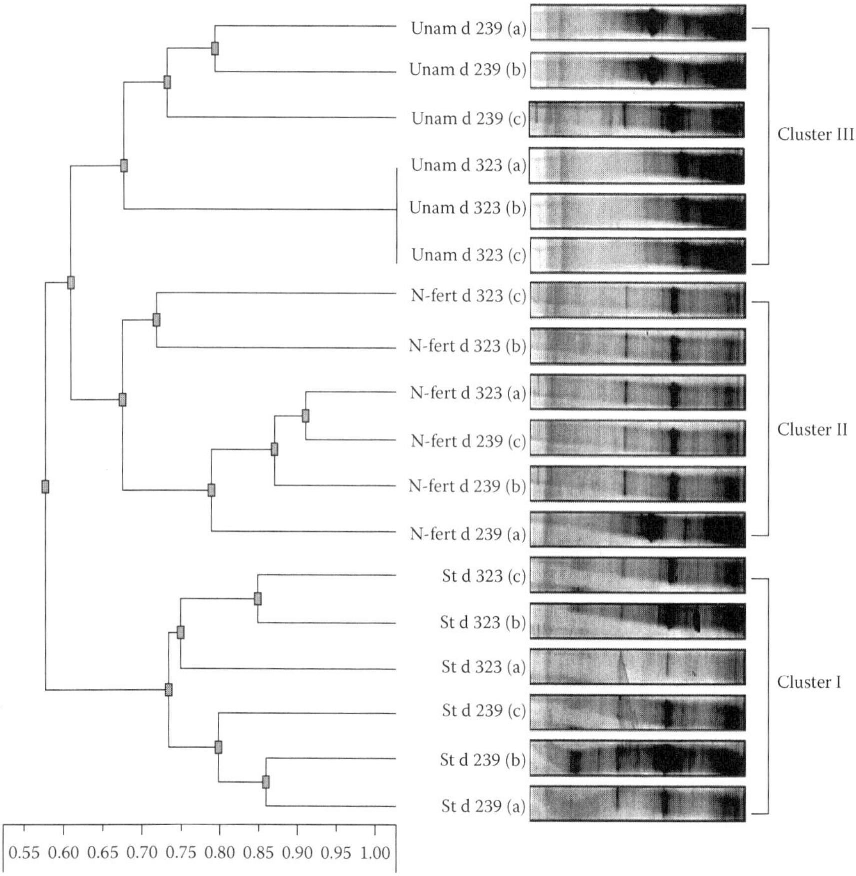


**Figure 6.7** UPGMA dendogram constructed from similarity matching data (Dice–Sorensen index) produced from DGGE profiles of 16S rDNA amplified from soil samples (collected at d 239 and d 323) amended with straw (represented as St) and N fertiliser (represented as N-fert) and unamended soil (represented as Unam). The scale bar represents similarity (as a proportion of 1.0) among the triplicate samples.

and analysis using the Phoretix ID Advanced software (Non Linear Dynamics, Newcastle, United Kingdom). The band intensity is usually a reflection of the relative abundance of a ribotype in the environment. This band intensity was then expressed as a proportion of the total intensity of all of the bands comprising a particular community profile. The software eliminates background and automatically detects peaks when noise levels and minimum peak thresholds are set; it was used as described in detail in Girvan et al. (2003). In this case the analysis shows that the land management practice, the addition of straw, and the addition of N fertiliser or unamended soil led to the clustering of the soil microbial community profile. The sampling of soil at different times only showed intracluster changes.

**Table 6.1 Measurement of Heterogeneity for the Nucleic Acid Profiles of the Soil Bacterial Communities under Different Treatment Regimes during the Study[a]**

| Julian days | Shannon index[b] of diversity derived from DGGE profiles (n = 3) | Equitability for DGGE profiles derived from DNA[c] | Total no. of bands |
|---|---|---|---|
| Straw d 239 | 3.1 ± 0.04 | 0.7 ± 0.02 | 45 |
| Straw d 323 | 3.7 ± 0.90 | 0.9 ± 0.03 | 51 |
| N fertiliser d 239 | 3.9 ± 0.01 | 0.6 ± 0.03 | 33 |
| N fertiliser d 323 | 3.1 ± 0.05 | 0.8 ± 0.03 | 37 |
| Unamended d 239 | 2.3 ± 0.04 | 0.5 ± 0.03 | 20 |
| Unamended d 323 | 2.7 ± 0.03 | 0.5 ± 0.06 | 28 |

*Note:* Treatment regimes = straw-amended soil, N-fertiliser-amended soil, and unamended soil.

[a] Data are reported as means ± SE ($n = 3$).

[b] Shannon index of diversity for DGGE profiles generated from amplified 16S rDNA from unamended N fertiliser- and straw-treated soil at d 239 and d 323.

[c] Equitability for the total number of DGGE profiles generated from amplified 16S rDNA.

Further analysis of the DGGE profile can be carried out and useful measurements, such as the Shannon index for diversity and index of equitability (Shannon and Weaver 1949), can be used:

$$H' = -\sum_{i=1}^{s} p_i \ln p_i$$

$H'$ is the value of the Shannon index, $p_i$ is the number of individuals of species (ribotypes) $i$, and $s$ is the number of species (ribotypes) found in the community profile. Relative comparison of diversity indices for DGGE data is used (Table 6.1), as these indices do not represent absolute measures of diversity that are the norm in classical macro-ecology studies (Girvan et al. 2003). The results show differences in microbial diversity. Equitability or evenness indices can also be calculated in order to deduce a relationship between intensity values between bands in a lane on a gel. Other analyses, such as multidimensional scaling and principal component analysis, can also be carried out on the gel (Nakatsu 2007).

The analysis shown in Figure 6.7 was carried out using DGGE, which is an efficient method for the detection of DNA sequence differences and a convenient tool for analysing changes in a community through analysis of only a small fragment (200–300 bp) of the 16S rRNA gene.

### *DGGE, T-RFLP, and Forensic Science*

The application of molecular fingerprinting techniques such as DGGE and T-RFLP in forensic science is a growing scientific front and, as such, is not as widely used as other DNA-based methods, such as the use of short tandem repeats (STRs) (Horswell et al. 2002). This is because these fingerprinting techniques are traditionally used for detecting mutations in genes and for studying soil microbial communities from an ecological point of view. However, growing interest in microbial forensics and the development of newer and simpler equipment and software for these fingerprinting techniques have made potential utilisation for forensic purposes easier and more feasible. In fact, when accurately used for soil biotic characterisation, molecular fingerprinting techniques can be a credible alternative to or support for other forensic techniques, such as STRs and geological fingerprinting, and provide suitable and consistent supporting evidence of a crime, which can aid a criminal conviction.

As human beings (and criminals) invariably spend a huge amount of time on land (soil), it follows that a substantial number of crimes are committed on land. Evidence derived from a crime scene (soil) can play either a major or a supporting role in securing a criminal conviction (Lerner et al. 2006). Surface soils teem with microbial life, which can be extracted, amplified, and analysed in such a way to generate a microbial profile of soil based on extracted nucleic acids. As different microbial groups abound in different soils and soil types, it is not unusual to generate different DGGE or T-RFLP profiles from different soils and soil types. These profiles may have features that are peculiar to specific ecosystems; this would potentially allow for matching of samples to different ecological sites and for forensic application such as matching of samples taken from a suspect to a crime site (Heath and Saunders 2006).

Investigations of the use of DGGE of bacterial populations in forensic science have shown that, provided adequate precautions and standardisations are carried out, DGGE can generate evidence that could tie a suspect to a crime scene or exonerate him. These standardisations would involve the use or development of simple DNA extraction methods that are fast, highly reproducible, and sensitive and can be used for a wide variety of organisms and soils (Lerner et al. 2006). Substantial research is still required in order to reduce the impacts of inhibitors found in soil, which may affect the DNA amplification process prior to DGGE or T-RFLP. Despite these challenges, Horswell et al. (2002, 2006) have carried out successful forensic comparison of soils by bacterial DNA where specific ribotypes were matched to specific ecosystems using terminal restriction fragments (TRFs) and DGGE. Theoretically, any crime committed in any of these ecosystems may be connected to a suspect, provided soil samples collected from the suspect's clothing or shoes showed the same unique profile as determined in samples from the crime scene and unrelated to the suspect's alibi scene. Apart from soil forensics application, DGGE has been used to analyse blood genotypes and successfully used to identify the victim of a fatal car accident and to connect the victim to the culprit's car tyre (Mukaida et al. 2003).

### *Single-Stranded Conformation Polymorphism Analysis (SSCP)*

Single-stranded conformation polymorphisms (SSCPs), like DGGE, distinguish DNA molecules of the same size, but with different nucleotide sequences. Separation is based on the unique three-dimensional features of single-stranded DNA, which allow small changes in the nucleotide sequence to be detected through conformational changes in the DNA. In the absence of a complementary strand, single-stranded DNA may undergo intrastrand base pairing, resulting in loops that give the DNA its unique three-dimensional structure. In turn, this unique structure imparts a specific motility through a polyacrylamide electrophoresis gel (Melcher 2004). SSCP analysis requires that DNA is extracted from an environmental sample; PCR is then conducted using one phosphorylated and one nonphosphorylated primer specific for the target gene (usually the 16S rRNA gene). Double-stranded amplicons are then converted to single strands through lambda exonucleases, which digest the phosphorylated strand. The primary steps associated with SSCP are:

- DNA extraction from an environmental sample;
- amplification of DNA fragments by PCR using a phosphorylated and a nonphosphorylated primer for the target gene;
- denaturation of the amplicons to a single-stranded form by exonuclaese digestion;
- separation of the denatured amplified fragments using polyacrylamide gel electrophoresis; and
- visualisation of the separated fragments by silver staining or autoradiography.

SSCP community patterns can be obtained from a range of environmental samples. Following visualisation, bands can be excised, reamplified by PCR, and sequenced directly. SSCP therefore represents a low-cost methodology for the analysis of the diversity of a microbial community in an environmental sample. However, single-stranded DNA mobility is dependent not only on the unique three-dimensional structure of the amplicon, but also on temperature and pH. Therefore, it is better to run gels at constant temperature and under low pH. Also, for optimal results, DNA fragment size should fall within the range of 150–300 bp (Petrisor et al. 2006).

SSCP represents a simple, low-cost technique for the analysis of the diversity of a microbial community in environmental samples, based on PCR-amplified small subunit (SSU) rRNA gene sequences from DNA extracted from environmental samples.

### *Terminal-Restriction Fragment Length Polymorphism (T-RFLP)*

TRFLP represents one of the most common DNA fingerprinting techniques in environmental forensic applications. T-RFLP is a microbial community profiling method usually based around the 16S rRNA gene. TRFLP is again based on PCR amplification of the target gene, but uses a fluorescent end-labelled primer. Following amplification, amplicons are digested using restriction endonucleases with high specificity. Fragments (different sizes) are separated by electrophoresis with the visualisation of only the terminal fragments, as they contain the fluorescent label. These amplified fragments of DNA originate from different organisms and consequently have sequence variations, ensuring that terminal restriction sites for different species in a community are unique. T-RFLP uses the differences in length from different DNA terminal fragments to differentiate between profiles of microbial communities.

The main steps of T-RFLP analysis are (Figure 6.8):

- DNA extraction from an environmental sample;
- amplification of DNA fragments by PCR using fluorescently labelled primers;
- digestion of amplicons with one or more restriction endonucleases; and
- separation and visualisation of fluorescently labelled terminal fragments.

T-RFLP patterns are routinely used to characterise the microbial communities from sites contaminated with a pollutant and comparison with the microbial community profile from uncontaminated areas of the site. Figure 6.9 illustrates how T-RFLP profiles distinguish between the microbial communities of groundwater samples contaminated with different levels of benzene (Fahy et al. 2005). The profile of the microbial community in the clean ground water sample (a) contains a number of peaks throughout the profile indicative of a complex community. However, in the presence of benzene (b and c), profiles become simpler with fewer peaks, which is indicative of a community dominated by fewer organisms, presumably as a result of the benzene contamination. The relative intensity of any single peak provides some information regarding the concentration of the particular micro-organism present in the community, with large peaks indicating the greater prevalence of that organism.

T-RFLP is a technique that is capable of rapidly analysing large amounts of information through automation, resulting in the production of large quantities of reproducible data. T-RFLP enables communities to be moni-

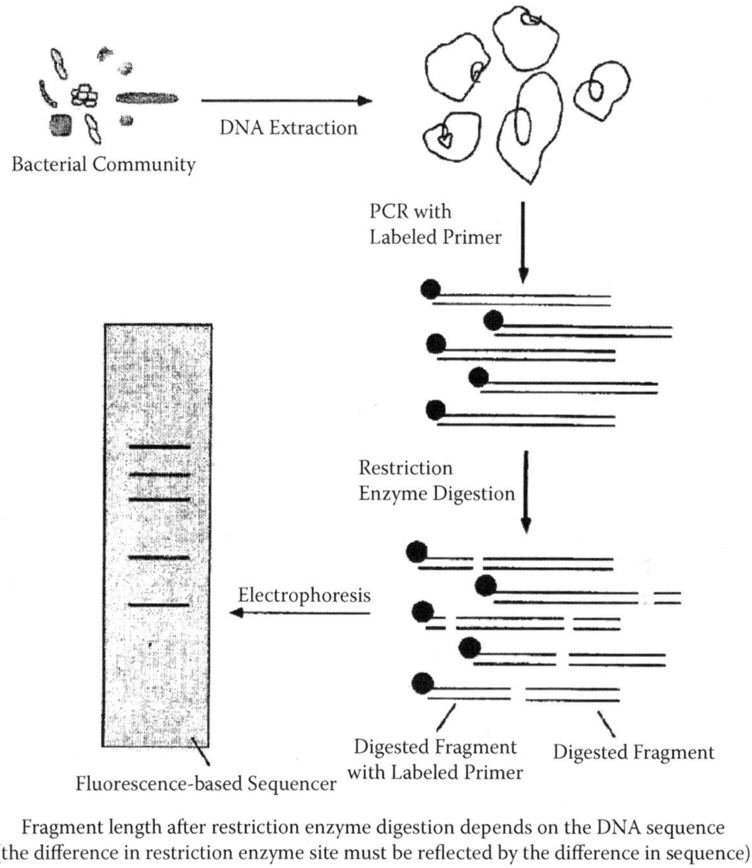


**Figure 6.8** The principle of T-RFLP. (Iwamoto, T., and M. Nasu, *Journal of Bioscience and Bioengineering*, 92: 1–8, 2001.)

tored to a high resolution, especially when T-RFLP is linked to a capillary electrophoresis sequencer, enabling increased throughput and greater reproducibility (Osborn, Moore, and Timmis 2000). One specific disadvantage of T-RFLP lies in the complexity associated with identification of organisms responsible for a particular element in a profile. This is because T-RFLPs are destructively sampled; the DNA cannot be recycled. This is in contrast to DGGE technology, which allows for either direct cloning of bands or the direct sequencing of bands excised from the gel.

## Limitations of PCR-Based Methodologies

PCR technology has limitations and, although some of these limitations are specific to the technique used (Table 6.2), there are some general caveats that can be made regarding the use of PCR:

- PCR primers are only able to amplify genes that match the primer sequence.
- PCR amplification can be uneven; some sequences are readily amplified while others are less so.
- Underamplification of sequences from abundant taxa is possible, skewing the community fingerprint.
- The results of studies based on these techniques may be regarded as an inventory of the community rather than a quantitative measurement of abundant taxa.
- The outcome illustrates the PCR-amplifiable community in the sample.

Another perceived limitation in terms of environmental forensics with PCR based on DNA is the fact that this technology does not differentiate between DNA from living or dead organisms. DNA from any organisms present in the sample may be amplified. This may be important when inves-

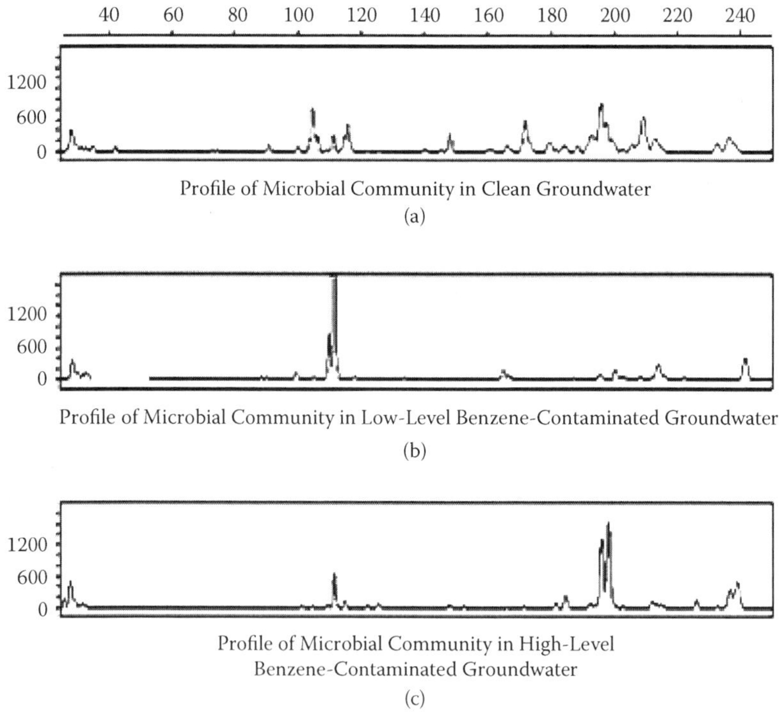


**Figure 6.9** Electropherograms showing bacterial 16S rDNA T-RFLP profiles of in situ communities from three benzene-contaminated groundwater wells containing clean groundwater (a), low levels of benzene contamination (b), or high levels of benzene contamination (c). The horizontal scale represents the T-RF length in nucleotides and the vertical scale the relative fluorescence. (Redrawn from Fahy, A. et al., *Environmental Microbiology*, 7: 1192–1199, 2005.)

**Table 6.2 Summary of the 16sRNA Community Fingerprinting Techniques Commonly Used in Environmental Forensics**

| Technique | Mode of differentiation | Advantages | Disadvantages | Ref. |
|---|---|---|---|---|
| Denaturing gradient gel electrophoresis | Melting temperature of DNA in a denaturing gradient | Bands can be excised from gel and sequenced<br>Probes can be used to hybridise to profile | Different sequences can have similar melting properties<br>Only small PCR products (2–300 bp) can be separated efficiently | Muyzer et al. 1993 |
| Temperature gradient gel electrophoresis | Melting temperature of DNA in a temperature gradient | Bands can be excised from gel and sequenced<br>Probes can be used to hybridise to profile | Different sequences can have similar melting properties<br>Only small PCR products (2–300 bp) can be separated efficiently | Heuer et al. 1997 |
| Single-stranded conformation polymorphism (SSCP) | Secondary structure of single-stranded DNA | Bands can be excised from gel and sequenced | Electrophoretic conditions are important variables<br>Only small PCR products (150–400 bp) can be used | |
| Terminal-restriction length fragment polymorphism | Terminal restriction fragment length | Automation is well established<br>Reproducible technology allowing comparisons to other data | Fragments cannot be sequenced directly | Osborn et al. 2000 |

tigating organisms capable of living in a contaminated sample—for example, when considering the application of bioremediation. In this instance, it is possible to extract RNA, rather than DNA, from an environmental sample. RNA is only present in living organisms. In terms of tracking pollution, non-discrimination between living and dead micro-organisms may be useful if the microbial community of the original material is to be traced through varying communities (e.g., faecal pollution). In contrast, examination of a contaminant plume through an environment may require the identification of living micro-organisms that are used to track the contaminant plume.

### Forensic Interpretation of Profiles

Ideally, the various DNA techniques that are used to generate community profiles should be analysed similarly. As we have seen, similarity indices such as Sorenson's similarity index can be used (Horswell et al. 2002; Blackwood et al. 2003). However, Horswell et al. suggested that such an index may not be robust enough to be used as evidence in a court of law. Clearly, high similarity indices between replicates and low similarity indices for different samples are preferred. One further factor to consider is the heterogeneity of soils even over very short distances (Prosser 1997). However, little is known about the variability in the microbial community over such distances. Felske and Akkermans (1998) showed through the use of TGGE that soil samples collected 1 m apart contained the same prominent bacteria, although these researchers did not investigate minor variations. Spatial variation is important when comparing soils for forensic purposes because, if the variability is great between sites in close proximity, then the sampling of the exact place in question is required. However, if variability in the profile of the microbial communities over short distances is low, but distinct in geographically distinct areas, then forensic analysis will be more useful (Girvan et al. 2003).

## Conclusions

The requirement for analysis of both aquatic and terrestrial samples in cases involving environmental forensics is becoming increasingly appreciated. However, because of the limitations of the currently available techniques, this analysis is rarely used as evidence. The recent developments in the application of molecular biology for the first time have provided effective tools to examine and compare the microbial community (through DNA fingerprinting) of environmental samples. These DNA profiling techniques are based around the extraction of DNA (and RNA) directly from the environmental samples. The three techniques described here (DGGE/TGGE, SSCP, and T-RFLP) profile PCR-amplified genes from microbial community (mainly bacterial) DNA through targeting of the 16S rRNA gene, resulting in the generation of a fingerprint of the microbial community. Each of the methods described has specific advantages and disadvantages, although all three are subject to the limitations associated with PCR. However, DNA fingerprinting of environmental samples offers great potential as a tool in environmental forensics.

## References

Arias, M. E., J. A. Gonzalez-Perez, F. J. Gonzalez-Vila, and A. S. Ball. (2005) Soil health—A new challenge for microbiologists and chemists. *International Microbiology,* 8: 13–21.

Ball, A. S. (1997) *Bacterial cell culture—Essential data,* 100. Chichester, England: Wiley.

Ball, A. S. (2004) Bacterial cell culture. In *Encyclopaedia of molecular cell biology and molecular medicine.* Chichester, England: Wiley.

Blackwood, C. B., T. Marsh, S. H. Kim, and E. U. Paul. (2003) Terminal restriction fragment length polymorphism data analysis for quantitative comparison of microbial communities. *Applied and Environmental Microbiology,* 69: 926–932.

Budowle, B., S. E. Schutzer, A. Einseln, L. C. Kelley, A. C. Walsh, J. A. Smith, B. L. Marrone, J. Robertson, and J. Campos. (2003) Building microbial forensics as a response to bioterrorism. *Science,* 301: 1852–1853.

Fahy, A., G. Lethbridge, R. Earle, A. S. Ball, K. N. Timmis, and T. J. McGenity. (2005) Effects of long-term benzene pollution on bacterial diversity and community structure in groundwater. *Environmental Microbiology,* 7: 1192–1199.

Fahy, A., T. J. McGenity, K. N. Timmis, and A. S. Ball. (2006) Heterogeneous aerobic benzene-degrading communities in oxygen-depleted groundwaters. *FEMS Microbiology Ecology,* 58: 260–270.

Felske, A., and A. D. Akkermans. (1998) Spatial homogeneity of abundant bacterial 16S rRNA molecules in grassland soils. *Microbial Ecology,* 36: 31–36.

Girvan, M. S., J. Bullimore, A. S. Ball, J. N. Pretty, and A. M. Osborn. (2004) Monitoring of seasonal trends in the soil microbial community of an agricultural field. *Applied and Environmental Microbiology,* 70: 2692–2701.

Girvan, M. S., J. Bullimore, J. N. Pretty, A. M. Osborn, and A. S. Ball. (2003) Soil type is the primary determinant of the composition of the total and active bacterial communities in arable soils. *Applied and Environmental Microbiology,* 69(3): 1800–1809.

Heath, L. E., and V. A. Saunders. (2006) Assessing the potential of bacterial DNA profiling for forensic soil comparisons. *Journal of Forensic Sciences,* 51: 1062–1068.

Heuer, H., M. Krsek, P. Baker, K. Smalla, and E. M. Wellington. (1997) Analysis of actinomycete communities by specific amplification of genes encoding 16S rRNA and gel-electrophoretic separation in denaturing gradients. *Applied and Environmental Microbiology,* 63: 3233–3241.

Horswell, J., S. J. Cordiner, E. W. Mass, B. W. Sutherland, T. W. Speier, B. Nogales, and A. M. Osborn. (2002) Forensic comparison of soils by bacterial community DNA profiling. *Journal of Forensic Sciences,* 47: 350–353.

Iwamoto, T., and M. Nasu. (2001) Current bioremediation practice and perspective. *Journal of Bioscience and Bioengineering,* 92: 1–8.

Lerner, A., Y. Shor, A. Vinokurov, Y. Okon, and E. Jurkevitch. (2006) Can denaturing gradient gel electrophoresis (DGGE) analysis of amplified 16s rDNA of soil bacterial populations be used in forensic investigations? *Soil Biology and Biochemistry,* 38: 1188–1192.

Madigan, M.T., and J. M. Martinko. (2006) *Brock biology of microorganisms,* 11th ed. London: Pearson Education International.

Melcher, U. SSCPs. http://opbs.okstate.edu/~melcher/MG/MGW1/MG11129.html. Accessed February 17, 2004.

Mudge, S. M., and A. S. Ball. (2006) Sewage. In *Environmental forensics—Contaminant specific guide,* ed. R. D. Morrison and B. L. Murphy, 36–53. New York: Academic Press.

Mukaida, M., Y. Takada-Matuzaki, T. Masuda, and H. Kimura. (2003) The identification of a victim using DGGE method for trace deposits collected on adhesive film. *Forensic Science International,* 132: 157–160.

Muyzer, G., E. C. Dewaal, and A. G. Uitterlinden. (1993) Profiling of complex microbial populations by denaturing gradient gel electrophoresis analysis of polymerase chain reaction amplified genes coding for 16S ribosomal RNA. *Applied and Environmental Microbiology,* 59: 695–700.

Nakatsu, C. H. (2007) Soil microbial community analysis using denaturing gradient gel electrophoresis. *Soil Science Society of America Journal,* 71: 562–571.

Osborn, A. M., E. R. Moore, and K. N. Timmis. (2000) An evaluation of terminal restriction fragment length polymorphism (T-RFLP) analysis for the study of microbial community structure and dynamics. *Environmental Microbiology,* 2: 39–50.

Petrisor, I. G., R. A. Parkinson, J. Horswell, J. M. Waters, L. A. Burgoyne, D. E. A. Catcheside, W. Dejonghe, N. Leys, K. Vanbroekhoven, P. Pattnaik, and D. Graves. (2006) Microbial forensics. In *Environmental forensics—Contamination specific guide,* ed. R. D. Morrison and B. L. Murphy, 227–257. New York: Academic Press.

Prosser, J. I. (1997) Microbial processes within the soil. In *Modern soil microbiology,* ed. J. D. van Elsas, J. T. Trevors, and E. M. Wellington. New York: Marcel Dekker.

Shannon, C. E., and W. Weaver. (1949) *The mathematical theory of communication.* Champaign: University of Illinois Press.

Truper, H. G. (1992) Prokaryotes—An overview with respect to biodiversity and environmental importance. *Biodiversity and Conservation,* 1: 227–236.

Tuma, R. S., M. P. Beaudet, X. K. Jin, L. J. Jones, C. Y. Cheung, S. Yue, and V. L. Singer. (1999) Characterization of SYBR gold nucleic acid gel stain: A dye optimized for use with 300-nm ultraviolet transilluminators. *Analytical Biochemistry,* 268: 278–288.

Turpeinen, R., T. Kairesalo, and M. M. Häggblom. (2004) Microbial community structure and activity in arsenic-, chromium-, and copper-contaminated soils. *FEMS Microbial Ecology*, 47: 39–50.

Woese, C. R., and G. E. Fox. (1977) Phylogenetic structure of the prokaryotic domain: The primary kingdoms. *Proceedings of the National Academy of Sciences USA*, 74: 5088–5090.

# 7

# Biological Communities as a Forensic Tool in Marine Environments

ANGEL BORJA
IÑIGO MUXIKA

## Contents

Introduction ........ 220
Ecological Tools for Assessing Impacts in Marine Systems ........ 221
Univariate Indices ........ 221
Multimetric Indices ........ 222
Multivariate and Modelling Approaches ........ 223
Approaches Using Indicator Species in Assessing Ecological Quality ........ 224
Examples of Detecting Environmental Impact Gradients in Marine Systems ........ 226
Case 1: Detecting Spatial and Temporal Changes in an Estuarine System ........ 227
Introduction ........ 227
Methodology ........ 227
Results ........ 229
Discussion ........ 231
Case 2: Detecting Spatial and Temporal Gradients in Relation to a Submarine Outfall ........ 233
Introduction ........ 233
Methodology ........ 233
Results ........ 234
Discussion ........ 237
Case 3: Detecting Spatial Gradients in Oilfield Exploitation ........ 238
Introduction ........ 238
Results ........ 238
Discussion ........ 240
General Discussion ........ 241
Conclusions ........ 243
References ........ 243

## Introduction

Marine environmental investigations have different goals ranging from assessing the ecological quality status of the ecosystems to investigating the response of such ecosystems to global change or any other source of disturbance; these include impacts from hazardous substances or human activities upon biological communities and ecosystems. Recently, significant efforts have been directed also towards resolving various legal disputes over environmental matters related mainly to hazardous substance releases, oil spills, etc. In these particular cases, the ability to obtain clear and defensible answers to some basic questions has been outlined by different authors (Ram 2000; Lundegard, Sweeney, and Ririe 2000). Hence, Ram highlights several issues that occur typically in environmental disputes:

When and how did the release occur?
What was the source of release?
Who contributed to the problem?
What historical industry practices and regulatory practices were in place at the time?
Did the release occur during an insurance coverage period?
How much will it cost to clean up?
If the cleanup is already completed, were the costs necessary and appropriate?
How should the costs be allocated among the responsible parties?

The response to such complex questions can be addressed by different environmental experts and technical professionals; these, in relation to environmental litigation support teams, include hydrogeologists, toxicologists, microbiologists, chemists, engineers, and safety professionals (Ram 2000). Surprisingly, no reference is made to ecologists, who are able to evaluate the biological damage produced to communities and ecosystems. Probably, such an absence is due to the necessity (on a legal basis) of (1) assessing the fingerprints of the chemicals released, (2) using biomarkers to detect the microbiological pathways in which the damage was produced, or (3) determining the ecotoxicological damage produced at species level, which requires only the previously mentioned experts. However, on the basis that benthic communities are used widely as indicators of environmental change (Bellan 1967; Pearson and Rosenberg 1978; Diaz, Solan, and Valente 2004; Hewitt, Anderson, and Thrush 2005), the question posed is: Could and should benthic communities be used as a forensic tool in marine environments? Subsequently, are ecologists able to respond to some, or all, of the previously mentioned questions in marine environmental disputes?

The response to both of the questions depends upon the tools and approaches that ecology can provide to forensic science. The first step towards understanding the environmental issues at an impacted site is to characterise the extent of the contamination; this is followed by fate and transport analyses to determine migration pathways and timing (Ram 2000; Seguel et al. 2001) and, further, the assessment of impact over marine benthic species and communities (Hewitt and Mudge 2004; Hopkins and Mudge 2004). Hence, methodologies and tools suitable for assessing the impact gradient and the environmental damage produced should be used in such an investigation. This chapter provides further information and examples on this particular issue.

## Ecological Tools for Assessing Impacts in Marine Systems

Marine benthic communities show dramatic spatial and temporal changes in species richness, diversity, relative abundance, and biomass. These variations are produced by the interactions of biotic (competence, depredation, reproduction, feeding, etc.) and abiotic (grain size, organic matter, depth, salinity and temperature changes, etc.) processes occurring at multiple spatial and temporal scales. Hence, the study of such variations following a human disturbance (e.g., an oil spill, dredged sediment dumping) can be very complicated in the absence of any previous monitoring design or a posterior adequate sampling strategy. Similarly, there is a need for powerful and appropriate analysis tools allowing natural and man-induced changes to be distinguished before and after the disturbance.

Several metrics or approaches have been developed in order to explain and reveal the impact of stressors on marine benthic communities; these can be used as a forensic tool. Following ICES (2004), these metrics can be grouped into three classes, based upon their complexity and information content: (1) univariate individual-species data or community structure measures, (2) multimetric indices combining several measures of community response to stress into a single index, and (3) multivariate methods describing the assemblages pattern, including modelling. These methodologies are described next.

### Univariate Indices

These approaches are the oldest used in marine ecology and have experienced several developments in order to improve their suitability in assessing impacts or determining gradients. Some of the most important or most used univariate indices for assessing impacts are the Shannon–Wiener Diversity Index (Shannon and Weaver 1949), the Benthic Pollution Index (BPI) (Leppäkoski 1975), the Infauna Trophic Index (ITI) (Word 1979, 1980), the

Abundance-Biomass Curves (ABC) (Warwick and Clarke 1994), the Annelid Index of Pollution (Bellan 1980), the Shannon–Wiener Evenness Proportion Index (McManus and Pauly 1990), the Taxonomic Diversity Index and Taxonomic Distinctness (Warwick and Clarke 1995), and the Ecological Evaluation Index (EEI) (Orfanidis, Panayotidis, and Stamatis 2001).

Normally, the differences between groups of sites (before and after impact) for each univariate index are analysed using one-way analysis of variance (ANOVA) or pairwise comparisons. A recent example of this application can be seen in Hewitt et al. (2005). All these methods have a simple derivation, with apparent robust responses to species changes' composition. However, its individual use leads to a loss of information, (i.e., the ecological value of a community dominated by opportunistic species may be similar to another dominated by species sensitive to disturbance).

## Multimetric Indices

Some multimetric indexing tools have been developed in recent years; as such, they incorporate some of the univariate approaches, together with the use of indicator species, in an attempt to increase the efficacy in assessing the marine benthic quality. Some of the approaches respond to legal requirements in the United States and Europe, such as the Clean Water Act (CWA), 1972 and the Water Framework Directive (WFD 2000/60/EC), respectively. Among the most extended in terms of their use are the Pollution Coefficient (CoP) (Satsmadjis 1982, 1985), the Biological Quality Index (BQI) (Jeffrey et al. 1985), the Infauna Ratio-to-Reference of Sediment Quality Triad (RTR) (Chapman, Dexter, and Long 1987), the Biotic Index (Hily 1984; Hily, Le Bris, and Glémarec 1986; Majeed 1987; Grall and Glémarec 1997), the Benthic Index of Estuarine Condition (BIEC) (Weisberg et al. 1993; Schimmel et al. 1994; Strobel et al. 1995), the Benthic Condition Index (BCI) (Engle, Summers, and Gaston 1994; Engle and Summers 1999; Paul et al. 2001), the Benthic Index of Biotic Integrity (B-IBI) (Ranasinghe et al. 1994; Weisberg and Ranasinghe 1997; Van Dolah et al. 1999; Llansó, Scott, Dauer, et al. 2002; Llansó, Scott, Hyland, et al. 2002), the AMBI (AZTI Marine Biotic Index) (Borja, Franco, and Pérez 2000; Borja, Muxika, and Franco 2003; Borja, Franco, and Muxika 2004; Muxika, Borja, and Bonne 2005), the Bentix (Simboura and Zenetos 2002), the Ecofunctional Quality Index (EQI) (Fano, Mistri, and Rossi 2003), the Indicator Species Index (Rygg 2002), and the Benthic Quality Index (Rosenberg et al. 2004). A review and comparison of 64 such indices can be found in Diaz et al. (2004).

Some of the preceding methods are based upon the paradigm of Pearson and Rosenberg (1976), in which increasing levels of organic matter and pollutants produce a disturbance gradient on the benthic communities, changing the species composition and the structural parameters. Normally, this produces an increasing dominance of opportunistic species (a species indicator of pollution or disturbance) and a decreasing dominance in sensitive species.

The growth in the number of these tools has been promoted by the need of the legislator for a reductionist approach to the environmental quality assessment. Basically, they integrate multivariate data into a single score, which can be interpreted by a nonspecialist within a 'good–bad' continuum, according to Diaz et al. (2004). These authors state that such indices have the following merits:

- Multiple benthic attributes are combined into a single measure designed to maximise the ability to distinguish between degraded versus nondegraded benthic conditions.
- They are developed with an appropriate methodology that accounts for biological variability associated with natural controlling factors, such as latitude, salinity, and sediment particle size.
- The indices permit the comparison of values that reflect the degree to which component measures of key biological attributes at one location deviate from corresponding optimum values expected under undisturbed or reference conditions.

However, other authors highlight the fact that they are not ideal for monitoring estuarine areas, which have highly variable natural conditions (Engle et al. 1994); these, in turn, are difficult to summarise into a single value.

## Multivariate and Modelling Approaches

Probably, in some cases, such approaches are the most powerful, incorporating multiple different population and community variables; these can be analysed, together with environmental variables, avoiding the loss of information. Some of the approaches are the Benthic Response Index (Smith et al. 2001), the Estuarine Trophic status (Bricker, Ferreira, and Simas 2003), the Principal Response Curves (PRCs) (Pardal et al. 2004), or M-AMBI (Borja, Franco, Valencia, et al. 2004; Muxika, Borja, and Bald 2007), also including several software packages, such as Multi-Dimensional Scaling (MDS) (Warwick and Clarke 1991), Canonical Correspondence Analysis (CANOCO) (ter Braak and Šmilauer 1998), and PRIMER (Clarke and Ainsworth 1993; Clarke and Gorley 2001).

Some of the most robust multivariate methods allow any dissimilarity measure to be used as the basis for the analysis (such as MDS) (Hewitt et al. 2005). Following these authors, such ordinations are useful in providing low-dimensional visualisations of the patterns of greatest variability in the more complex multidimensional data cloud. However, the direction of variation due to anthropogenic disturbance may not be the same as the direction of greatest variability (Hewitt et al. 2005). Hence, in an estuary in which the strongest axis of variation is provided by salinity, the human disturbance axis can be different—for example, due to a source of impact in the mouth of the estuary (see, for example, Muxika, Borja, and Franco 2003 and this chapter). For this reason, some ordination methods may not be helpful for modelling disturbances, especially when they are diffuse or chronic disturbances. On the other hand, its derivation used to be complex and the interpretation was very difficult for nonexperts—similarly, with some degree of subjectivity, even for experts.

## Approaches Using Indicator Species in Assessing Ecological Quality

Man-induced changes influence benthic communities in both a direct and indirect way through alterations to the environmental properties. Therefore, evaluation of disturbance by using indicator species could be a valid strategy. A biological indicator is described as

> an organism or a group of organisms whose biochemical, cytological, physiological or ecological response can characterise, in a practical, yet sound way, the health of an ecosystem or an ecocomplex (considered as a localised set of interdependent ecosystems with a common ecological history) highlighting, as early as possible, its alterations. (Blandin 1986)

Therefore, depending upon the situation, a bioindicator can be a community, a population, a single species, or a portion of an organism (Occhipinti-Ambrogi and Forni 2004).

The tools derived from the list described previously that include the presence of indicator species as a determination criterion can be grouped into four 'families' (ICES 2004):

1. Tools using the ecological adaptive strategies. These approaches are based upon the ecological adaptive strategies of the *r, k,* and *T* and the progressive response of biota to the gradient in stressed environments. The species should be classified into several ecological groups, based upon sensitivity to and tolerance of pollution (or disturbance). These metrics are calculated based upon the proportions between the ecological groups, with the most representative being the BPI, the Biotic Index, the AMBI, and the Bentix.

   Recently, the AMBI has experienced increasing use, both as a tool in assessing the ecological quality status within the WFD (Borja, Franco, Valencia, et al. 2004; Borja, Franco, Muxika 2004; Borja and Muxika 2005; Muxika et al. 2007) and as a tool in detecting different sources of human-induced disturbance (Borja et al. 2003; Salas et al. 2004; Muniz et al. 2005; Muxika et al. 2005). There exist some guidelines for its use in environmental studies, together with free software and a regularly updated species list (Borja and Muxika 2005).

2. Empirical relationships between abundance and environmental parameters. The most representative is the CoP, which is based upon the empirical relationships between the number of individuals and species in unpolluted macrobenthic communities, together with sediment grain size and water depth.
3. Tools including several structural and environmental variables. The most important of these approaches are the BIEC, the BCI, the B-IBI, and the M-AMBI, which can consider species diversity, total abundance, total biomass, percentage of pollution-indicative taxa density, percentage of pollution-sensitive taxa density, percentage of pollution-sensitive taxa biomass, and percentage of biomass > 5 cm below the sediment–water interface. These metrics combine structural parameters from the community with physicochemical substrate conditions.
4. Tools using diversity. These approaches are based upon the assumption that increased disturbance leads to decreased diversity; the most representative are the Indicator Species Index and the Benthic Quality Index. Hurlbert's rarefaction index is used to calculate sensitivity and tolerance.

When selecting an indicator (applicable also, in our case, to the selection of a tool for detecting impact in forensic science), Rice (2003) suggests that many different criteria should be taken into account:

meaning—should be reported and interpreted meaningfully by decision makers and stakeholders;
measurement—the ideal situation is when it can be derived from monitoring data, as it is widely applicable, inexpensive, etc.;
accuracy/precision—should reflect the actual state of the environment;
representativeness—the seasonal and geographic variation in properties should be adequately calibrated;
availability of historic data—needed to calibrate or check the indicator;
specificity—sensitivity to environmental variation, especially of human disturbances;
ability to set reference points—associated with serious or irreversible harm;
sensitivity—the best indicators have smooth, monotonic relationships of high slope, with changes to the ecosystem that they represent;
responsiveness—should respond to ecosystem change, on time scales useful in management decision making;
legal considerations—the requirement could arise from international agreements to community by-laws; and

theoretical basis—should be based upon broadly accepted ecological theories.

## Examples of Detecting Environmental Impact Gradients in Marine Systems

In recent years, the use of indices involving indicator species in assessing benthic environmental impacts has experienced increasing use on a worldwide basis (Diaz et al. 2004). This is probably due to the need to provide advice to politicians and stakeholders in a comprehensive and simple, yet scientific, way.

In Europe, one of the indices that is most extended in use throughout estuarine and coastal waters (including Atlantic, Baltic, North Sea, and Mediterranean) is AMBI (Borja et al. 2000). This particular index has been applied in environmental impact assessment and in the WFD implementation (Borja, Franco, and Muxika 2004; Muxika et al. 2005). Moreover, its use has recently been extended to South America (Muniz et al. 2005) and it is being applied in other seas worldwide.

The power of the AMBI in assessing impacts and gradients arises from the ecological models on which it is based, such as the ecological adaptive strategies of the *r*, *k*, and *T* (McArthur and Wilson 1967; Pianka 1970; Gray 1979) and the ecological succession in stressed environments (Bellan 1967; Pearson and Rosenberg 1978). These are of worldwide application under different impacts. These impacts include drill cutting discharges, submarine outfalls, harbour and dike construction, heavy metal inputs, eutrophication, engineering works, diffuse pollutant inputs, recovery in polluted systems under the impact of sewage schemes, oxygen depletion, dredging processes, mud disposal, sand extraction, oil spills, and fish farming (Borja et al. 2000, 2003; Muxika et al. 2005). AMBI is sensitive to human impacts; it reveals low changes in its value in the absence of those impacts, together with a response to seasonal variability (Borja et al. 2003; Salas et al. 2004; Reiss and Kröncke 2005).

AMBI is based upon the proportions of five ecological groups (EGs) to which the benthic species are allocated (based upon Leppäkoski 1975; Glémarec and Hily 1981; Grall and Glémarec 1997):

EG I corresponds to the disturbance-sensitive species.
EG II corresponds to the disturbance-indifferent species.
EG III corresponds to the disturbance-tolerant species.
EG IV corresponds to the second-order opportunistic species.
EG V corresponds to the first-order opportunistic species (see Borja et al. 2000).

Finally, AMBI accomplishes all the criteria mentioned by Rice (2003) in selecting indicators. Hence, for the application in forensic use, we have selected this index and three examples showing the response of AMBI to several impact gradients (see Figure 7.1). The selected case studies were (1) the detection of spatial and temporal changes in an estuarine system with different weak impact sources, (2) the impact assessment of a submarine outfall discharging urban and industrial wastes, and (3) the detection of spatial gradients in oilfields exploitation affected by mud-drilling discharges.

The corresponding AMBI values were calculated, using freely available software, on www.azti.es (AMBI 3.0 version); this includes the EG of more than 3400 taxa, updated in October 2005. The AMBI was calculated for each of the replicates and then averaged for the entire station, as recommended by Borja, Franco, and Muxika (2004) and in the guidelines of the index (Borja and Muxika 2005).

The disturbance assessment was undertaken according to the thresholds established over a scale of 0–7 for the AMBI based upon the proportions among the various ecological groups: 0 to ≤1.2 values correspond to undisturbed communities, 1.2 to ≤3.3 correspond to slightly disturbed, 3.3 to ≤5 correspond to moderately disturbed, 5 to ≤6 correspond to heavily disturbed, and 6–7 are extremely disturbed (for details of the scale, see Borja et al. 2000; Muxika et al. 2005).

## Case 1: Detecting Spatial and Temporal Changes in an Estuarine System

### *Introduction*

In this particular case, AMBI is used in an estuarine system (Plentzia, northern Spain) and its adjacent coastal area (Figure 7.1a) to determine the spatial gradient induced by several contamination sources (riverine inputs, an aquaculture farm, several urban discharges within the estuary until 1998, and a small submarine outfall of treated domestic sewage) within the coastal area (Borja et al. 2005, 2006), together with the temporal evolution in the quality status due to changes in these sources (Muxika et al. 2003).

### *Methodology*

Five sampling stations, with three biological replicates obtained at each of them, were used to analyse the temporal trends of the benthic communities: St. 1 (a coastal location, near the submarine outfall) was sampled from 1996 to 2003; St. 2 (estuarine) was sampled from 1995 to 2003; and St. 3, St. 4, and St. 6 (estuarine) were sampled from 1997 to 2003. Each station was sampled annually, in autumn and winter.

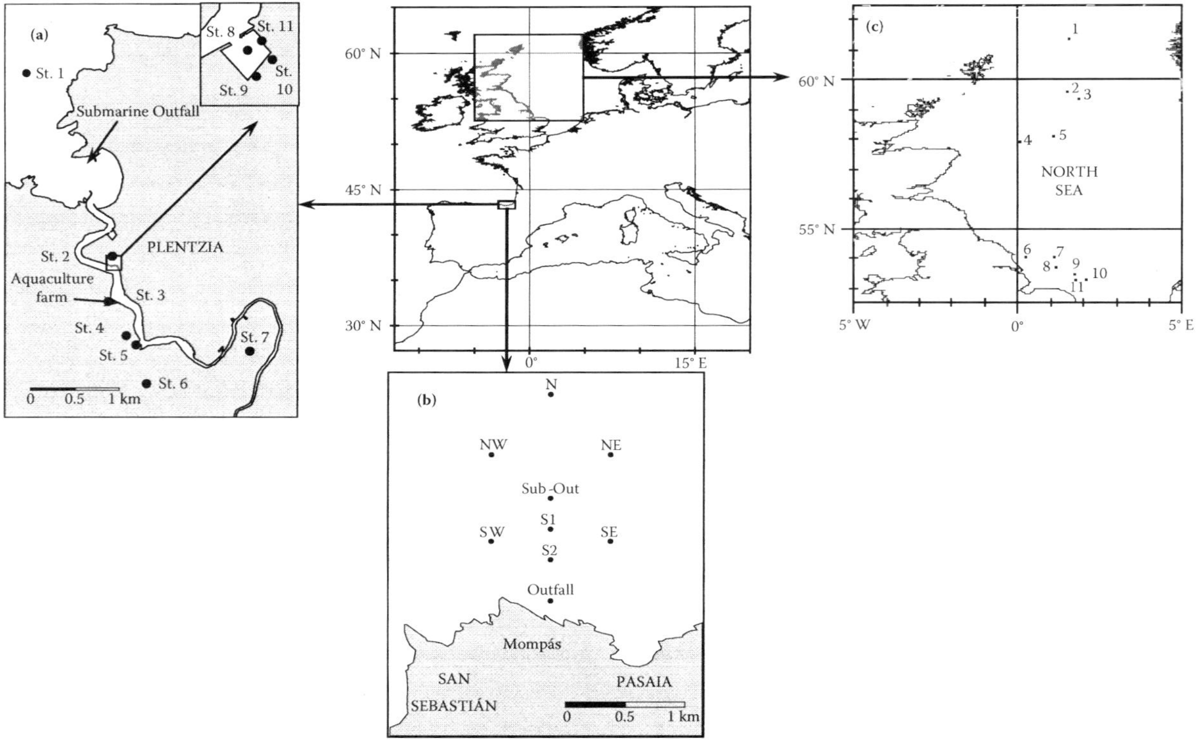


**Figure 7.1** Location of the three case studies analysed in this chapter, together with the position of each sampling station in the study areas: (a) sampling stations in Plentzia area (northern Spain); (b) sampling stations in Mompás coastal area (northern Spain); and (c) oil platforms from which data were obtained for the North Sea (1 = Thistle A; 2 = Beryl B; 3 = Beryl A; 4 = Buchan A; 5 = Miller; 6 = Cleeton 'P/Q'; 7 = Ravenspurn North 'CPP'; 8 = West Sole; 9 = Barque PB; 10 = Audrey 'A'; 11 = Clipper).

In autumn 2002 and winter 2003, 11 stations were sampled to establish the spatial gradient: St. 1 (coastal); St. 2, St. 3, St. 4, St. 5, St. 6, and St. 7 (estuarine); and St. 8, St. 9, St. 10, and St. 11 (located within the harbour) (Figure 7.1a). Differences between sampling years for each of the sampling stations, together with differences between the sampling stations in 2002, were analysed by one-way ANOVA. Post hoc comparisons were made by means of a Fisher's least significant difference (LSD) multiple range test.

### *Results*

The time evolution of the AMBI for each station is shown in Figure 7.2. Hence, St. 1 was slightly disturbed in 1997, 1999, 2000, and 2002 and undisturbed in 1996, 1998, and 2003. The results from this station show a narrow range of variability in the mean AMBI (0.7–1.7) without any clear trend and with low standard errors (0.0–0.6).

At St. 2, the mean AMBI shows high variability between years (1.0–5.2), with high standard errors in some cases (up to 1.8). The station improved in quality from 1995, when it was moderately disturbed, to 1998 (slightly disturbed throughout the years). In 1999 it was classified as heavily disturbed; however, it improved to undisturbed in 2000. Afterwards, the AMBI

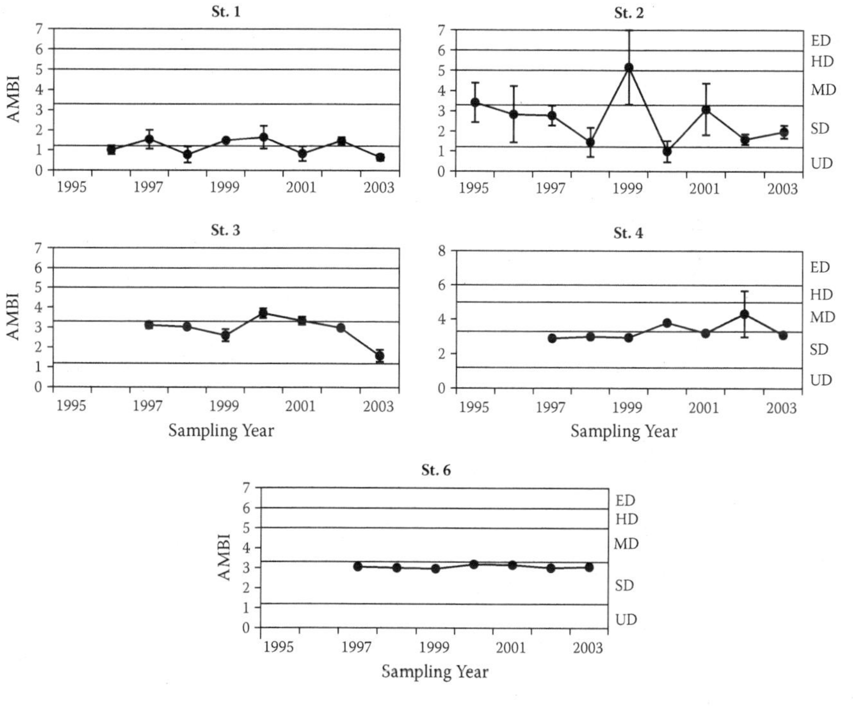


**Figure 7.2** Temporal evolution of the AMBI for selected stations, with the standard error as vertical error bars. Key: UD = undisturbed; SD = slightly disturbed; MD = moderately disturbed; HD = heavily disturbed; ED = extremely disturbed.

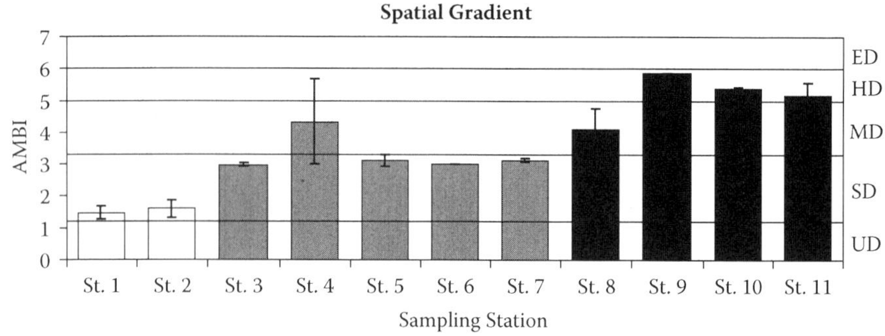


**Figure 7.3** Spatial gradient of the AMBI values in 2002, with the standard error as vertical error bars. The sampling stations close to the mouth of the estuary are shown in white; inner estuarine sampling stations are in light gray; and the sampling stations located in the harbour are shown in dark gray. Key: UD = undisturbed; SD = slightly disturbed; MD = moderately disturbed; HD = heavily disturbed; ED = extremely disturbed.

increased in 2001 and decreased again in 2002, with a similar value in 2003. The station was classified as slightly disturbed during the latter years.

At St. 3, the AMBI ranged from 1.6 in 2003 to 3.7 in 2000, but standard errors are small (0.0–0.3), indicating a high degree of similarity between replicates. The station was classified as slightly disturbed in all samplings, except in 2000 and 2001, when it was moderately disturbed. However, AMBI values were very similar along all the series (with a decreasing trend from year 2000), except for 2003 when AMBI was much lower than in previous years.

At St. 4, AMBI ranges from 2.9 to 4.3, with low standard errors (0.0–0.1), except in 2002 (1.3), coinciding with the highest AMBI value for the station. This station was classified as slightly disturbed, from 1997 to 1999. Afterwards, its situation worsened and it was classified as moderately disturbed in 2000 and 2002. In 2001 and 2003, it improved to slightly disturbed. However, the AMBI was still higher than in the first 3 years of the series.

St. 6 is the most homogeneous, both in terms of replicate and temporal variability in AMBI, which ranged from 3.0 to 3.2 and presents low standard errors (0.0–0.1). This station has been classified throughout as slightly disturbed (at the boundary with moderately disturbed).

On the other hand, the sampling stations could be divided into three groups in terms of their spatial gradient classification (Figure 7.3):

The coastal stations or stations close to the mouth of the estuary (St. 1 and St. 2) were slightly disturbed, with AMBI ≈ 1.5.

The inner estuarine stations (stations 3 to 7) were also slightly disturbed, with AMBI ≈ 3 (except St. 4, which was moderately disturbed; however, it showed a high standard error of 1.3).

The stations within the harbour (stations 8 to 11) were moderately (the station located at the mouth of the harbour) or heavily (the remaining stations) disturbed.

## *Discussion*

There are no important pollution sources within the estuary (Borja et al. 2005). The most important pollutant inputs are transported by the river from several small metallurgical and chemical industries. The main source of organic matter is agricultural farms within the basin. Further, the channel and the harbour were dredged between October 2002 and May 2004, except during summer. Overall, the Plentzia estuary cannot be considered highly stressed (Borja et al. 2006).

In spite of the preceding points, some sewage works were undertaken and two water-treatment plants were constructed. The main plant commenced working in 1998, with only primary treatment processes. It discharges almost $1500 \times 10^3$ $m^3 \cdot y^{-1}$ of treated, mainly urban, waste waters through a submarine outfall located near St. 1 (Figure 7.1a). The other plant discharges $21 \times 10^3$ $m^3 \cdot y^{-1}$ of urban waste waters through a collector located near St. 7 (Figure 7.1a), after primary treatment.

No clear trend has been detected by the AMBI in the benthic communities throughout the time series, except over several years at St. 2 and St. 3, which were more affected by the small discharges before the diversion. The absence of AMBI temporal trends in the remainder of the stations could be explained in terms of the absence of important impact sources in the surroundings, with almost all the stations being slightly disturbed. This pattern has been detected also in previous works using the AMBI (Salas et al. 2004; Reiss and Kröncke 2005). This demonstrates the sensitivity of this particular tool in the presence of an impact (or changes in the impacts, such as in St. 2 and St. 3) and the absence of response without impact (St. 1, St. 4, and St. 6).

The narrow range in variability of the AMBI at St. 1 denotes high homogeneity between replicates and temporal variation (i.e., low differences between the sampling years). In spite of the presence of a small submarine outfall located within the area, after 1998 it can be assumed that its impact on the benthic communities is negligible (note the small increase in AMBI between 1998 and 2000 in Figure 7.2).

St. 2 is located near the harbour (and near the main outfall until 1997 and 1998). Following diversion of the discharges to the submarine outfall, the quality improved considerably (see Figure 7.2). However, some dysfunctionalities in water treatment, together with occasional discharges through the old outfall, can explain the AMBI values established in 1999 and 2001. On the other hand, heterogeneity between replicates in the AMBI can be explained by an important heterogeneity in the sediment; this is probably due to the proximity of the harbour, together with changes in current speed near the estuary mouth.

St. 3 is located close to some of the collectors that are used only as stormwater runoff. This use implies that significant differences ($p < 0.05$) have been found between sampling years, depending upon differences in precipitation, with the AMBI significantly lower in 2003 than in the remaining sampling years. This sampling station shows a slight decreasing trend following the diversion of discharges within the estuary; this is in spite of a peak in 2000 that could be related to a particularly frequent use of the collectors that year. Changes in benthic fauna associated with differences in stormwater runoff inputs have been detected in estuaries elsewhere (e.g., Morrisey et al. 2003).

There were no significant differences in the AMBI over the years for St. 4. The unexpectedly high AMBI value in 2002 may be attributed to the influence of the aquaculture farm, located near the sampling station. The homogeneity observed at St. 6 indicates that there are no significant differences between sampling years and that no temporal trend can be detected; this is probably due to the absence of pollution inputs in the surrounding area.

Taking into account the spatial gradient, all the sampling stations within the estuary are classified as slightly disturbed, except St. 4, which is moderately disturbed being situated near the outfall of an aquaculture farm (Figure 7.1a). Further, there is no significant difference between it and the remainder of the inner estuarine sampling stations (Figure 7.3). Conversely, there are significant differences between the next sampling stations: (1) the outer sampling stations (St. 1 and St. 2) are near to the undisturbed limit (AMBI = 1.2) and there is no significant difference between them, while (2) the inner sampling stations (St. 3, St. 5, St. 6, and St. 7) lie closer to the moderate disturbance limit (AMBI = 3.3); likewise, there is not any significant difference between them. This difference between stations is expected, as the inner stations are more stressed by changes in salinity and pollutant inputs carried by the river and the aquaculture farm.

Within the harbour, all the sampling stations (with the exception of St. 8) are heavily disturbed due to the enclosure effect, implying (1) slow water renewal, (2) an increase in pollutant retention, and (3) increasing levels of organic matter. These effects have been detected elsewhere (Je et al. 2004; Guerra-García and García-Gómez 2005). The differences in water renewal rate explain the differences in AMBI, as follows:

St. 8 is located in the mouth of the harbour and is moderately disturbed, with a high standard error (0.7). This high value indicates an unstable environment, as tidal water movements can carry pollution from the harbour and clean water from the estuary. However, there is no significant difference with any of the other sampling stations, except the two outermost stations (St. 1 and St. 2) and the innermost harbour station (St. 9).

St. 9 is located within the innermost part of the harbour and shows less water renewal and hence higher stress.

St. 10 is very similar to St. 9; the AMBI is lower as it is located closer to the mouth of the harbour.

St. 11 is located near the mouth, presenting a lower value of AMBI and a higher standard error (0.4); this indicates also an unstable environment, but not as unstable as at St. 8. There is no significant difference between the last three sampling stations (St. 9, St. 10, and St. 11) and St. 4.

The magnitude of the AMBI standard errors could be used as a measure of the degree of stability of the system, both in terms of the physicochemical nature of the waters or on the dynamics of the substrate. However, this hypothesis needs to be investigated.

## Case 2: Detecting Spatial and Temporal Gradients in Relation to a Submarine Outfall

### *Introduction*

In the spring of 2001, as a transitory solution until complete water cleaning, within the context of the sewage scheme, the initial discharges (the old outfall, functioning since 1970) from the town of San Sebastián and the Pasaia area (both in northern Spain) were diverted into a submarine outfall. This outfall is located approximately 1.2 km from the coast, in a water depth of around 47 m, with combined discharges $45{,}727 \times 10^3$ $m^3 \cdot y^{-1}$ of untreated urban and industrial waste water (Figure 7.1b).

### *Methodology*

The benthic communities were studied: (1) 5 months before the diversion (in 2000), (2) 4 and 16 months after the diversion (in 2001 and 2002, respectively), and (3) subsequently in 2003 and 2004. The benthos was sampled with a box-corer grab at 9 sampling stations (see the sampling pattern shown in Figure 7.1b). Three replicates were obtained at each sampling site. All of the samples were sorted, identified, and counted; subsequently, species richness (number of species), Shannon's diversity (based upon abundance), and the AMBI (calculated as described earlier) were derived. Benthic communities at two separate stations (sampled before the diversion in sea bed areas in 50 (R-50) and 160 (R-160) m water depth, some 7 km apart) were used as a proxy to reference (R-) conditions over the area (data obtained from Martínez and Adarraga 2001).

Differences between sampling years (before–after impact) for each of the sampling stations, together with differences between the various sampling

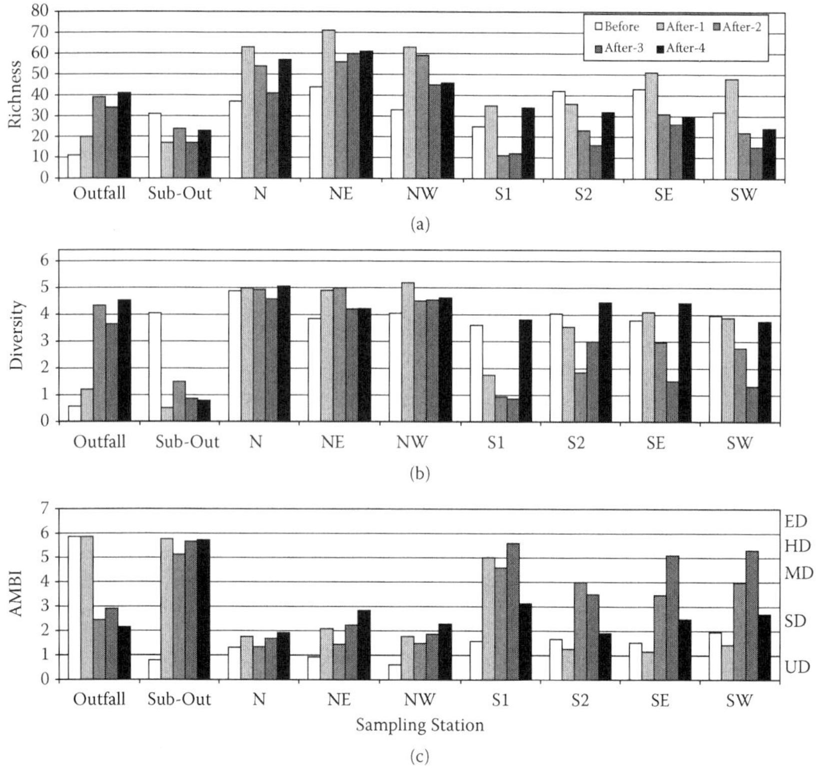


**Figure 7.4** Structural parameters measured in the San Sebastián–Pasaia area before (before: 2000) and after (after-1: 2001; after-2: 2002; after-3: 2003; and after-4: 2004) the diversion of discharges: (a) richness (number of species); (b) Shannon's diversity; and (c) AMBI. Key: UD = undisturbed; SD = slightly disturbed; MD = moderately disturbed; HD = heavily disturbed; ED = extremely disturbed.

stations, were analysed using a two-way ANOVA approach. Post hoc comparisons were made by means of an LSD multiple range test.

## *Results*

Before the discharge diversion, the highest richness (>70 species) was observed at the reference stations (R-50 and R-160); the lowest value (11 species) was reached near the old outfall and in its immediate surroundings (Figure 7.4). The old outfall area is more affected by polluted water discharged since 1970 containing metals and organic compounds. After the diversion in 2001, there is a progressive improvement in the richness values near the old outfall (approaching 40 species after 16 months). However, the new submarine outfall, together with those stations to the south, experiences some deterioration in richness following the diversion. No clear trends are observed in the stations located to the north of the submarine outfall.

Before the diversion, the highest diversities were found at the reference stations (between 5.2 and 5.65 $bit.individual^{-1}$) and the nonaffected stations (Figure 7.4). After the discharge diversion, the benthic community at the old outfall station improved significantly in its diversity, from <1 to >4 values. The area near the new submarine outfall reduced from 4 to 1 $bit.ind^{-1}$. In terms of diversity, the most affected area is that situated to the south of the new impact source point, especially at S1; in comparison, changes to the north are indistinguishable.

The same pattern was detected in the AMBI values (Figure 7.4). The reference stations presented low AMBI values (around 1.1) equivalent to an undisturbed situation (following the terminology of Borja et al. 2000 and Muxika et al. 2005) before the diversion. Those stations located in the northern area (far from the outfall source) were also undisturbed before the diversion. In contrast, the station near the outfall presented high AMBI values; it was the area that was heavily disturbed. Likewise, the immediate area (stations S1, S2, SE, and SW) was slightly disturbed. Following the diversion, there was a rapid improvement (after 16 months) in terms of AMBI at the outfall station; now, it can be considered as slightly disturbed. However, the area located near the new submarine discharge worsened rapidly, reaching a heavy disturbance. The stations located to the south of the submarine outfall presented an increasing gradient in terms of AMBI values, with some of them lying between the limit of moderate and heavily disturbed situations. In general, well-marked gradients, both spatial and temporal (Figure 7.4), can be detected by means of the AMBI.

There is an increase in the pollution load within the surroundings of the submarine outfall (Cu, following the diversion of discharges, and Hg and polyaromatic hydrocarbons [PAHs], without data before the diversion but with higher values around the submarine outfall than in the remaining stations after the diversion) (Table 7.1). Conversely, Cd, Pb, and Zn have reduced, homogeneously, the whole concentration in the area after diversion.

The most important change in the area is the progressive decrease in the redox potential values, especially around the submarine outfall and over the southern part of the area (Table 7.1). The mean values for stations Sub-Out, S1, SE, and SW together changed from oxygenated sediments before diversion (with a mean redox value of 132 mV) to reduced sediments after the diversion, with progressively decreasing values (–58, –126, –226, and –194 mV, in 2001, 2002, 2003, and 2004, respectively).

On the basis of the ANOVA analysis, there is no significant ($p > 0.05$) interaction between stations and the values before and after the discharge diversion in Cu, Pb, Zn, organic matter, and redox potential. Among these parameters, those showing only significant ($p < 0.05$) differences before and after the impact are Pb, Zn, and redox potential. Conversely, no differences between stations have been detected at a significant level ($p < 0.05$).

**Table 7.1 Mean Values of the AMBI, Metals,[a] PAH,[b] Organic Matter,[c] and Redox Potential[d] in Sediments within Different Areas**

| | AMBI | | Cadmium | | Copper | | Mercury | | Lead | | Zinc | | PAH | | OM | | Eh | |
|---|---|---|---|---|---|---|---|---|---|---|---|---|---|---|---|---|---|---|
| Area | B | A | B | A | B | A | B | A | B | A | B | A | B | A | B | A | B | A |
| Outfall | 5.86 | 2.49 | 0.88 | 0.12 | 23.4 | 29.2 | | 0.52 | 61.5 | 33.8 | 501.0 | 158.9 | | 64.8 | 3.78 | 4.49 | 35 | −128 |
| Sub-out | 0.79 | 5.51 | 0.63 | 0.09 | 20.9 | 30.1 | | 0.80 | 64.2 | 32.6 | 385.6 | 113.2 | | 2493.4 | 5.85 | 4.85 | 210 | −212 |
| South | 1.68 | 3.81 | 0.32 | 0.11 | 18.8 | 31.1 | | 0.57 | 65.9 | 33.7 | 398.4 | 165.7 | | 113.3 | 5.28 | 3.29 | 112 | −166 |
| North | 0.95 | 1.90 | 0.25 | 0.11 | 18.6 | 23.2 | | 0.25 | 66.0 | 34.6 | 388.3 | 175.0 | | 683.5 | 4.13 | 3.56 | 156 | 52 |

*Notes:* South includes stations S1, S2, SE, and SW; north includes stations N, NE, and NW before (B, year 2000) and after (A, years 2002–2004) the diversion of the discharges.

[a] $mg.kg^{-1}$.

[b] $\mu g.kg^{-1}$.

[c] OM, %.

[d] Eh, mV.

The ANOVA analysis shows also significant ($p < 0.05$) interaction between stations, together with the values before and after the discharge diversion, in AMBI and Cd. Hence, there is a negative interaction in the AMBI, decreasing in value for the old outfall station and then increasing for the remainder of the stations following the diversion of the discharge. The submarine outfall shows the highest increase in the AMBI value, followed by those stations in the southern part of the area (S1, SW, SE, and S2). This pattern can be identified also in Table 7.1. In the case of Cd, there is a general decrease in the concentration, with positive interaction between stations and the values before and after the discharge diversion; the most important is that in the old outfall location (see also Table 7.1).

### *Discussion*

It should be noted that the deterioration of the benthic communities in the area affected by the new discharge is much more rapid (less than 6 months) than the recovery of the communities in the areas positively affected by the waste elimination (more than 1 year). In the other areas, the impact of the new outfalls can be detected over time periods of 3 years (Ferraro et al. 1991). Similar patterns of response to outfall impacts, in terms of richness and diversity decrease and increase of opportunistic species, have been found worldwide (Bellan and Bourcier 1990; Solís-Weiss et al. 2004).

The pattern of impact coincides with the general water circulation pattern over the area (González et al. 2004), in which the mean transport is towards the south. In general, the effects on the benthic communities are related inversely to the distance from the outfalls, but in response to the prevalent current direction (Ros and Cardell 1991), as detected in this study. Other investigations have found impacted zones near submarine outfalls between 100 and 500 m (Anderlini and Wear 1992; Chapman et al. 1996) and >3,000 m (Bellan and Bourcier 1990), depending upon the discharge flow and the local prevailing hydrodynamic characteristics. In this particular case study in San Sebastián–Pasaia, the affected area extends over 1 km to the south.

Changes in benthic communities near submarine outfalls are a response to an increase in organic matter, together with an increase in pollutants (metals, PAH, etc.), as shown by Ferraro et al. (1991), Chapman et al. (1996), and Kress, Herut, and Galil (2004). However, very few studies have highlighted the relationship between changes in redox potential (as a response to increasing organic matter loads and oxygen consumption within the bottom water layers) and changes in benthic communities, such as mentioned here.

The new discharge has led to an overall worsening of the soft-bottom benthic communities compared to the previous situation. This is because now the discharge is injected directly to the bottom (previously it was discharged in the surface). When the biological water treatment is completed, an improvement in the benthic communities is expected.

Most of the previously mentioned studies utilised multivariate methods (MDS, cluster analysis, canonical correlations, etc.) and ABC curves, together with univariate methods, in assessing environmental impacts produced by the outfall. However, the results obtained from these methods, in terms of spatial and temporal impact or response to pollutants, are similar to those obtained by using AMBI in this particular case study.

## Case 3: Detecting Spatial Gradients in Oilfield Exploitation

### *Introduction*

The North Sea incorporates numerous oilfields, which have been exploited over many years (Figure 7.1c). Such exploitation results in the disposal of hydrocarbon-polluted drilling muds together with circumstantial oil spills and other impacts. Many authors have studied the impacts produced by these kinds of platforms on the benthic communities worldwide. For this particular case, based upon the results of Muxika et al. (2005), biological and physicochemical data were obtained from the Marine Environmental Surveys Database on the UKCS-UK Benthos, provided by the U.K. Offshore Operators Association (UKOOA). From this database, the 11 areas sampled in 1988 were selected for this chapter; 5 of them were situated in the northern North Sea (Beryl A, Beryl B, Buchan, Miller, and Thistle), and 6 in the southern/central North Sea (Audrey, Barque, Cleeton, Cilpper, Ravenspurn, and Sole). Moreover, complementary biological and chemical information was obtained from Davies et al. (1984), Shimmield et al. (2000), and Breuer et al. (1999, 2004).

### *Results*

The AMBI values show a clearly decreasing gradient, away from the stations located near the platform wells in all of the cases studied (Figure 7.5a). However, the regression between the distance (ranging from 0 to 12,000 m) and the AMBI relating to stations in the prevailing current direction shows only a weak correlation ($F = 0.389$; $p = 0.000$; $r = 0.150$). On the other hand, the regression for distances ranging between 0 and 1200 m (Figure 7.5b) is AMBI $= -0.004 \times \text{distance} + 5.354$, with the correlation being strong and highly significant ($F = 168.31$; $p = 0.000$; $r = 0.928$).

From this analysis, the following gradient pattern can be detected (Figure 7.5b, Table 7.2): (1) from 0 to 100 m, where the AMBI values lie between 4.89 and 5.97 (heavily disturbed) and the benthic community is dominated by first-order opportunistic species (EG V); (2) from 100 to 500 m, where the AMBI values lie between 3.2 and 5.82 (moderately to heavily disturbed), with the increasing dominance of EG IV and III and the presence of EG I and II; (3) from 500 to 1000 m, where the AMBI values lie between 0.84 and 3.13

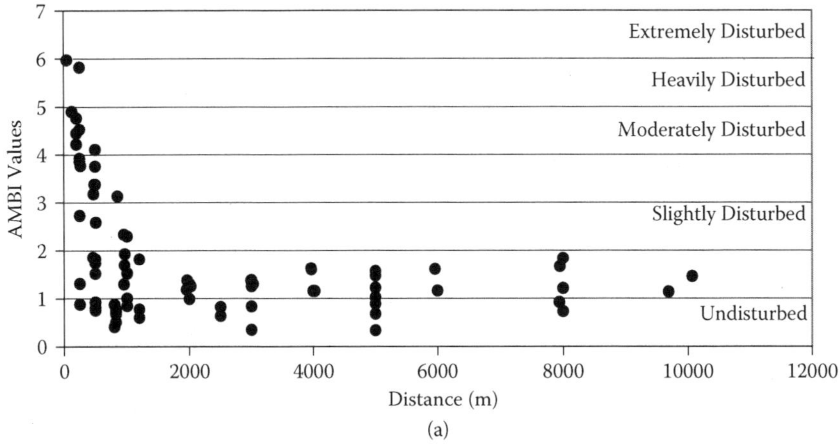


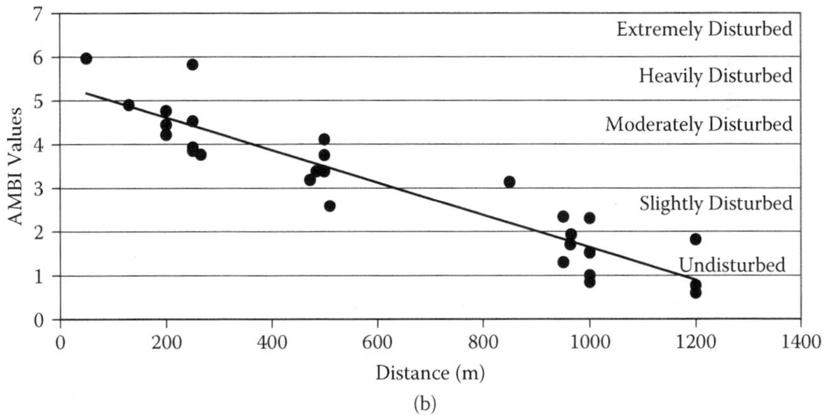


**Figure 7.5** (a) Variation of the AMBI values, with distance from oil platforms ranging from 0 to 12,000 m. (b) Regression between AMBI values and the distance from oil platforms (from 0 to 1,200 m).

(undisturbed to slightly disturbed), with low values of opportunistic species and the dominance of EG III; and (4) over 1000 m, with the AMBI values < 1.83 (undisturbed or slightly disturbed), with the EG I and II dominating. However, this gradient depends upon the regional prevailing current direction and, probably, upon the current speed.

Conversely, an excellent and highly significant correlation was found between the total hydrocarbons in the sediment and the AMBI values, following a logarithmic model ($F = 157.02$; $p = 0.000$; $r = 0.914$) (see Muxika et al. 2005). Hence, at the farthest stations, sensitive species are dominant in all cases. Approaching the oil platforms, they are progressively substituted by indifferent, tolerant, and second- and first-order opportunistic species. These changes are related to the high hydrocarbon values in the underlying

**Table 7.2 Mean and the Range of Values of the AMBI at Different Distances from Oil Platforms in the North Sea in the Prevalent Current Direction**

| Distance | AMBI | | THC (ppm) | |
|---|---|---|---|---|
| (m) | Mean | Range | Mean | Range |
| 0–100 | 5.43 | 4.89–5.97 | 15,854 | 5069–26,639 |
| 100–500 | 4.08 | 3.18–5.82 | 3,933 | 808–6440 |
| 500–1000 | 1.83 | 0.84–3.13 | 191 | 7.2–1063 |
| 1000–5000 | 1.12 | 0.33–1.82 | 82 | 9.8–316 |
| >5000 | 1.30 | 0.73–1.83 | 17 | 7.2–30.8 |

*Note:* Mean and range values of total hydrocarbons (THCs) were calculated from the previous AMBI ranges (for details, see text).

sediments (Table 7.2). Likewise, Muxika et al. (2005) have demonstrated that, in these particular locations, correlations between grain size and AMBI values, together with those between organic matter and AMBI, were only moderate ($p = 0.000$; $r < 0.50$). Clear relationships can be established between the hydrocarbon enrichment near the oil platforms and the increasing levels of the AMBI values (Table 7.2). Hydrocarbon values < 1000 ppm (with mean values < 200 ppm) correspond to undisturbed or slightly disturbed areas. Hydrocarbon values between 1,000 and 6,000 ppm (with a mean near to 4,000 ppm) correspond mainly to moderately disturbed areas. Finally, hydrocarbon values over 6,000 ppm (with mean values approaching 16,000 ppm) produce highly to extremely disturbed areas.

## *Discussion*

Oil platforms can produce several environmental impacts (Frascari et al. 1992): (1) physical impacts, such as the generation of turbulence, erosion, changes in sediment grain size, and (2) biological impacts, such as community changes and pollutant incorporation. These impacts are in response to the platform itself and to the discharge of drilling muds and cuttings. For example, the amount of diesel oil discharged (associated with drill cuttings, used in drilling operations) in 1981 into U.K. continental shelf waters was estimated to be 7,000 t (Davies et al. 1984). Drilling chemicals discharged in the same area, up until 1989, were 39,902 t $y^{-1}$ (Breuer et al. 2004).

In this particular case study, the highest AMBI values (therefore, the highest disturbance) under all circumstances are reached near the oil platforms. Clear gradients are evident in all directions, but preferential currents are indicated by the smoothest gradients (Muxika et al. 2005). The impact of the oil platforms reached up to 500–1,000 m, as detected by the AMBI (e.g., in the northern North Sea, some of the stations located some 10,000 m from the platform are still slightly disturbed).

The same pattern has been detected by Davies et al. (1984) for the same oilfields, using community structure parameters, as well as by Borja et al. (2003), using AMBI on ester-based muds in the Dutch area of the North Sea. Davies et al. (1984) detected oil concentrations of between 1,000 and 10,000 times the background levels within 250 m of the platforms; this explains the high correlation obtained in this contribution between the AMBI values and the total hydrocarbons. The pattern of the distribution of the pollutants (together with its impact on the benthic community, detected by means of the AMBI) coincides with the axis of the most persistent current, often producing an ellipsoidal distribution (Davies et al. 1984). Further, Shimmield et al. (2000) found high disturbances in cores obtained from sediments retrieved at a distance of 65 m from the drilling cutting piles. Higher depletion of interstitial dissolved oxygen concentrations was found in these cores in comparison with those obtained at 165 m and 300 m, as well as higher heavy metal concentrations in the superficial layer and higher Ba concentrations.

## General Discussion

Occhipinti-Ambrogi and Forni (2004) synthesise the main advantages in using benthic fauna in assessing the environmental quality, outlined as follows:

Benthic organisms are sedentary and therefore most likely to respond to local environmental conditions.
Benthic organisms are sensitive to different kinds of pollutants, accumulating typically in the sediments.
Many benthic species have relatively long life spans and, as such, they provide an integrated response, over time, to variations of water and sediment quality changes.
Benthos include many species characterised by different life cycles, trophic roles, and tolerance of stress.
Benthic species are important links for nutrients and material exchanges between the underlying sediment and the overlying water column.
The relationships between benthos and the main environmental variables (substratum type, oxygen, depth, etc.) are well known.

Such capacity of response to disturbance, together with its integration over time and space, could lead to benthic communities being used as a suitable forensic tool in marine environments (see Hewitt and Mudge 2004; Hopkins and Mudge 2004). One of the problems could be the reduction of such complexity into a single tool, suitable in detecting impact gradients or clear responses, in terms of typical issues in environmental disputes; these have been highlighted by Ram (2000) and have been compiled in the introduction to this chapter. Hence, some univariate variables, such as diversity, evenness, or species richness, can change minimally, despite obvious differences in impacts between sites, thus making them sometimes unsuitable as indicators of specific contamination sources (Hewitt and Mudge 2004). In other cases, these structural parameters have varied in response to changes in environmental conditions (grain size, sediment texture, etc.) rather than to a contamination gradient (Hopkins and Mudge 2004). Under such circumstances, an index such as AMBI can help in identifying these gradients (see the case studies in this chapter). The potential in assessing impacts probably is likely to increase by using these tools in a multivariate analysis, as recommended by Borja, Franco, Valencia, et al. (2004) and demonstrated by Hewitt et al. (2005).

Not all of the tools available are applicable in all cases and under all circumstances. The index selected here (AMBI) works under many different impact sources, as demonstrated in this chapter and by several publications (Borja et al. 2000, 2003; Salas et al. 2004; Muniz et al. 2005; Muxika et al. 2005). However, in some particular cases, such as physical disturbance (sand extraction, fish trawling, etc.), the AMBI is not applicable (Muxika et al. 2005; Borja and Muxika 2005).

These tools can contribute to two types of error when used within a decision-making context. For example, the tool could fail to provide information in relation to the events that have occurred in the real world or can provide false alarms about events that did not occur (Rice 2003). However, in the case of the AMBI, when the cases in which it can fail or not are known, it can be used in forensic science more accurately than other tools. Likewise, this risk can be minimised when previous data exist or very good reference points are taken after the impact to be studied, as shown in the previous case studies.

In this way, the best conditions under which to use the AMBI (or any other metric) as a forensic tool in assessing impacts is when it is possible to examine a system before and after that particular impact, as shown previously. This approach is defined as BACI (before/after control/impact) (Green 1979); it permits the quantitative definition of an impact, such as the statistical interaction between the impacted and control locations, from before to after the disturbance. This approach acknowledges the existence of natural variability among locations in the studied environment, allowing interpretation of the human disturbance separately. The use of benthic communities, within this approach, can permit the tracking of the source of an impact; this is through the real effect on the biota by monitoring the footprint of a chemical, which can be concentrated by a particular indicator species or a community. Therefore, the existence of monitoring networks or previous data permits natural trends to be discarded and focus put on the real effects of the studied impact.

## Conclusions

On the basis of the large amount of published literature, it should be considered that benthic communities can be used as a forensic tool in understanding marine environments. Hence, ecologists should be able to respond to some of the questions referred to in the introduction to this chapter in marine environmental disputes. Despite the large number of ecological tools and approaches available, not all of them can be used in forensic science. In our case, the AMBI has demonstrated a logical scientific approach to characterising the extent of contamination and assessing the impact on marine benthic species and communities. Moreover, the index is suitable in assessing the impact gradient, at spatial and temporal scales, over a broad geographical area.

## References

Anderlini, V. C., and R. G. Wear. (1992) The effect of sewage and natural seasonal disturbances on benthic macrofaunal communities in Fitzroy Bay, Wellington, New Zealand. *Marine Pollution Bulletin,* 24: 21–26.

Bellan, G. (1967) Pollution et peuplements benthiques sur substrat meuble dans la région de Marseille. 1 Partie. Le secteur de Cortiu. *Revue Internationale d'Océanographie Medicale,* VI–VII: 53–87.

______. (1980) Relationships of pollution to rocky substratum polychaetes on the French Mediterranean coast. *Marine Pollution Bulletin,* 11: 318–321.

Bellan, G., and M. Bourcier. (1990) Les enseignements d'une étude sur dix ans (1976–1986) des peuplements de substrats meubles au large d'un émissaire d'eaux usées, Marseille-Cortiu. *Cahiers de Biologie Marine,* 31: 225–249.

Blandin, M. (1986) Bioindicateurs et diagnostic des systèmes écologiques. *Bulletin d'Ecologie,* 17(4): 215–307.

Borja, A., J. Bald, M. J. Belzunce, J. Franco, I. Muxika, V. Pérez, O. Solaun, et al. (2005) *Red de seguimiento del estado ecológico de las aguas de transición y costeras de la Comunidad Autónoma del País Vasco.* Report from AZTI-Tecnalia for the 'Dirección de Aguas del Departamento de Ordenación del Territorio y Medio Ambiente' Basque Government. 14 volumes, 718 pp.

Borja, A., J. Franco, and I. Muxika. (2004) The Biotic Indices and the Water Framework Directive, the required consensus in the new benthic monitoring tools. *Marine Pollution Bulletin,* 48: 405–408.

Borja, A., J. Franco, and V. Pérez. (2000) A marine biotic index to establish the ecological quality of soft-bottom benthos within European estuarine and coastal environments. *Marine Pollution Bulletin,* 40: 1100–1114.

Borja, A., J. Franco, V. Valencia, J. Bald, I. Muxika, M. J. Belzunce, and O. Solaun. (2004) Implementation of the European Water Framework Directive from the Basque Country (northern Spain), a methodological approach. *Marine Pollution Bulletin,* 48: 209–218.

Borja, A., I. Galparsoro, O. Solaun, I. Muxika, E. M. Tello, A. Uriarte, and V. Valencia. (2006) The European Water Framework Directive and the DPSIR, a methodological approach to assess the risk of failing to achieve good ecological status. *Estuarine, Coastal and Shelf Science,* 66: 84–96.

Borja, A., and I. Muxika. (2005) Guidelines for the use of AMBI (AZTI's Marine Biotic Index) in the assessment of the benthic ecological quality. *Marine Pollution Bulletin,* 50: 787–789.

Borja, A., I. Muxika, and J. Franco. (2003) The application of a Marine Biotic Index to different impact sources affecting soft-bottom benthic communities along European coasts. *Marine Pollution Bulletin,* 46: 835–845.

Breuer, E., J. A. Howe, G. B. Shimmield, D. Cummings, and J. Carroll. (1999) *Contaminant leaching from drill cuttings piles of the northern and central North Sea: A review.* Center for Coastal & Marine Sciences and Scottish Association for Marine Science, Dunstaffnage Marine Laboratory, Scotland, United Kingdom, 49 pp.

Breuer, E., A. G. Stevenson, J. A. Howe, J. Carroll, and G. B. Shimmield. (2004) Drill cutting accumulations in the northern and central North Sea: A review of environmental interactions and chemical fate. *Marine Pollution Bulletin,* 48: 12–25.

Bricker, S. B., J. G. Ferreira, and T. Simas. (2003) An integrated methodology for assessment of estuarine trophic status. *Ecological Modeling,* 169: 39–60.

Chapman, M., D. Paine, A. D. Arthur, and L. A. Taylor. (1996) A triad study of sediment quality associated with a major, relatively untreated marine sewage discharge. *Marine Pollution Bulletin,* 32: 47–64.

Chapman, P. M., R. N. Dexter, and E. R. Long. (1987) Synoptic measures of sediment contamination, toxicity and infauna community composition (the Sediment Quality Triad) in San Francisco Bay. *Marine Ecology Progress Series,* 37: 75–96.

Clarke, K. R., and M. Ainsworth. (1993) A method of linking multivariate community structure to environmental variables. *Marine Ecology Progress Series,* 92: 205–219.

Clarke, K. R., and R. N. Gorley. (2001) *PRIMER v5, user manual/tutorial.* PRIMER-E Ltd, Plymouth Marine Laboratory, UK.

Davies, J. M., J. M. Addy, R. A. Blackman, J. R. Blanchard, J. E. Ferbrache, D. C. Moore, H. J. Somerville, A. Whitehead, and T. Wilkinson. (1984) Environmental effects of the use of oil-based drilling muds in the North Sea. *Marine Pollution Bulletin,* 15: 363–371.

Diaz, R. J., M. Solan, and R. M. Valente. (2004) A review of approaches for classifying benthic habitats and evaluating habitat quality. *Journal of Environmental Management,* 73: 165–181.

Engle, V. D., and J. K. Summers. (1999) Refinement, validation and application of a benthic condition index for Northern Gulf of Mexico estuaries. *Estuaries,* 22: 624–635.

Engle, V. D., J. K. Summers, and G. R. Gaston. (1994) A benthic index of environmental condition of Gulf of Mexico estuaries. *Estuaries,* 17: 372–384.

Fano, E. A., M. Mistri, and R. Rossi. (2003) The ecofunctional quality index (EQI), a new tool for assessing lagoonal ecosystem impairment. *Estuarine, Coastal and Shelf Science,* 56: 709–716.

Ferraro, S. P., R.C. Swartz, F. A. Cole, and D. W. Schults. (1991) Temporal changes in the benthos along a pollution gradient, discriminating the effects of natural phenomena from sewage-industrial wastewater effects. *Estuarine, Coastal and Shelf Science,* 33: 383–407.

Frascari, F., G. Rosso, G. Bortoluzzi, A. Barbanti, A. M. Bonvicini, R. Pagliai, A. Crema, et al. (1992) Environmental impact of water-based drilling muds and cuttings in a northern Adriatic Sea site. *Bulletin de l'Institut Océanographique, Monaco,* no. spécial 11, 305–324.

Glémarec, M., and C. Hily. (1981) Perturbations apportées à la macrofaune benthique de la baie de Concarneau par les effluents urbains et portuaires. *Acta Oecologica,* 2:139–150.

González, M., A. Uriarte, A. Fontán, J. Mader, and P. Gyssels. (2004) Marine dynamics. In *Oceanography and marine environment of the basque country,* ed. A. Borja and M. Collins. Elsevier Oceanography Series, 70, 133–157.

Grall, J., and M. Glémarec. (1997) Using biotic indices to estimate macrobenthic community perturbations in the bay of Brest. *Estuarine, Coastal and Shelf Science,* 44: 43–53.

Gray, J. S. (1979) Pollution-induced changes in populations. *Philosophical Transactions of the Royal Society of London Series B,* 286: 545–561.

Green, R. H. (1979) *Sampling design and statistical methods for environmental biologists.* Chichester, England: Wiley Interscience, 257 pp.

Guerra-García, J. M., and J. C. García-Gómez. (2005) Oxygen levels versus chemical pollutants. Do they have similar influence on macrofaunal assemblages? A case study in a harbour with two opposing entrances. *Environmental Pollution,* 135: 281–291.

Hewitt, E. J., and S. M. Mudge. (2004) Detecting anthropogenic stress in an ecosystem. 1. Meiofauna in a sewage gradient. *Environmental Forensics,* 5: 155–170.

Hewitt, J. E., M. J. Anderson, and S. F. Thrush. (2005) Assessing and monitoring ecological community health in marine systems. *Ecological Applications,* 15: 942–953.

Hily, C. (1984) *Variabilité de la macrofaune benthique dans les milieux hypertrophiques de la Rade de Brest.* Thèse de Doctorat d'Etat, Univ. Bretagne Occidentale. Vol. 1, 359 pp., Vol. 2, 337 pp.

Hily, C., H. Le Bris, and M. Glémarec. (1986) Impacts biologiques des émissaires urbains sur les écosystèmes benthiques. *Oceanis,* 12: 419–426.

Hopkins, F. E., and S. M. Mudge. (2004) Detecting anthropogenic stress in an ecosystem. 2. Macrofauna in a sewage gradient. *Environmental Forensics,* 5: 213–223.

ICES. (2004) *Report of the study group on ecological quality objectives for sensitive and for opportunistic benthos species,* 22–24 March, 2004 ICES, Copenhagen; ICES Advisory Committee on Ecosystems, ICES CM 2004/ACE, 01 Ref. E, WGECO, BEWG, 41 pp.

Je, J.-G., T. Belan, C. Levings, and B. J. Koo. (2004) Changes in benthic communities along a presumed pollution gradient in Vancouver Harbour. *Marine Environmental Research,* 57: 121–135.

Jeffrey. D. W., J. G. Wilson, C. R. Harris, and D. L. Tomlinson. (1985) *The application of two simple indices to Irish estuary pollution status. Estuarine management and quality assessment,* 147–165. London: Plenum Press.

Kress, N., B. Herut, and B. S. Galil. (2004) Sewage sludge impact on sediment quality and benthic assemblages off the Mediterranean coast of Israel—A long-term study. *Marine Environmental Research,* 57: 213–233.

Leppäkoski, E. (1975) Assessment of degree of pollution on the basis of macrozoobenthos in marine and brackish-water environments. *Acta Academiae Aboensis, Ser. B,* 35: 1–89.

Llansó, R. J., L. C. Scott, D. M. Dauer, J. L. Hyland, and D. E. Russell. (2002) An estuarine benthic index of biotic integrity for the mid-Atlantic region of the United States. I. Classification of assemblages and habitat definition. *Estuaries,* 25: 1219–1230.

Llansó, R. J., L. C. Scott, J. L. Hyland, D. M. Dauer, D. E. Russell, and F. W. Kutz. (2002) An estuarine benthic index of biotic integrity for the mid-Atlantic region of the United States. II. Index development. *Estuaries,* 25: 1231–1242.

Lundegard, P. D., R.E. Sweeney, and G. T. Ririe. (2000) Soil gas methane at petroleum contaminated sites, forensic determination of origin and source. *Environmental Forensics,* 1: 3–10.

Majeed, S. A. (1987) Organic matter and biotic indices on the beaches of North Brittany. *Marine Pollution Bulletin,* 18: 490–495.

Martínez, J., and I. Adarraga. (2001) Distribución batimétrica de comunidades macrobentónicas de sustrato blando en la plataforma continental de Guipúzcoa (Golfo de Vizcaya). *Boletín del Instituto Español de Oceanografía,* 17(1–2): 33–48.

McArthur, R. H., and E. O. Wilson. (1967) *The theory of island biogeography.* Princeton, NJ: Princeton University Press.

McManus, J. W., D. and Pauly. (1990) Measuring ecological stress, variations on a theme by R. M. Warwick. *Marine Biology,* 106: 305–308.

Morrisey, D. J., S. J. Turner, G. N. Mills, R. B. Williamson, and B. E. Wise. (2003) Factors affecting the distribution of benthic macrofauna in estuaries contaminated by urban runoff. *Marine Environmental Research,* 55: 113–136.

Muniz, P., N. Venturini, A. M. S. Pires-Vanin, L. R. Tommasi, and A. Borja. (2005) Testing the applicability of a Marine Biotic Index (AMBI) to assessing the ecological quality of soft-bottom benthic communities, in the South America Atlantic region. *Marine Pollution Bulletin,* 50: 624–637.

Muxika, I., Á. Borja, and J. Bald. (2007) Using historical data, expert judgment and multivariate analysis in assessing reference conditions and benthic ecological status, according to the European Water Framework Directive. *Marine Pollution Bulletin,* 55: 16–29.

Muxika, I., A. Borja, and W. Bonne. (2005) The suitability of the marine biotic index (AMBI) to new impact sources along European coasts. *Ecological Indicators,* 5(1): 19–31.

Muxika, I., A. Borja, and J. Franco. (2003) The use of a biotic index (AMBI) to identify spatial and temporal impact gradients on benthic communities in an estuarine area. *ICES CM 2003/Session J-01, Tallinn (Estonia),* 24–28 September, 2003.

Occhipinti-Ambrogi, A., and G. Forni. (2004) Biotic indices. *Biologia Marina Mediterranea,* 11(suppl. 1): 545–572.

Orfanidis, S., P. Panayotidis, and N. Stamatis. (2001) Ecological evaluation of transitional and coastal waters, a marine benthic macrophytes-based model. *Mediterranean Marine Science,* 2: 45–65.

Pardal, M. A., P. G. Cardoso, J. P. Sousa, J. C. Marques, and D. Raffaelli. (2004) Assessing environmental quality, a novel approach. *Marine Ecology Progress Series,* 267: 1–8.

Paul, J. F., K. J. Scott, D. E. Campbell, J. H. Gentile, C. S. Strobel, R. M. Valente, S. B. Weisberg, A. F. Holland, and J. A. Ranasinghe. (2001) Developing and applying a benthic index of estuarine conditions for the Virginian biogeographic province. *Ecological Indicators,* 1: 83–99.

Pearson, T., and R. Rosenberg. (1976) A comparative study of the effects on the marine environment of wastes from cellulose industries in Scotland and Sweden. *Ambio,* 5: 77–79.

______. (1978) Macrobenthic succession in relation to organic enrichment and pollution of the marine environment. *Oceanography and Marine Biology Annual Review,* 16: 229–311.

Pianka, E. R. (1970) On r- and K-selection. *American Naturalist,* 104(940): 592–597.

Ram, N. (2000) The tools of environmental litigation support, how environmental litigation support teams employ a unique set of skills and tools to improve the outcome and reduce the cost of legal disputes. *Environmental Forensics,* 1: 25–30.

Ranasinghe, J. A., S. B. Weisberg, D. M. Dauer, L. C. Schaffner, R. J. Diaz, and J. B. Frithsen. (1994) *Chesapeake Bay Benthic Community Restoration Goals. CBP/TRS 107/94.* Annapolis, MD: Chesapeake Bay Program Office, USEPA, 49 pp.

Reiss, H., and I. Kröncke. (2005) Seasonal variability of benthic indices. An approach to test the applicability of different indices for ecosystem quality assessment. *Marine Pollution Bulletin,* 50: 1490–1499.

Rice, J. (2003) Environmental health indicators. *Ocean & Coastal Management,* 46: 235–259.

Ros, J. D., and M. J. Cardell. (1991) Effect on benthic communities of a major input of organic matter and other pollutants (coast off Barcelona, western Mediterranean). *Toxicological and Environmental Chemistry,* 31-32: 441–450.

Rosenberg, R., M. Blomqvist, H. C. Nilsson, H. Cederwall, and A. Dimming. (2004) Marine quality assessment by use of benthic species-abundance distributions, a proposed new protocol within the European Union Water Framework Directive. *Marine Pollution Bulletin,* 49: 728–739.

Rygg, B. (2002) *Indicator species index for assessing benthic ecological quality in marine waters of Norway.* Norwegian Institute for Water Research, report no. 40114, 1–32.

Salas, F., J. M. Neto, A. Borja, and J. C. Marques. (2004) Evaluation of the applicability of a marine biotic index to characterise the status of estuarine ecosystems, the case of Mondego estuary (Portugal). *Ecological Indicators,* 4: 215–225.

Satsmadjis, J. (1982) Analysis of benthic data and measurement of pollution. *Revue Internationale d'Océanographie Medicale,* 66–67: 103–107.

______ (1985) Comparison of indicators of pollution in the Mediterranean. *Marine Pollution Bulletin,* 16: 395–400.

Schimmel, S. C., B. D. Melzian, D. E. Campbell, S. J. Benyi, J. S. Rosen, and H. W. Buffum. (1994) *Statistical Summary, EMAP—Estuaries Virginian Province, 1991.* Narragansett, RI: Office of Research and Development, Environmental Research Laboratory, USEPA, 77 pp.

Seguel, C. G., S. M. Mudge, C. Salgado, and M. Toledo. (2001) Tracing sewage in the marine environment, altered signatures in Concepción Bay, Chile. *Water Research,* 32: 4166–4174.

Shannon, C. E., and W. Weaver. (1949) *The mathematical theory of communication.* Urbana: the University of Illinois Press, 115 pp.

Shimmield, G. B., E. Breuer, D. G. Cummings, O. Peppe, and T. Shimmield. (2000) Contaminant leaching from drill cuttings piles of the northern and central North Sea, field results from the Beryl 'A' cuttings pile. Scottish Association for Marine Science, 28 pp.

Simboura, N., and A. Zenetos. (2002) Benthic indicators to use in ecological quality classification of Mediterranean soft bottom marine ecosystems, including a new biotic index. *Mediterranean Marine Science,* 3: 77–111.

Smith, R. W., M. Bergen, S. B. Weisberg, D. Cadien, A. Dalkey, D. Montagne, J. K. Stull, and R. G. Velarde. (2001) Benthic response index for assessing infaunal communities on the southern California mainland shelf. *Ecological Applications,* 11: 1073–1087.

Solís-Weiss, V., F. Aleffi, N. Bettoso, P. Rossin, G. Orel, and S. Fonda Umani. (2004) Effects of industrial and urban pollution on the benthic macrofauna in the Bay of Muggia (industrial port of Trieste, Italy). *The Science of the Total Environment,* 328: 247–263.

Strobel, C. J., H. W. Buffum, S. J. Benyi, E. A. Petrocelli, D. R. Reifsteck, and D. J. Keith. (1995) *Statistical Summary. EMAP—Estuaries Virginian Province—1990 to 1993.* Narragansett, RI: National Health Environmental Effects Research Laboratory, Atlantic Ecology Division, USEPA, 72 pp.

ter Braak, C. J. F., and P. Šmilauer. (1998) *CANOCO reference manual and user's guide to Canoco for Windows. Software for canonical community ordination (version 4).* Center for Biometry Wageningen (Wageningen, NL) and Microcomputer Power (Ithaca, NY), 352 pp.

Van Dolah, R. F., J. L. Hyland, A. F. Holland, J. S. Rosen, and T. R. Snoots. (1999) A benthic index of biological integrity for assessing habitat quality in estuaries of the southeastern USA. *Marine Environmental Research,* 48: 269–283.

Warwick, R. M., and K. R. Clarke. (1991) A comparison of some methods for analysing changes in benthic community structure. *Journal of the Marine Biological Association of the UK,* 71: 225–244.

______. (1994) Relating the ABC, taxonomic changes and abundance/biomass relationship in disturbed benthic communities. *Marine Biology,* 118(4): 739–744.

______. (1995) New 'biodiversity' measures reveal a decrease in taxonomic distinctness with increasing stress. *Marine Ecology Progress Series,* 129: 301–305.

Weisberg, S. B., J. B. Frithsen, A. F. Holland, J. F. Paul, K. J. Scott, J. K. Summers, H. T. Wilson, D. G. Heimbuch, J. Gerritsen, S. C. Schimmel, and R. W. Latimer. (1993) *Virginian Province Demonstration Project Report, EMAP-Estuaries, 1990.* EPA/620/R-93/006. Washington, D.C.: Office of Research and Development, USEPA.

Weisberg, S. B., and J. A. Ranasinghe. (1997) An estuarine benthic index of biotic integrity (B-BY) for Chesapeake Bay. *Estuaries,* 20: 149–158.

Word, J. Q. (1979) The Infaunal Trophic Index. *6th California Coastal Water Research Project Annual Report*, El Segundo, California, 19–39.

______. (1980) Classification of benthic invertebrates into Infaunal Trophic Index feeding groups. In *Coastal Water Research Project Biennial Report 1979–1980.* SCCWRP, Long Beach, California, pp. 103–121.

# Normalisation Techniques in the Forensic Assessment of Contaminated Environments

# 8

GAVIN F. BIRCH
ANDREW T. RUSSELL
STEPHEN M. MUDGE

## Contents

Introduction ........ 251
Commonly Used Normalisation Techniques ........ 254
Physical Segregation ........ 254
Granulometric Normalisation ........ 255
Elemental Normalisation ........ 256
Enrichment Factors ........ 259
Choice of Normalising Elements ........ 260
Postextraction Normalisation (PEN) Technique ........ 263
Application of Normalisation Techniques ........ 265
Marine Sediments from the Continental Shelf Off New South Wales Shore ........ 265
Estuarine Sediments from Port Jackson ........ 267
Fluvial Sediments from the Catchment ........ 269
Conclusions ........ 271
References ........ 272

## Introduction

Sediments are frequently used in forensic science for identifying contaminant sources, determining dispersion pathways, and locating areas of pollutant deposition and accumulation (Matthai and Birch 2000c). Sediments are also used to assess the magnitude of anthropogenic impacts, determine the timing of onset of contamination, and provide a history of adverse environmental change (Birch and Taylor 2002). Sediments provide this type of information by faithfully recording and time-integrating the environmental status of aquatic systems (i.e., sediments have a 'memory'). This attribute is a major advantage sediments have over other media (e.g., hydrologic, atmospheric, and biotic environments, which are spatially and temporally more variable). Being less dynamic, sedimentary systems require smaller sample sizes for meaningful interpretation (Bubb, Rudd, and Lester 1990; Birch and Taylor 2000a, 2000b) and sediment does not require such high levels of expertise to collect and analyse (compared to the water column), thus further reducing costs. Sediments are, therefore, economically attractive as a forensic tool and are being increasingly employed in environmental assessment of aquatic systems.

Historically, water and biota have been used as preferred forensic tools in environmental assessment due to the lack of uniformity in analysis and difficulties in interpretation of sedimentary data. With more advanced understanding of partitioning, speciation, and mobilisation of contaminants in sedimentary systems, interpretation has become clearer; with the establishment of accepted analytical protocols, these issues have been largely overcome. More problematic, however, is the difficulty in interpreting sediment-derived contaminant data due to the confounding effects of variable grain size. Grain size (or, more correctly, grain surface area) is the dominant parameter controlling contaminant concentrations in sedimentary systems, and variability in grain size imposes considerable spatial and temporal variance on sediment-derived chemical data (Forstner and Wittmann 1979; Loring 1991). Size partitioning of chemicals in sediments is due to an affinity of contaminants for the fine fraction (<4 μm) and is related to the exponential increase in surface area with decreasing particle size and increased surface charge of these materials (Forstner 1982; Sakai, Kojima, and Saito 1986; Bubb et al. 1990). A normalising procedure for reducing the confounding caused by variable sediment size is essential for interpretation of sediment contaminant data.

Intensive studies have been conducted to determine the environmental impact of contaminants discharged in estuarine environments. These studies often try to establish if concentrations of contaminants are increasing or decreasing over time. It has been long established that contaminants are intrinsically controlled by the physical and chemical parameters of the sediments with which they are associated. Therefore, it is important to remove the effect of these parameters on the trends in the data; this is generically known as normalisation.

The sediment composition in an estuary is dependent on the nature of the source materials (i.e., the catchment rock types, the terrestrial and marine material input, and the anthropogenic contaminants). As a result of changes in erosion and particle transport largely controlled by the energy in the system, concentrations in the sediments may vary over time. Therefore, it is important to establish if changing sediment characteristics are influencing the trend in the anthropogenic contamination of the sediments. If the concentration of contaminant is varying with time (i.e., seasonally), it is important to standardise the results so as to remove the variability.

There are several factors controlling the variability in sediments:

- grain size (change in erosion and deposition patterns—energy in the system);
- lithology (catchment erosion and coastal processes);
- biological activity (e.g., bioturbation mixing old with new sediments deposited under different energy regimes);
- Other physicochemical parameters (e.g., pH, Eh, % TOC, which may affect speciation and solubility); and
- contaminant input.

It is important to understand the interaction and behaviour of the contaminants with the environmental processes so that the natural variation can be removed and the anthropogenic variation can be determined. Although some studies have been able to establish trends in concentrations over a large time scale, there may still be problems associated with these studies. For example, Harland, Taylor, and Wither (2000) showed that between 1974 and 1998 the concentration of anthropogenic mercury and other trace metals in the River Mersey, United Kingdom, had decreased. For mercury, the normalised data indicated a decrease of >70% during this period. However, when the data were studied from year to year, there was a deviation from this trend. During the early 1990s the concentrations in sediments of the River Mersey appeared to be rising and, for three successive years, the concentration increased to levels not seen for a decade. The first impression was that increased discharges from industrial sources had occurred. However, after detailed review of the estuary using a geographical information system (GIS), the authors suggested the increase was due to a change in the estuarine dynamics and erosion processes leading to the reworking of historical deposits. Therefore, the normalisation method used—mercury to grain size ratio—failed to account for the reworking of historical contamination. It is difficult to interpret such problems until further data have been collected.

Correlation between some contaminant concentrations and physical properties of the sediment has been established (Loring 1990; Bothner, Buchholtz ten Brink, and Manheim 1998; Ackermann 1980). This correlation is due to the chemical interactions between the appropriate bonding sites. Sediments from most natural waters consist of a variety of components, including clay minerals, carbonates, quartz, feldspar, and organic matter. These fractions are usually 'coated' with hydrous manganese and iron oxides with nonspecific organic substances (humic materials). Together these materials provide active surfaces that can scavenge metals and organic compounds from the water column and accumulate them on the surface of sediments.

Therefore, a degree of the variability seen in sediments may be due to changes in the sediment lithology and environmental history.

Tessier, Campbell, and Bisson (1979) identified the importance of different sediment coatings on the partitioning of contaminants in particles. The environmental history will have a key role in the accumulation of these active coatings and therefore will affect the number of binding sites for different contaminants. Typical sequential leaching schemes have been based on the early work by Tessier and generate fractions divided into classes according to the reagents used. These reagents become more chemically stringent through the process, and the class fractions are entitled exchangeable, bound to carbonate, bound to iron/manganese oxides, bound to organic matter and residual. Metals bind to each of these phases depending upon their availability and the elemental chemistry.

The trace elements that reach the marine environment can be retained by the sediments through the following processes:

- fixation on suspended matter with subsequent deposition;
- direct precipitation of colloidal forms;
- direct fixation by adsorption; and
- deposition of organic matter that previously had incorporated the trace element.

It can be deduced that the interactions of the trace elements with the suspended matter, which also depend on the chemical state of the element, play an important role in fixation to sediments. Some investigations have shown a dependence of trace elements, including radionuclides, on grain surface area, magnetic susceptibility, lithology, and/or organic content. Furthermore, metal concentrations are influenced by other factors dependent on the medium, such as pH or redox potential and the physical or chemical state of the trace element (Ligero et al. 2001).

In this chapter, normalising methodologies are reviewed and case studies from the fluvial, estuarine, and marine environments are discussed to demonstrate the improved interpretative capability provided by these techniques.

## Commonly Used Normalisation Techniques

### Physical Segregation

The most common form of normalisation of sedimentary contaminant data is achieved by wet sieving at 62.5 μm using MilliQ water (Forstner and Wittmann 1979; Klamer, Hegeman, and Smedes 1990; Martincic, Kwokal, and Branica 1990; Birch 1996; Birch and Taylor 2000b). Analysing only the <62.5

μm fraction of the sediment reduces the diluent effect of the variable, contaminant-poor coarse component, thereby improving the compatibility of the data. Advantages of using the <62.5 μm fraction are that it produces adequate material for analysis and is rapid and therefore more economical than using smaller sediment sizes. Because calcareous material has minor adsorptive properties, contaminant data for sediment comprising shelly material are frequently expressed on a calcium carbonate-free basis to further reduce confounding effects introduced by a variable calcareous content (Jickells and Knap 1984; Leoni and Sartori 1991).

Size fractionation greatly assists in identification of contaminant sources using geochemical fingerprinting and in determination of dispersion pathways by tracing concentration gradients. Normalisation is essential when conducting forensic work in widely different environments (e.g., marine, estuarine, and especially fluvial environments where sediments of different types and textures are being considered; Matthai and Birch 2001). Variable ambient energy in a system or location will result in a range of sediment size distributions, necessitating a normalisation procedure to reduce confounding in the chemical data. Size normalisation is essential in determining preanthropogenic background metal concentrations and is important in temporal studies, which require comparable data (Birch and Taylor 2004). Disadvantages of the physical separation technique are that the data do not represent total sediment concentrations and therefore cannot be used in sediment quality assessment and also that the process is time consuming.

## Granulometric Normalisation

A widely used method of normalisation is the granulometric method. This is a procedure that involves the mathematical normalisation of the contaminant concentration data to a weight percent of various size fractions removed from the sediments, such as <62.5 μm (Ackermann, Bergmann, and Schleichert 1983), <20 μm (Ackermann 1980; Ackermann et al. 1983), or <16 μm (De Groot 1964).

Taylor (1986) and Harland et al. (2000) have proposed granulometric normalisation techniques to compare data from yearly sampling programs. This involves determining correlation and the regression equations for each individual year; from this, a concentration can be calculated for a given percentage grain size (e.g., 40% < 62.5 μm). Using this actual concentration value, an overall trend can be determined. This has disadvantages in that the method requires many samples from a large range of grain sizes to determine an accurate regression equation and thus additional expense.

Boust (1999) states that the abundance of radionuclides in sediments depends, at least partially, upon grain size—that is, the efficiency of the surface-controlled nonspecific sorption processes and clay abundances—as

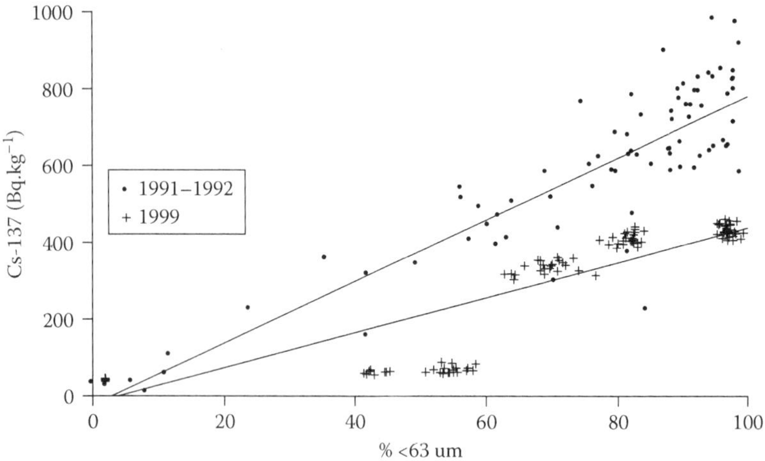


**Figure 8.1** The activity of $^{137}$Cs in sediments of the Ribble Estuary in relation to the percentage of grains < 63 µm.

these minerals have much higher concentrations of most naturally occurring radionuclides than any other lithological phase present. Using data from Ribble Estuary collected in 1991 and 1992 (Mudge, Assinder, and Bourne 1994), data from a microscale variation project collected in 1999 and 2000 (Mudge, Assinder, and Russel 2002), and the method described by Harland et al. (2000), the $^{137}$Cs data were plotted against percent < 62.5 µm (Figure 8.1). Linear regression was performed on these data sets. These data were collected from many sediment types in the estuary, exhibiting a range of grain sizes; in total, each data set comprised about 125 samples.

The correlation for both sets is highly significant with $r^2 = 0.78$ and $r^2 = 0.80$ for 1991/1992 and 1999, respectively. Using the linear regression equation, the concentration of $^{137}$Cs was calculated for a given percentage <62.5 µm (e.g., 40%). The activity in 1991/1992 was calculated to be 298.8 Bq kg$^{-1}$ and in 1999 equalled 165.4 Bq kg$^{-1}$, a decrease of approximately 45%. During this time, the $^{137}$Cs discharge from Sellafield had decreased by approximately 50% (BNFL 1993, 2000). In this case, this normalisation procedure adequately reflects the change in contaminant availability between these two sampling times.

## Elemental Normalisation

The elemental normalisation technique involves the use of 'conservative' trace elements, which are nonanthropogenic and act as a proxy for the fine fraction in the sediment, particularly clay minerals. Total metal data are normalised to (divided by) the total concentration of a conservative element, which reduces variance due to an irregular, metal-poor coarse component. Many elements (Co, Li, Sc, Cs, Rb) have been used for normalising, but the most common are Al (Hirst 1962; Windom et al. 1989; Daskalakis and O'Connor 1995; Matthai and Birch 2001; Bertine and Goldberg 1977; Bruland et al. 1974; Carral et al. 1995; Sharma, Borole, and Zingde 1994), lithium (Loring 1990; Aloupi and Angelidis 2001), scandium (Ackerman 1980; Grousset et al. 1995), and Fe (Heltz et al. 1985; Hornung, Krom, and Cohen 1989; Daskalakis and O'Connor 1995; Blomqvist, Larsson, and Borg 1992). For this geochemical approach, it is important that the normalising element represents the natural background of the lattice structure within the binding minerals or the availability of surface binding sites (e.g., Fe).

In most cases, the mineralogy of the sediment has been assumed to be constant so that emphasis has been placed on accounting for the physical effect, such as grain size, as a proxy for surface area and therefore surface-bound contaminants. Often, analyses have been carried out on a specific size fraction to correct for natural variability. However, this approach requires a separation step and results are often confusing since concentrations in a certain size fraction may not reflect the concentration in the total sediment (Loring 1991).

Elemental normalisation utilises total sediment analysis and hence the resulting metal values include the anthropogenic and natural adsorbed phases, as well as the natural matrix component. For elemental normalisation to be effective, it is necessary for the relationship between the normalising element and the fine fraction to be positive and linear. Caution must be exercised in the use of normalising elements as the nonanthropogenic content may vary regionally depending on local geology. The Fe content of marine sediment has also been shown to vary due to diagenetic mobilisation and Fe staining (Grant and Middleton 1990), whereas nonanthropogenic Al may be highly variable in marine and fluvial sediments due to an inconsistent presence of Al-rich minerals, especially feldspar (Loring 1991).

Advantages of the elemental normalisation technique are that it is a single chemical analysis eliminating the time-consuming size separation procedure. The method also provides total sediment chemical data, which are assessable for sediment quality and toxicity assessment. Disadvantages of the method are that no single normalising element may be used globally, severely reducing the compatibility and value of the data. Normalisation results are expressed as a ratio, which has less discriminative power and is difficult to comprehend in assessing magnitude of impact. Also, a significant correlation must first be established between the normalising element and grain size before the chemical may be used with confidence; therefore, textural analyses are required in any event.

Normalisation to organic carbon content has also been used in some cases (Horowitz 1991; Ruiz and Saiz-Salinas 2000). However, it has been shown to be variable due to variation in its origin, including residues of sewage, planktonic production, and terrestrial components.

When normalising by a geochemical approach, the basic criteria for the normalising element are that

- it is a lattice component of fine grained, important trace element–bearing minerals such as the phyllosilicates and clay minerals or
- it reflects the grain surface area in the case of surface-bound elements;
- it reflects the granular or phase variability of its host component;
- it is conservative; and
- it is cost effective to analyse.

In addition, it should be present in easily detectable concentrations and it is usually not present in feldspars or influenced by anthropogenic inputs. Therefore, these criteria substantially limit the number of elements that can be used for normalisation.

It is assumed that the relationship between a normalising constituent, such as aluminium, and another metal like zinc will form a positive, linear trend (i.e., as the concentration of the normalising element varies with changing mineralogy and particle size, the concentration of the other metal will change proportionally). The normalising element must, therefore, be a constituent of one or more of the major trace metal–bearing minerals and reflect the granular variability in the sediments. As a result, the spatial trace metal/normaliser ratio remains constant with changes in grain size and sedimentation rates (Loring 1990). For linear geochemical normalisation to be of value, the following requirements should be met:

- Significant granular variations should occur between sediment samples.
- A strong, significant relationship, at least at the 95% confidence level (99% level would be preferred), should exist between (1) the normaliser and grain size distribution and (2) the metal concentration and the normalising element.
- It should be possible to provide accurate and precise analysis of the metal and the relevant normaliser to minimise errors of measurements that may mask some of the natural variability of the parameters.

Aloupi and Angelidis (2001) suggest that geochemical normalisation uses metal data from noncontaminated sediments of the study area to calculate the regression line of the metal on the normaliser and then tests the metal/normaliser ratio at other, possibly contaminated, stations. For the production of such a figure, it is necessary to remove outlier values and to delineate a confidence band of 95% of the regression line of the metal on the normaliser. Then, the data points from the possibly contaminated areas are projected on the diagram. All points found inside the 95% confidence band can be characterised as uncontaminated, while all points above this area should be considered as contaminated (Loring 1990).

### *Enrichment Factors*

An alternative way of expressing the results is by standardising the measurements to a reference material and defining an enrichment factor. For example, an enrichment factor (EF) for zinc relative to aluminium is

$$EF = \frac{\left(\frac{Zn}{Al}\right)_{sample}}{\left(\frac{Zn}{Al}\right)_{reference}}$$

Enrichment factors have been used to establish a value for the increased concentration from anthropogenic sources that correct for the natural variability. This has the benefit of allowing comparison between sites if a regional background reference is used. There have been two methods proposed for the calculation of enrichment factors:

- to compare the ratio in the sample with a value from an average abundance in the Earth's crust; or
- to compare with a value from a more local 'background' sample.

Each method has respective problems. There have been many values given in the literature for the average abundance, but there has not been a clearly agreed-upon value. The second method uses a value of a local background: this may be calculated from a sediment core before any anthropogenic interference (Figure 8.2).

In this example, lead data have been normalised to iron with the pre-Industrial Revolution ratios used as the regional/local background. By dividing all ratios by this background value, the degree of enrichment (EF) is apparent. Here, EF values reach ~7, indicating a seven-times enrichment of lead relative to iron in the 1970s.

The other method of determining the background is identifying an area close to the site of interest that is uncontaminated and using this as natural background. This is more difficult as most areas are proving to be contaminated. The validity of such an enrichment factor varies with the values used for the reference material. Most authors have used concentrations for crustal abundance that are of little applicability because they represent neither regional background levels nor the analytical uncertainties associated with their measurement. To be of any relevance, the reference ratio must represent the natural sediment for which a comparison is required (Loring 1991).

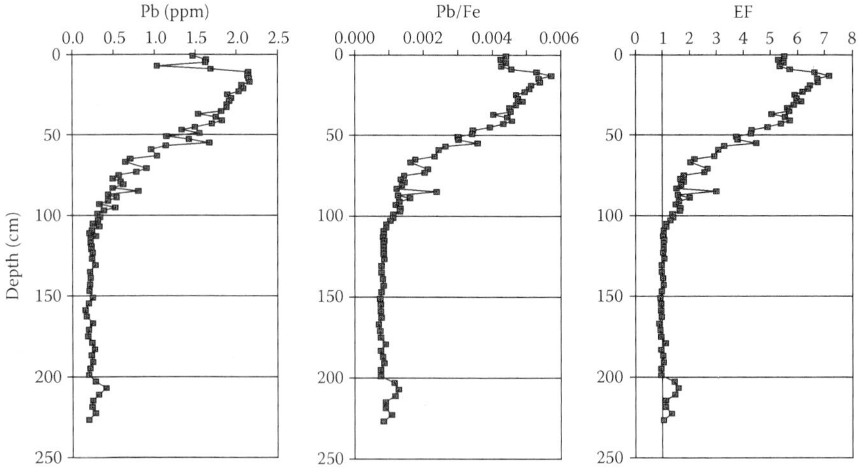


**Figure 8.2** Example of EF calculations for Pb using Fe as the normaliser. (Treadwell, unpublished data.)

## *Choice of Normalising Elements*

**Radionuclides.** Within the literature there has been little use of natural radionuclides (e.g., $^{226}$Ra and $^{40}$K) as normalisers. The following have been established:

- Radionuclide concentrations increase when grain size of the materials decreases.
- Concentration depends on the composition. Thus, the radionuclides uranium and thorium are associated mainly with heavy minerals, while their concentrations in the less dense fractions of the materials are relatively low. This light fraction is generally quartz and feldspar and can contain high concentrations of potassium associated with the feldspar.

In the process of mineral formation, selected radionuclides may be incorporated into the crystal lattice. The concentrations of these elements depend on the type of mineral and the abundance in the parent magma. Thus, the activity depends on the mineral and the provenance. Subsequently, through erosion, these minerals are transported and may reach the coast or the marine environment, becoming part of the sediments. There are also some radionuclides incorporated in sediments through anthropogenic contamination, such as potassium in phosphate-containing fertilisers (Ligero et al. 2001). $^{40}$K is found in both clays and sometimes in coarser grained sediments containing K-feldspar, so it does not show a consistent relationship with grain size. Also, $^{40}$K is found in agricultural fertiliser, so large areas of arable farming may be dominated by $^{40}$K from fertiliser use on land. However, there is also the effect of which chemical state $^{40}$K is in, as the potassium in fertilisers usually has a high solubility in the water column. Therefore, only a small proportion may be fixed to the sediment.

**Aluminium.** The most basic geochemical approach has been to use aluminium as a normaliser of trace metal concentrations since it has high natural abundance and is not commonly associated with anthropogenic inputs. Hirst (1962), one of the first to use aluminium as a normaliser, found that regional metal/aluminium ratios remained relatively constant with changes in the sediment grain size. Aluminium is associated with the fine-grained fraction of sediments, due to clay minerals being rich in alumino-silicates.

Windom et al. (1989) suggested that the covariance of metals with aluminium provides a useful basis for identification and comparison of anthropogenic inputs to the southeastern United States. From sediments collected off the coast of Florida, Georgia, and South Carolina, the correlation between metals and aluminium proved to be consistently significant. However, the basic geochemical assumptions used in the metal/aluminium ratio approach are such that it cannot be applied universally. One of the main problems is that aluminium concentrations do not always vary significantly with grain size in high-latitude sediments composed mainly of material derived from the glacial erosion of igneous rocks (Loring 1991). This is due to significant aluminium concentrations that are also present in aluminium-rich feldspars (Bothner et al. 1998).

Bruland et al. (1974) stated that aluminium is assumed to have had a uniform flux to the sediments over the past century from crustal rock sources. Bertine and Goldberg (1977) assumed that aluminium has had a uniform flux to the deposits over the past 10,000 years. Consequently, changes in the water, salt, calcium carbonate, or organic matter content, especially in the surface sediments, can be compensated for through normalisation of the metal concentrations to that of aluminium.

Normalising the radionuclide concentrations to aluminium content is used in trace element geochemistry to compare data obtained from sediments with a wide range of clay abundances. The method gives good results in muddy areas, but fails in coarser sediments in which rock debris may also occur (Boust 1999).

**Lithium.** Aloupi and Angelidis (2001) also identified the need for a conservative element when normalising. Elements of natural origin, which are structurally combined to one or more of the major fine-grained trace metal carriers, are considered conservative. These researchers used lithium to normalise core data from the Aegean Sea. Loring (1990) suggested that the use of lithium is equal or superior to aluminium as a reference element for normalisation purposes, with lithium giving consistently stronger positive covariance for chromium, copper, and zinc.

Working in the St. Lawrence Estuary and the Gulf of St. Lawrence, Loring (1990) states that lithium concentrations vary directly with the particle size and mineralogy. Statistical analyses indicate that the increase in lithium concentrations was significantly correlated with increasing amounts of fine-grained material (<53 μm) in the estuary ($r = 0.88$) and open gulf ($r = 0.84$) sediments. In contrast, the relationship between aluminium and mud content was weaker ($r = 0.41$) in the St. Lawrence Estuary and in the open gulf ($r = 0.71$). In addition, the lithium/organic carbon correlations, unlike the aluminium/organic carbon correlations, remain very strong ($r = 0.81$ and 0.86, respectively) with the changes in sediment texture. This is a consequence of lithium being incorporated in fine-grained alumino-silicate metal-bearing minerals, but not in the aluminium-rich but metal-poor feldspars that occur throughout the grain size spectrum of such sediments. Lithium has had limited use as a normaliser in European sediments (Loring et al. 1998).

**Scandium.** Ackermann (1980) showed that scandium increased significantly ($p < 0.01$) with the amount of the <63 and <20 μm fractions in sediments from the Ems Estuary and that metal/scandium ratios could be used effectively to normalise for the grain size effect in natural and contaminated sediments.

Loring (1990) identified that scandium has the potential as a reference metal because it is associated with the fine-grained clay minerals, but not with feldspars. Scandium's ionic radius (0.83Å) permits its substitution for magnesium (0.78Å) and iron$^{2+}$ (0.83Å) during crystallisation, but it cannot replace calcium, aluminium, or iron$^{3+}$ in igneous minerals. As a result, scandium resides in the ferro-magnesium minerals, such as the pyroxenes, amphiboles, and biotite mica, but the calcium and aluminium minerals, such as plagioclase feldspars, contain little or no scandium (Loring 1991). Since this early work, scandium has been used very seldom as a normaliser in the literature.

**Iron.** Iron has been used with differing degrees of success as concentrations may have large variations within intertidal sediments. There are studies that suggest a good correlation between iron and grain size < 63 μm (Herut et al. 1993; Lee, Fang, and Hsieh 1998). Bothner et al. (1998) selected iron as a normalising constituent because it was determined in all samples and because its behaviour was empirically observed to reduce erratic variation in key contaminant trends better than aluminium. However, Bothner et al. (1998) acknowledged that iron is itself a constituent of sewage effluent, although it is far less enriched than other contaminants. Iron is used as a normaliser to measure the surface adsorption component of sediments: the iron (oxy-) hydroxide coating (Whalley et al. 1999). Iron, however, is not always suitable as a normaliser because secondary accumulation of iron compounds can distort the total iron concentration of many near-shore sediments. In addition, iron concentrations are also affected by the redox potential of the sediment. Iron$^{3+}$ hydroxides are easily reduced to iron$^{2+}$ sulphides, which are less water soluble. The redox conditions can directly influence the migration of iron in the sediment layers. This reaction is also affected by the pH of the sediment.

**Cobalt.** Matthai and Birch (2001) used cobalt to identify accumulation of contaminants on the continental shelf off the shore of central New South Wales, Australia. The spatial distribution of cobalt-normalised concentrations of copper, lead, and zinc was similar to the distribution of trace metals in size-normalised material. The study identified higher concentrations of metals in adjacent Sydney, possibly due to the discharge of large volumes of sewage offshore.

**Organic carbon.** Organic matter in sediments has highly variable origins, including sewage discharges, phytoplankton production, and terrestrial materials, such as humic acid, pollen, and plant matter. The relationship of contaminants to these sources of organic carbon may vary as a consequence of different organic matter types (Bothner et al. 1998). Langston (1982) reports that there is better correlation between metals and total organic material, measured by a loss on ignition method, compared to that between metals and total organic carbon measured by CHN analysis. Balls et al. (1997) highlighted the variability of organic carbon as a normaliser; sediments from the Clyde and Forth estuaries were studied and the correlation coefficients for trace metals to organic carbon were consistently higher for the Forth Estuary than the Clyde Estuary. However, in this study the authors identify aluminium as being a more effective normaliser. For some radionuclides, Ligero et al. (2001) identified the organic content of sediment as important for the binding and fixation of these elements. They suggested that organic content is more important for the binding of $^{137}Cs$ and $^{40}K$ ($r^2 = 0.9$ and 0.8, respectively) than for $^{232}Th$ and $^{226}Ra$ ($r^2 = 0.5$ and 0.4, respectively).

The correlation coefficients for selected radionuclides with organic carbon were calculated in a study of microscale variability (Mudge et al. 2002) (Figure 8.3). Although samples were from both the Ribble Estuary and the Mersey Estuary and they were sampled across a 2-year time span, the data described a single regression equation. Similar results were seen with $^{241}Am$. Therefore, using TOC as the normaliser would suggest that $^{137}Cs$ is intrinsically linked to the organic carbon in some manner. With further research, TOC could be used for the normalisation of caesium and americium at these sites.

In summary, several potential normalising elements could be used, although they all have apparent advantages and disadvantages. These are summarised in Table 8.1.

## Postextraction Normalisation (PEN) Technique

The PEN technique is a relatively new method and utilises the total sediment for analysis. After weak-acid digestion (e.g., *aqua regia* [50:50 $HNO_3$: HCl]) (Birch 2003) (US EPA Method 200.8), the solute is decanted into a

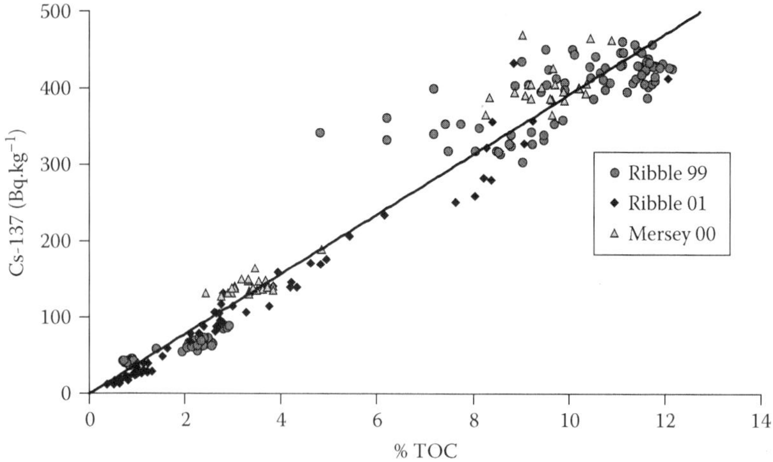


**Figure 8.3** The activity of $^{137}Cs$ as a function of total organic carbon in Ribble and Mersey estuaries' sediments.

**Table 8.1 Summary of Advantages and Disadvantages of Several Normalising Approaches**

| Method | Advantages | Disadvantages |
|---|---|---|
| Geochemical | Provides understanding of lithology | All require separate analysis |
| Aluminium | Extensively used in literature<br>Large concentrations in sediments | Not always as useful in higher latitudes due to K-feldspar dominated sediments |
| Lithium | Superior to aluminium in higher latitudes<br>Large concentrations in sediments<br>Not found in feldspar minerals | May be more expensive to analyse |
| Scandium | Not found in feldspar minerals | Low concentrations in sediments<br>May be more expensive to analyse |
| Iron | Large concentrations in sediments<br>Sometimes good covariation with surface bound elements | Large variation in intertidal sediments<br>Secondary accumulation can distort the total iron concentrations |
| Cobalt | Good relationship to size-normalised data<br>Not found in feldspar minerals | Must establish a significant relationship with fine sediment first |
| Organic carbon | Simple and cheap to analyse | Highly variable<br>Many origins, some also from same source at contaminant |
| Grain size | Extensively used in literature | Wide range of grain sizes required to define relationship |

scintillation vial for analysis and the residue is washed and dried before being weighted. The normalised metals data (N) are calculated by:

$$N = \frac{\text{total digestion weight}}{\text{residue weight}} \times \text{measured concentration } (\mu g \cdot g^{-1})$$

The technique assumes that no metals are associated with the residue material after the digestion process. Microscopic analysis and x-ray diffraction tests of residue show it to comprise quartz and other stable silicious minerals exclusively (Birch 2003).

Advantages of the PEN method are that it is quicker and therefore cheaper to analyse because wet sieving is not required in the procedure. Because the method requires digestion and analysis of the total sediment fraction, the results include metals associated with the coarse fraction, which is not the case with size-normalising techniques. Also, the technique provides normalised data for identification of contaminant sources, as well as total metal concentrations for sediment quality and toxicity assessment in a single procedure. A disadvantage of the technique is that very fine-grained, quartzitic sediment may be difficult to manage and measure.

## Application of Normalisation Techniques

Size-, elemental-, and PEN-normalisation techniques have been applied to detect contaminant sources, establish dispersion pathways, and determine preanthropogenic metal concentrations in fluvial, estuarine, and offshore marine systems in central New South Wales, Australia (Birch and Taylor 1999, 2000b; Birch and Davey 1995). The textural and compositional characteristics of sediments in these environments are highly variable, ranging from very immature gravelly sands through quartzitic muddy sands to organic-rich muds. This magnitude of textural variability within these systems necessitates some form of normalisation procedure to minimise the confounding effects of variable size and to improve comparability of metals data from these environments.

### Marine Sediments from the Continental Shelf Off New South Wales Shore

Detailed sampling ($n = 309$) has been carried out on the central New South Wales continental shelf to determine the source and impact of anthropogenic materials discharged to the marine environment (Figure 8.4). The sediments on this shelf are slightly muddy quartzitic inner shelf sands, midshelf sandy muds, and calcareous outer shelf gravelly sands (Matthai and Birch 2001). The mud content of these sediments varies from <1 to 73%.

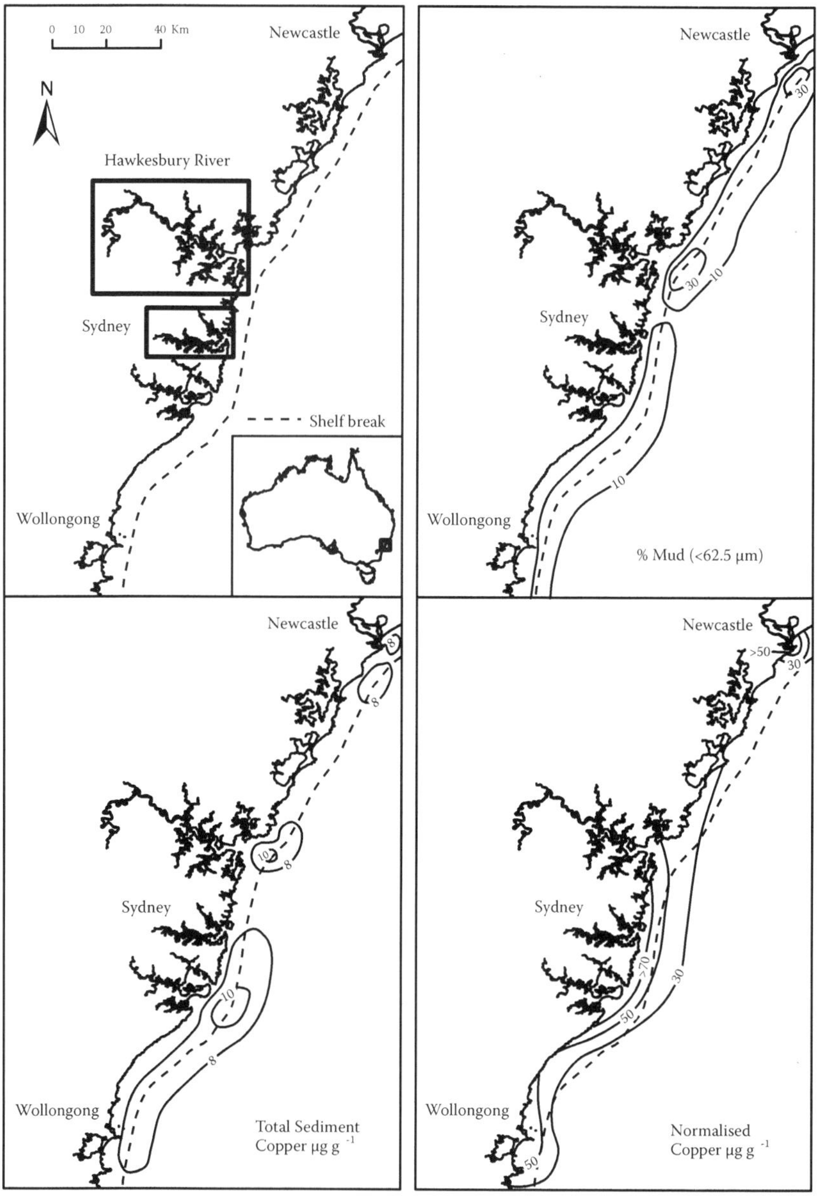


**Figure 8.4** Sample locations on the shelf off New South Wales, Australia.

On a total-sediment basis, heavy metal concentrations (depicted in this work by copper) are highest coincident with the midshelf sandy mud belt and decline rapidly both seawards and landwards (Birch and Davey 1995; Matthai and Birch 2000a). Total metal analyses clearly reflect the textural characteristics of the surficial sediments on the shelf becoming more enriched with increasing mud content. This metals distribution pattern implies an unlikely contaminant dispersion from the central shelf area seawards and landwards. Size-normalised heavy metal concentrations, however, decrease strongly and regularly seawards from a maximum close to the coast. The fine-fraction heavy metal distribution, in contrast, indicates a strong seawards dispersion of contaminants from sources close to the coast (Matthai and Birch 2000b).

The aim of this study was to determine the impact of sewage discharged adjacent to Sydney and Wollongong and harbour waste disposed offshore Newcastle. Sewage disposal, at the time of this research, was via three major cliff-face outfalls off Sydney and into the near shore off Wollongong, whereas contaminated sediment dredged from the harbour had been dumped at a specific site off Newcastle for decades. Size-normalised data show highest heavy metal concentrations immediately adjacent to the coast opposite Sydney and Wollongong and as a point source in the near shore adjacent to Newcastle (Matthai and Birch 2000a, 2000b, 2000c). More detailed work on normalised data shows that the nature of enrichment adjacent to these major urban centres has characteristic chemical fingerprints (Matthai and Birch 2000c; Mudge, Birch, and Matthai 2003). The fine fraction of sediments offshore Newcastle is enriched in cadmium, copper, iron, manganese, nickel, lead, and zinc associated with dredged spoil from metalliferous industries; adjacent to Sydney, enrichment of copper, lead, and zinc in the fine fraction results mainly from disposal of large volumes of sewage. The use of total sediment chemistry to determine source and dispersion of particle-bound contaminants on the New South Wales continental shelf resulted in a false interpretation (i.e., that heavy metals were originating from the central shelf adjacent to all three centres). Other studies on the central New South Wales continental margin using total sediment chemistry also showed maximum metal concentrations on the middle shelf and resulted in erroneous interpretations of source, dispersion, and accumulation (de Forrest, Murphy, and Pettis 1978; Schneider and Davey 1995). Misinterpretation of environmental data, as demonstrated in this example, may result in erroneous management or judicial decisions being made.

## Estuarine Sediments from Port Jackson

The northeast arm of North Harbour, Port Jackson (inset to Figure 8.4), comprises a broad, shallow (<2 m) bayhead mantled in quartzitic sand (>95% sand) and a deep (10 m), muddy (83% mud) central basin. Total sediment heavy metal concentrations ($n$ = 20) (depicted by copper in Figure 8.5) increase rapidly towards the center of the bay, reflecting closely the overall finer in the textural characteristics of surficial sediment. Size-normalised metal concentrations increase strongly and regularly towards the bayhead and attain a maximum in the northwestern part of the embayment. Distribution of the PEN data is remarkably similar to the size-normalised pattern showing rapid decreasing concentrations toward the mouth of the embayment. Elemental

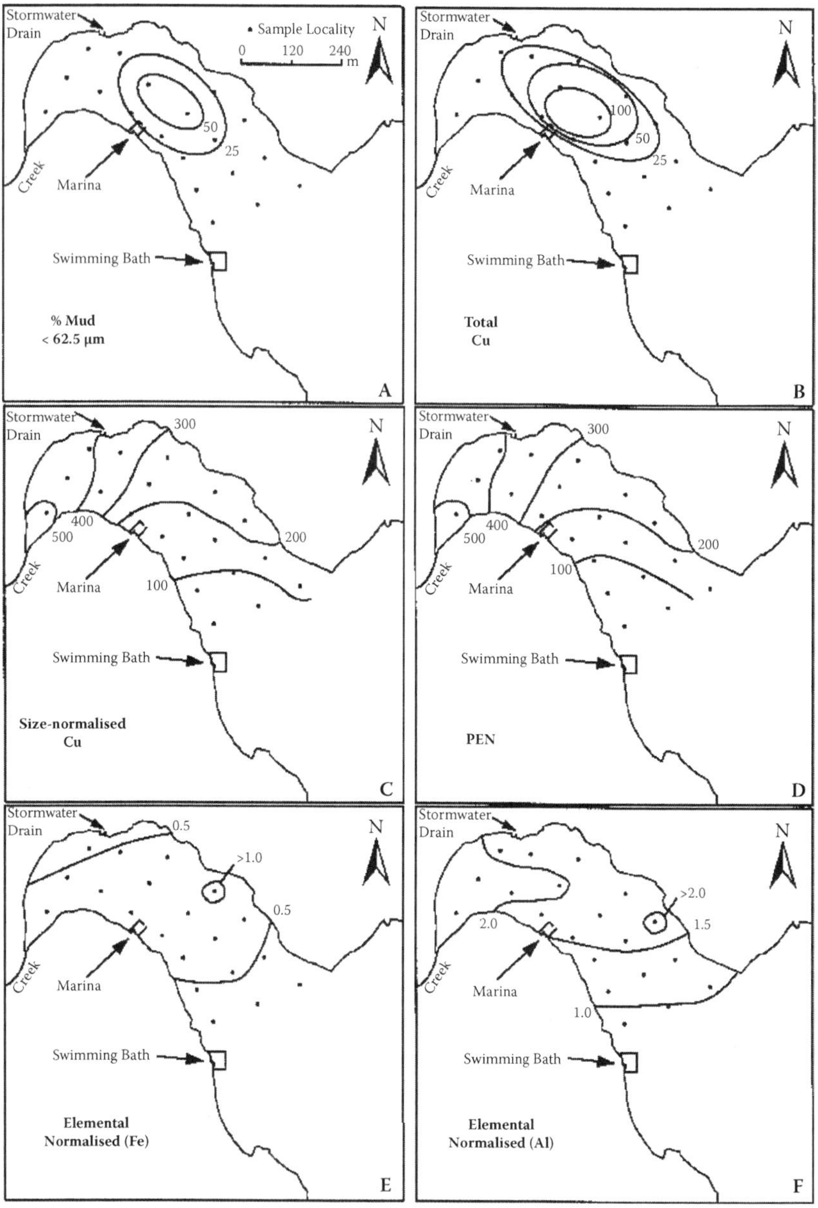


**Figure 8.5** (See Figure 8.4 for location.) Total sediment heavy metal concentrations increase rapidly towards the centre of the bay, reflecting closely the texture of surface sediment. Size-normalised metal concentrations and PEN values increase strongly and regularly towards the head of North Harbour in Port Jackson (Sydney). These distributions suggest the creek and stormwater drain are the main source of contamination. Elemental normalisation using iron presents a weak and diffuse distribution, whereas aluminium-normalised data indicate a weak but regular pattern of decreasing ratios towards the head of the bay.

normalisation using iron presents a weak and diffuse distribution and aluminium-normalised data indicate a weak but regular pattern of decreasing ratios towards the southeast.

There are four major possible sources of contaminants to this embayment: a stormwater drain and creek in the upper bay and a marina and swimming pool in the central east section of the location. The size-normalised and PEN data strongly indicate a major source in the upper reaches of the bay and probably favour the creek as the dominant source, with the stormwater drain providing an additional contribution of copper to the system. Contours depicting copper concentration are deflected towards the central basin in the east in both size-normalised and PEN distributions. This pattern may be explained by dispersion primarily down the eastern shore of the embayment. This subtle pattern in the distribution of copper in sediments of this bay is not uncommon in normalised metals data and demonstrates the power of this technique in detecting dispersion pathways in aquatic environments. The elemental normalisation techniques, in this case, do not provide strong evidence of source or dispersion. The aluminium-normalised data suggest the source is towards the north, and a weak declining gradient indicates dispersion towards the mouth of the embayment. There is no observable influence from the marina or the swimming pool on this environment by any of the normalised heavy metals data. The total-sediment metal concentration gradients imply an unlikely dispersion pattern from the central basin area outwards towards the bay margins.

Normalisation of organic contaminants contained in aquatic sediments is not commonly undertaken. However, a study of sediment-bound organochlorine pesticides in North Harbour revealed a similar distribution pattern to that of the fine fraction heavy metals presented in the current work (Birch 2003).

The distribution of both organic and inorganic contaminants in surficial sediments in North Harbour strongly indicates that stormwater discharged from the creek in the upper bay is the dominant source of toxicants to this environment. Additional contributions of copper to the bay from the drain are likely. Information on the source and dispersion of contaminants is important in the management of an environment such as North Harbour, as a major concern at the time was the influence of the marina on the embayment. The allocation of funds and effort to reduce the influence of the marina would be fruitless because a very different strategy is required to remediate the stormwater being discharged to the bay.

## Fluvial Sediments from the Catchment

Heavy metal characteristics of fluvial sediments in the Port Jackson catchment and other central New South Wales catchments are being determined to identify major sources of contaminants to estuaries in the region. Stream-bed

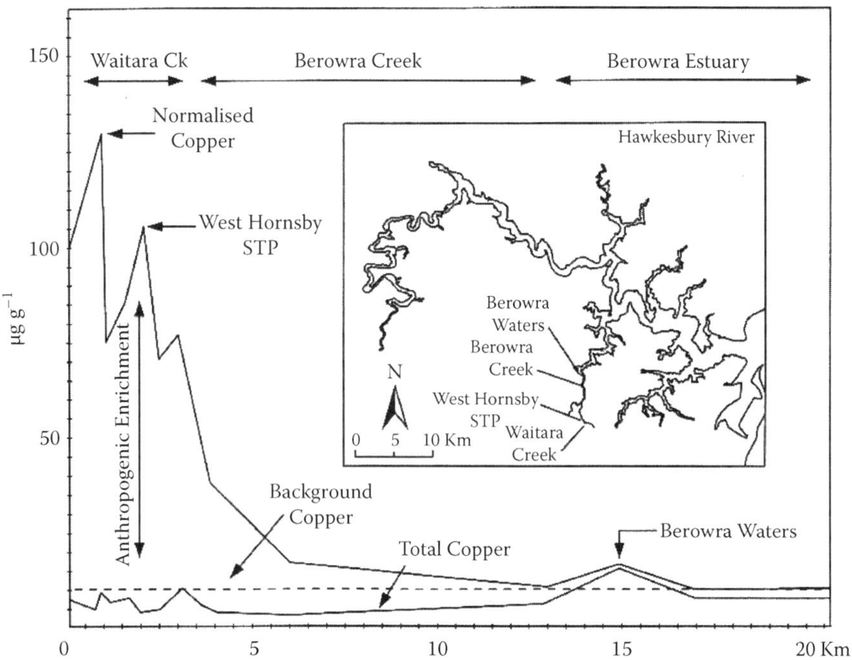


**Figure 8.6** (See Figure 8.4 for location.) Heavy metal concentrations of total sediment from above and below the sewage treatment plant (STP) on the Waitara Creek are low due to sediments being very coarse. This distribution suggests that the small township of Berowra Waters is a source of contamination with no obvious association with the STP. However, size-normalised copper distributions show a marked increase in concentration, not only below but also above the STP. High-normalised concentrations upstream of the STP are probably due to leaking sewage pipes higher in the valley.

sediments vary markedly in texture due to considerable fluctuations in ambient energy. These sediments are also generally immature, containing a high proportion of rock fragments and a range of mafic and opaque minerals. Sediments typically highly variable in texture and composition are materials that most require normalisation to provide compatible data for environmental assessment.

One of the many contaminant sources to fluvial systems in the Sydney region is effluent discharged by sewage treatment plants (STPs). Heavy metal concentrations (depicted in Figure 8.6 by copper, but also for other metals, e.g., lead and zinc) of total sediment ($n$ = 20) recovered from above and below the STP on the Waitara Creek (a tributary of the Berowra Creek on the Hawkesbury River) are low and indicate no obvious increase associated with the sewage treatment works (Birch, Evenden, and Teutsch 1996). This section of the river is in the headwaters of the catchment where the thalweg gradient is steep and ambient energy is high. Consequently, sediments in this reach of the creek are very coarse and variable with little fine material. However, size-normalised copper distributions of fluvial sediment from this part of Waitara Creek clearly show a marked increase in concentration, not only below but also above the STP. Normalised concentrations are high upstream of the STP, probably due to leaking of sewage pipes higher in the valley. This explanation was later confirmed by an independent study. Total copper concentrations increase in the estuarine section of the valley due to the sediment comprising predominantly mud-size material and thus the normalised and total metals data become similar and consistent. Normalised, preanthropogenic copper concentrations (16 $\mu g.g^{-1}$; Birch et al. 1998) determined for this fluvial system provide an estimate of the magnitude of change induced by the STP to the upper parts of Waitara Creek (i.e., five to nine times enrichment).

The important lesson for forensic scientists and environmental managers in this study is that if total contaminant concentrations were used as an indicator of human impact, the small community of Berowra Creek would be the focus of environmental scrutiny and perhaps unnecessary remedial actions, and the STP in the headwaters of the estuary would have escaped attention.

## Conclusions

Sediments are a key tool for forensic scientists and are particularly important in assessment of change related to human activity in aquatic environments. Sediments are the only media capable of distinguishing between natural, stress-related change and anthropogenic impact and estimating the magnitude of human-related influence. Sediments are also able to identify contaminant sources and determine dispersion pathways in aquatic systems. However, effective interpretation of sediment chemical data requires specialised sample treatment and a particular approach to analysis of the data.

Total sediment contaminant data are confounded by variable grain size distributions and cannot be used in determining anthropogenic impact or for identification of contaminant sources or dispersion. For information on these issues, a normalisation procedure must be employed to reduce confounding effects of irregular grain size. Of the three normalisation techniques described in the current chapter, size normalisation appears to provide the strongest and most reliable information on impact, source, and dispersion. A new postextraction technique (PEN) provides less detailed but nevertheless reliable information on these topics without the need for time-consuming sediment fractionation. Elemental normalisation is a rapid, single analysis technique, but the information provided is weak and interpretation is ambiguous in some cases.

The examples provided from the marine, estuarine, and fluvial environments of central New South Wales, Australia, in this chapter demonstrate how misinterpretation of data may result in inappropriate management decisions and a waste of valuable and scarce resources. In all three cases, misinterpretation of total sediment chemical data may lead to erroneous decisions on contaminant source and dispersion. The use of normalised contaminant data is essential in forensic investigative work in sedimentary aquatic environments.

## References


Ackermann, F. (1980) A procedure for correcting the grain size effect in heavy metal analyses of estuarine and coastal sediments. *Environmental Technology Letters,* 1: 518–527.

Ackermann, F., M. Bergmann, and G. U. Schleichert. (1983) Monitoring of heavy sediments in coastal and estuarine sediments—A question of grain size: <20 μm versus <60 μm. *Environmental Technology Letters,* 1: 518–527.

Aloupi, M., and M. O. Angelidis. (2001) Normalisation to lithium for the assessment of metal contamination in coastal sediment cores from the Aegean Sea, Greece. *Marine Environmental Research,* 52: 1–12.

Balls, P. W., S. Hull, B. S. Miller, J. M. Pirie, and W. Proctor. (1997) Trace metal in Scottish estuarine and coastal sediments. *Marine Pollution Bulletin,* 34: 42–50.

Bertine, K. K., and E. D. Goldberg. (1977) History of heavy metal pollution in southern California coastal zone—Reprise. *Environmental Science and Technology,* 11: 297–299.

Birch, G. F. (1996) Sediment-bound metallic contaminants in Sydney's estuaries and adjacent offshore, Australia. *Estuarine, Coastal and Shelf Science,* 42: 31–44.

______. (2003) A test of normalisation methods for marine sediments, including a new postextraction normalisation (PEN) technique. *Hydrobiologia,* 492: 5–13.

Birch, G. F., and S. Davey. (1995) Accumulation of metallic contaminants in surficial sediments on a high-energy continental shelf, Sydney, Australia. *Science of the Total Environment,* 170: 81–93.

Birch, G. F., D. Evenden, and M. E. Teutsch. (1996) Dominance of point source in heavy metal distributions in sediments of a major Sydney estuary (Australia). *Environmental Geology,* 28: 169–174.

Birch, G. F., and S. E. Taylor. (1999) Source of heavy metals in sediments of Port Jackson estuary, Australia. *Science of the Total Environment,* 227: 123–138.

______. (2000a) The distribution and possible sources of organochlorine residues in sediments of a large urban estuary, Port Jackson, Sydney. *Australian Journal of Earth Sciences,* 47: 749–756.

______. (2000b) The use of size-normalisation procedures in the analysis of organic contaminants in estuarine environments. *Hydrobiologia,* 431: 129–133.

______. (2002) Application of sediment quality guidelines in the assessment and management of contaminated surficial sediments in Port Jackson (Sydney Harbour), Australia. *Environmental Management,* 29: 860–870.

Blomqvist, S., U. Larsson, and H. Borg. (1992) Heavy metal decrease in the sediments of a Baltic Bay following tertiary sewage treatment. *Marine Pollution Bulletin,* 24: 258–266.

BNFL (British Nuclear Fuel). (1993) *Annual report on radioactive discharges and monitoring of the environment, 1992.* Warrington, U.K.: BNFL, 99 pp.

______. (2000) *Annual report on radioactive discharges and monitoring of the environment, 1999.* Warrington, U.K.: BNFL, 163 pp.

Bothner, M. H., M. Buchholtz ten Brink, and F. T. Manheim. (1998) Metal concentrations in surface sediments of Boston Harbour—Changes with time. *Marine Environmental Research,* 45: 127–155.

Boust, D. (1999) Distribution and inventories of some artificial and naturally occurring radionuclides in medium to coarse-grained sediments of the channel. *Continental Shelf Research,* 19: 1959–1975.

Bruland, K. W., K. Bertine, M. Koide, and E. D. Goldberg. (1974) History of metal pollution in southern California coastal zone. *Environmental Science and Technology,* 8: 425–432.

Bubb, J. M., T. Rudd, and J. N. Lester. (1990) Distribution of heavy metals in the River Yare and its associated broads. III. Lead and zinc. *Science of the Total Environment,* 102: 189–208.

Carral, E., R. Villares, X. Puente, and A. Carballeira. (1995) Influence of watershed lithology on heavy metal levels in estuarine sediments and organisms in Galicia (northwest Spain). *Marine Pollution Bulletin,* 30: 604–608.

Daskalakis, K. D., and T. P. O'Connor. (1995) Normalisation and elemental sediment contamination in the coastal United States. *Environmental Science Technology,* 29: 470–477.

de Forrest, A., S. P. Murphy, and R. W. Pettis. (1978) Heavy metals in sediments from the central New South Wales coastal region. *Australian Journal of Marine and Freshwater Research,* 29: 777–785.

De Groot, A.J. (1964) Mud transport studies in coastal waters from the western Scheldt to the Danish frontier. In *Deltaic and shallow marine deposits,* ed. L. M. J. U van Straaten. *Developments in Sedimentology,* 1: 93–103.

Forstner, U. (1982) Accumulative phases for heavy metals in limnic sediments. *Hydrobiologia,* 91: 269–284.

Forstner, U., and G. T. W. Wittmann. (1979) *Metal pollution in the aquatic environment.* New York: Springer–Verlag.

Grant, A., and R. Middleton. (1990) An assessment of metal contamination of sediments in the Humber Estuary, U.K. *Estuarine, Coastal and Shelf Science,* 31: 71–85.

Grousset, F. E., C. R. Quetel, B. Thomas, O. F. X. Donard, C. E. Lambert, F. Guillard, and A. Monaco. (1995) Anthropogenic vs. lithogenic origins of trace elements (As, Cd, Pb, Rb, Sb, Sc, Sn, Zn) in water column particles: Northwestern Mediterranean Sea. *Marine Chemistry,* 48: 291–310.

Harland, B. J., D. Taylor, and A. Wither. (2000) The distribution of mercury and other trace metals on the sediments of the Mersey Estuary over 25 years 1974–1998. *Science of the Total Environment,* 253: 45–62.

Heltz, G. R., S. Sinex, K. L. Ferri, and M. Nichols. (1985) Processes controlling Fe, Mn and Zn in sediments of northern Chesapeake Bay. *Estuarine, Coastal, and Shelf Science,* 21: 1–16.

Herut, B., H. Hornung, M. D. Krom, N. Kress, and Y. Cohen. (1993) Trace metals in shallow sediments from the Mediterranean coastal region of Israel. *Marine Pollution Bulletin,* 26: 675–682.

Hirst, D. M. (1962) The geochemistry of modern sediments from the Gulf of Paria I. *Geochimica Cosmochimica Acta,* 26: 309–334.

Hornung, H., M. D. Krom, and Y. Cohen. (1989) Trace metal distribution in sediments and Benthic fauna of Haifa Bay, Israel. *Estuarine, Coastal and Shelf Science,* 29: 45–56.

Horowitz, A. J. (1991) *A primer on sediment-trace element chemistry.* Chelsea, MI: Lewis Publishers, Inc.

Jickells, T. D., and A. H. Knap. (1984) The distribution and geochemistry of some trace metals in the Bermuda coastal environment. *Estuarine, Coastal and Shelf Science,* 18: 245–262.

Klamer, J. C., W. J. M. Hegeman, and F. Smedes. (1990) Comparison of grain size correction procedures for organic micropollutants and heavy metals in marine sediments. *Hydrobiologia,* 208: 213–220.

Langston, W. J. (1982) The distribution of mercury in British estuarine sediments and its availability to deposit feeding bivalves. *Journal of the Marine Biological Association U.K.,* 62: 667–684.

Lee, C., M. Fang, and M. Hsieh. (1998) Characterisation and distribution of metals in surficial sediments in southwestern Taiwan. *Marine Pollution Bulletin,* 36: 464–471.

Leoni, L., and F. Sartori. (1991) Trace element distributions in surficial sediments of the northern Tyrrhenian Sea: Contribution to heavy-metal pollution assessment. *Environmental Geology Water Science,* 2: 103–116.

Ligero, R. A., I. Ramos-Lerate, M. Barrera, and M. Casas-Ruiz. (2001) Relationships between sea-bed radionuclide activities and some sedimentological variables. *Journal of Environmental Radioactivity,* 57: 7–19.

Loring, D. H. (1990) Lithium—A new approach for the granulometric normalisation of trace metal data. *Marine Chemistry,* 29: 155–168.

______. (1991) Normalisation of heavy-metal data from estuarine and coastal sediments. *Journal of Marine Science,* 48: 101–115.

Loring, D. H., S. Dahle, K. Naes, J. Dos Santos, J. M. Skei, and G. Matishov. (1998) Arsenic and other trace metals in sediments from the Kara Sea and the Ob and Yenisey estuaries, Russia. *Aquatic Geochemistry,* 4: 233–252.

Martincic, D., Z. Kwokal, and M. Branica. (1990) Distribution of zinc, lead, cadmium and copper between different size fractions of sediments in Limski canal (North Adriatic Sea). *Science of the Total Environment,* 95: 201–215.

Matthai, C., and G. F. Birch. (2000a) Trace metals and organochlorines in sediments near a major ocean outfall on a high-energy continental margin (Sydney, Australia). *Environmental Pollution,* 110: 411–423.

______. (2000b) Dispersion of dredge spoil dumped on a high-energy continental margin (southeastern Australia). *Journal of Marine Environmental Engineering,* 6: 1–32.

______. (2000c) Effects of coastal cities on surficial sediments mantling an adjacent high-energy continental margin—Central New South Wales, Australia. *Marine and Freshwater Research,* 51, 565–576.

______. (2001) Detection of anthropogenic Cu, Pb, and Zn in continental shelf sediments off Sydney, Australia—A new approach using normalisation with cobalt. *Marine Pollution Bulletin,* 42(11): 1055–1063.

Mudge, S. M., D. J. Assinder, and G. S. Bourne. (1994) *A survey of radioactivity in the Ribble Estuary. Part 1: Activity concentrations and estuarine dynamics.* HMIP Environmental Series No. 2, 67 pp., HMSO.

Mudge, S. M., D. J. Assinder, and A. T. Russel. (2002) *Micro-scale variability of contaminants in surface sediments: The implications for sampling.* Technical Report P3-057/TR, 88 pp., Environment Agency.

Mudge, S. M., G. F. Birch, and C. Matthai. (2003) The effects of grain size and element concentrations in identification of contaminant source. *Journal of Forensic Science,* 4: 305–312.

Ruiz, J. M., and J. I. Saiz-Salinas. (2000) Extreme variation in the concentration of trace metals in sediments and bivalves from the Bilbao estuary (Spain) caused by the 1989–90 drought. *Marine Environmental Research,* 49: 307–317.

Sakai, H., Y. Kojima, and K. Saito. (1986) Distribution of heavy metals in water and sieved sediment in the Toyohira River. *Water Research,* 20: 559–567.

Schneider, P. M., and S. B. Davey. (1995) Sediment contaminants off the coast of Sydney, Australia: A model for their distribution. *Marine Pollution Bulletin,* 26: 262–272.

Sharma, P., D. V. Borole, and M. D. Zingde. (1994) $^{210}$Pb based trace element fluxes in the near shore and estuarine sediments off Bombay, India. *Marine Chemistry,* 47: 227–241.

Taylor, D. (1986) Changes in the distribution patterns of trace metals in sediments of the Mersey Estuary in the last decade (1974–83). *Science of the Total Environment,* 49: 257–295.

Tessier, A., P. G. C. Campbell, and M. Bisson. (1979) Sequential extraction procedure for the speciation of particulate trace metal. *Analytical Chemistry,* 51: 844–851.

Whalley, C., S. Rowlatt, M. Bennet, and D. Lovell. (1999) Total arsenic in sediments from the western North Sea and the Humber estuary. *Marine Pollution Bulletin,* 38: 394–400.

Windom, H. L., S. J. Schropp, F. D. Calder, J. D. Ryan., R. G. Smith, Jr., L. C. Burney, F. G. Lewis, and C. H. Rawlinson. (1989) Natural trace metal concentrations in estuarine and coastal marine sediments of the southern United States. *Environmental Science and Technology,* 23: 314–320.

# Multivariate and Geostatistical Methods in Environmental Forensics

# 9

STEPHEN M. MUDGE

## Contents

Introduction.......... 277
Statistical Approaches .......... 278
Comparative Statistics .......... 278
Systematic and Random Error.......... 279
Type 1 and Type 2 Errors .......... 279
Statistical Significance .......... 281
Multivariate Methods.......... 281
PCA.......... 282
Normality and Transformations .......... 287
Mean Centring to Unit Variance .......... 288
Using Proportions and Closure.......... 288
Zeros.......... 291
Summary: Checklist for PCA .......... 292
PLS .......... 292
Overlap between Signatures.......... 295
Summary: Checklist for PLS.......... 296
Geostatistics.......... 297
Variogram/Semivariogram Analysis.......... 298
Kriging .......... 301
Anisotropy.......... 302
Summary: Checklist for Geostatistics.......... 304
References.......... 304

## Introduction

> Lies, damned lies and statistics
>
> **Attributed to Benjamin Disraeli, a U.K. politician of the nineteenth century**

It has been intimated that you can make statistics show anything you want them to; politicians from opposing parties often use the same data to provide support for opposite standpoints. In an environmental forensic context, however, statistics can lend considerable support to other evidence and lead to the interpretation of complex data. The use of statistical methods must be conducted with utmost rigor and be applied only where relevant and necessary. Statistics for the sake of statistics may lead to doubt being raised and imply obfuscation. It has also been suggested that a well planned and executed analytical program should not need statistics to prove a point. This may be true when simple comparisons are all that is required; however, when examining complex signatures, statistical methods might be the only ones that can identify the underlying truth.

## Statistical Approaches

Statistical methods may be used to achieve many functions. These may be grouped into one of three categories: (1) comparative techniques, where it is simply necessary to demonstrate that one set of values is greater than another set; (2) multivariate methods, where many analytes are used to provide a form of signature analysis; and (3) geostatistical techniques that indicate a spatial distribution or pattern. The main thrust of this chapter will be in the use of multivariate statistical methods for signature analysis and geostatistics. Comparative techniques are briefly included as they help in the design and execution of sampling programs.

### Comparative Statistics

One function of statistical analysis may be to demonstrate that one set of values for an analyte is different from another set. The success of this rests upon many factors unrelated to the statistical treatment of the data; the quality of the values obtained is paramount. It is possible to summarise the steps that lead to data collection in the following nonexhaustive list:

- sampling program design
  - intrinsic spatial and temporal variability in the sample medium
  - number of replicates
  - method of collection
- sample storage and security
  - temperature
  - materials used
  - time of storage
  - security from deliberate interference or unintentional contamination
- analytical procedures
  - appropriateness of method used

blanks
instrumental detection limits
security of data

Each step of this procedure will have errors associated with it. These may mount up when each one is compounded by the next step in the process.

### *Systematic and Random Error*

Errors in analytical methods may arise due to several processes, which can be grouped into either systematic errors or random errors. In the case of the former, if an instrument consistently underrecords a value, this is a systematic error and affects all measurements made. If the same instrument fluctuates in its response, sometimes overreading and sometimes underreading, this becomes a random error. These errors are treated differently.

Systematic errors can be difficult to identify and require an audit of the method; this can best be achieved by use of accredited reference materials (ARMs) or standards where the value of the analyte is well known and characterised. Any deviation from this value in a system is an error and, if it is consistently in one direction, is a systematic error. The regular use of ARMs in analyses is recommended. One method to provide regular checks on performance is the use of Shewhart plots; this aspect of statistical process control (SPC) provides a check on the method's performance (Macii, Carbone, and Petri 2003). The process involves the analysis of several ARMs; some authors recommend 25, but others recommend significantly more (Quesenberry 1993). Comparison of the results with the published values provides a direct measure of the offset between the analyses and the accepted values. Corrections to methods can be made on the basis of these results. The second useful property of the Shewhart plot is monitoring individual batch performance.

An example of a Shewhart plot can be seen in Figure 9.1. In the upper pane is the development of the upper and lower acceptable values based on three standard deviations ($\pm 3\sigma$) about the mean value in a normal distribution. The lower pane shows a real example for $^{137}Cs$ measured in intertidal sediments (Mudge, Assinder, and Russell 2001); the confidence levels are based on analysis of an ARM. One standard was recounted in every 10 analyses to assess the continued performance of the analytical system with time.

### *Type 1 and Type 2 Errors*

There is a whole series of considerations that need to be given to statistical analyses including type 1 and type 2 errors. Type 1 errors are false positives where, due to natural variability in samples (and a small sample size), a positive relationship is indicated. If we set our confidence interval to be 95%, by definition 5 in 100 samples will be outside those limits. If we happened to

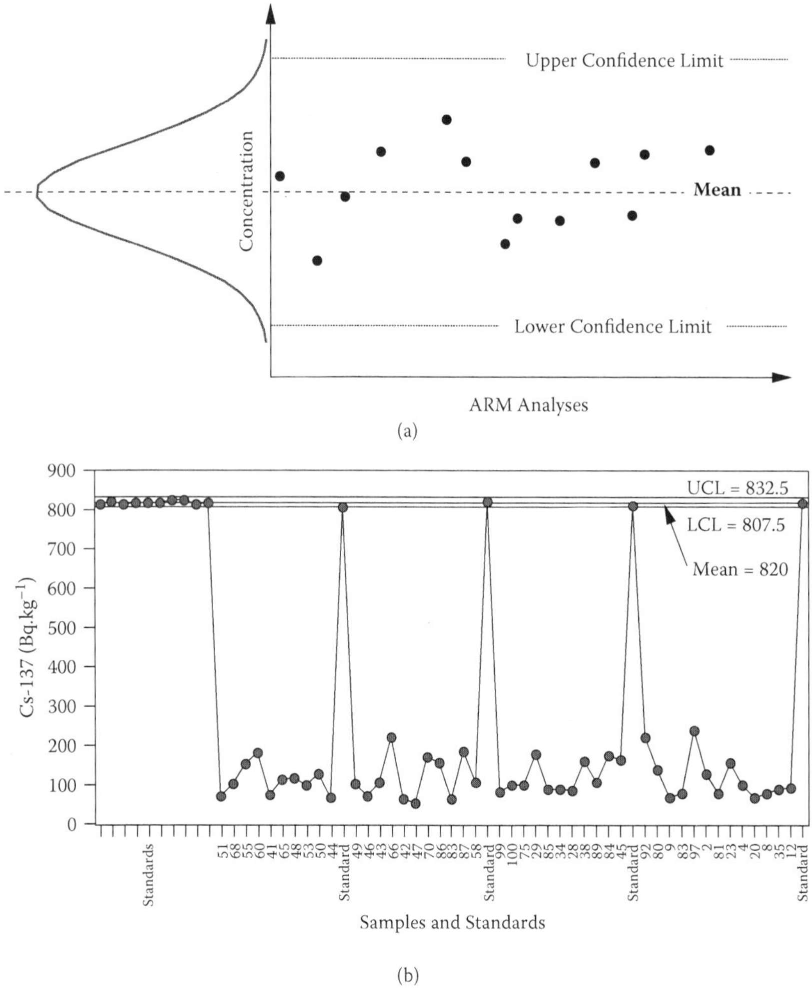


**Figure 9.1** (a) The development of the confidence intervals about the mean of several accredited reference values. (b) An example of a Shewhart plot for the analysis of $^{137}$Cs in environmental samples with a standard every 10.

collect one of those five during the sampling process, we might state that it is different from the others, while in reality it is part of the same distribution but we have not been able to define it well enough.

In contrast, type 2 errors are false negatives where a relationship is missed. This also tends to occur when small sample sizes are involved. The confidence interval is dependent on the number of samples used to generate it; if the number of samples is small, the interval may be large and overlap zero. However, if a larger number of samples were collected, the interval would get smaller and may indicate a significant difference from zero. Both of these errors depend upon the number of samples analysed, so it is sensible to make more than one measurement at a site, although there are cost implications.

### *Statistical Significance*

It is usual to take several samples at a site under investigation and compare these values to values obtained from other, unaffected or control sites. It is this comparison that needs to be conducted when making claims that one site has significantly more of something than another. Apart from the question of true controls, it is the 'significance' that usually requires definition. It is the extent to which a result deviates from that expected to arise simply from random variation or errors in sampling. Therefore, it is a measure of our confidence in a number. As such we can express the confidence as a chance of it happening through random effects (e.g., 1 in 100) or express it as a percentage (e.g., 99%). Most statistical methods in this class focus on providing an estimate of the likelihood that the two sets of values came from same population (or not). There are several methods to choose from, and the correct choice depends on the question being asked and the nature of the data.

There are many issues regarding the application of these relatively simple statistics and although the methods may be easily applied with computer applications, it is necessary to understand the limitations and relevance of each before using it. Other issues include normality of the data, distributions of the residuals, etc., which lead to the choice of parametric or nonparametric techniques. This is beyond the scope of this chapter, although there are many statistical books available that can assist in the choice of methods.

## Multivariate Methods

There are several multivariate statistical methods available to the environmental forensic investigator; many of them do the same thing but accomplish it in different ways with a range of possible outputs. The most useful are those that reduce the large number of potentially correlated variables measured (e.g., metals, organic compounds, bacterial species, etc.) to a small number of uncorrelated vectors that can be easily plotted in two dimensions. Typical methods for this purpose include principal components analysis (PCA), multidimensional scaling (MDS), and factor analysis (FA).

Other multivariate methods sort samples into groups and also produce graphical output that can be readily interpreted; examples of this type of analysis include cluster analysis (CA) and discriminant analysis (DA).

Another analysis that builds on PCA techniques is the 'projection to latent structure by means of partial least squares' (PLS; Wold et al. 1984). This method allows semiquantitative or better measures of the amount of variance explained by a chemical or biological 'signature' to be made, useful

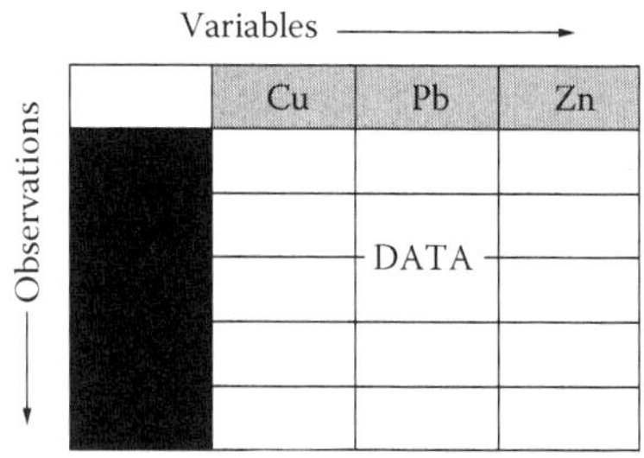


**Figure 9.2** Structure of the data matrix prior to PCA.

in apportioning the contribution from different sources in a complex environmental/multisource regime.

This section principally presents the use of PCA and PLS methods in environmental forensics as these have proven to be most useful in both visually and quantifiably presenting results (Hewitt and Mudge 2004; Mudge, 2002a, 2002b; Mudge, Birch, and Matthai 2003; Mudge and Duce 2005; Yunker et al. 1995). This does not mean that the other methods are less good, but rather reflects the preference of the author.

## PCA

One of the ways to investigate a source is to use the well established statistical technique known as principal components analysis. Data from a sampling campaign are arranged in a matrix in the form shown in Figure 9.2. Here the variables form the columns and the observations the rows; this is usually generated within a spreadsheet program.

In this technique, the data may be proportions or concentrations of each compound present, species numbers, etc. The data are subjected to matrix reduction to identify a small number of uncorrelated vectors in the matrix. The result of the matrix solution in Figure 9.2 can be visualised conceptually as if each variable were 'plotted' against each other and the best fit line to the data found. A schematic figure for three example compounds is shown in Figure 9.3.

The data may exhibit an underlying trend, which could be represented by the best fit straight line through the data; this is shown for the three-dimensional example in Figure 9.3 by the vector arrow labelled PC1. In statistics, it is usual to speak of *variance,* which is defined as the square of the standard deviation, which is a measure of how spread data are about a mean. In the situation here, variance is used to describe how much of the variability in the data can be explained by a model or line. When fitting lines to data, the proportion of variance explained can vary from 0 (does not fit the data at all) to 1.0 (explains everything). The significance of the amount of variance explained also changes according to the number of data points used;

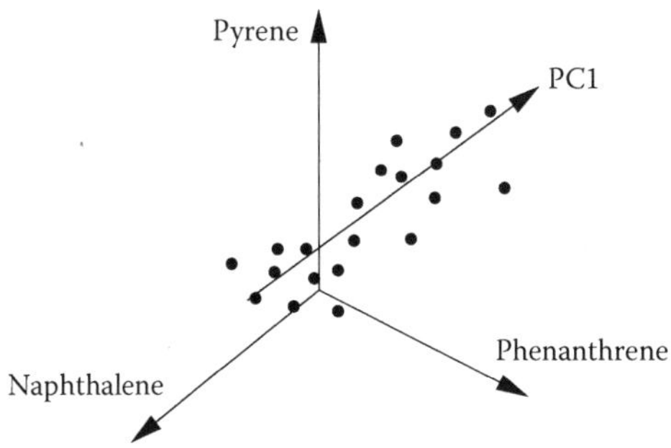


**Figure 9.3** Schematic figure showing how the first principal component (PC1) is fitted to data made up from three different PAHs.

the more data that are available, the more significant any explained variance will have.

This first line is known as principal component 1 (PC1) and explains the majority of the variance in the data. Typically, the amount of variance explained by the first component when using chemical compounds varies between 15 and 60%. In biological data, values towards the lower end of the range are more typical (e.g., Hewitt and Mudge 2004; Hopkins and Mudge 2004). A second component (PC2) is calculated orthogonally (at right angles) to the first so that there is no component of PC1 in PC2. All of these calculations are made using matrices in statistical software packages. If compounds co-vary (behave as if they had the same source) in this type of analysis, they will have similar loading factors. In the preceding example, if the pyrene and phenanthrene came from the same source, as the concentration of one went up, so would the other—they co-vary.

In a PCA, the numbers that describe the line labelled as PC1 in Figure 9.3 are called *loading factors*. Each compound (three in the preceding example) would have its own loading value. When compounds come from the same source, these loading factors are similar for each compound and, if these loadings were plotted against each other, would group closely together. It is possible to present this in a numerical form as:

$$PC1 = a[pyrene] + b[naphthalene] + c[phenanthrene]$$

where *a*, *b*, and *c* are the loading factors for each of the three PAHs. An example of the loading factors created from the PAH data from surface sediment samples collected in the Guanabara Bay, Brazil, can be seen in Figure 9.4 (Mudge and Neto, unpublished data).

The data in Figure 9.4 show several things: the PAHs with five or more rings cluster together to the left of the figure (load negatively on PC1) indicating that they co-vary. All the smaller PAHs are to the right of the figure

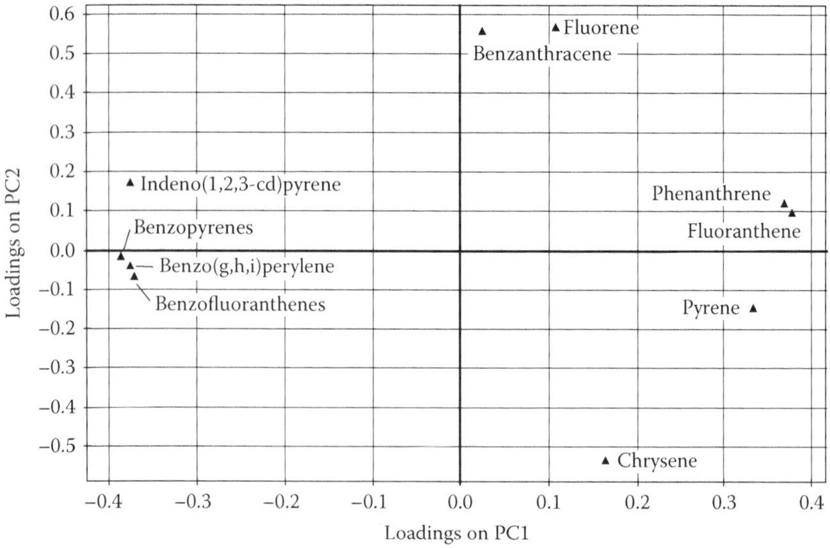


**Figure 9.4** Loadings in the PCA of example polyaromatic hydrocarbon (PAH) data from the Guanabara Bay, Brazil. PC1 explains 64.6% of the variance in the data and PC2 explains 16.8%. Those compounds that co-vary will lie close to each other in this two-dimensional diagram.

(load positively on PC1), indicating that they also co-vary to an extent and are mutually exclusive with regard to the 5+ ring PAHs. This implies that as the concentration of the 5+ ring compounds increases, the concentration of the smaller PAHs decreases. As this is PC1, the major factor, this also means that this distribution of these compounds has the most influence in the data. If we now look at the distribution along the second principal component (PC2), we see that the fluorene and benzanthracene co-vary and load positively, while chrysene loads negatively and thus behaves oppositely to the former two PAHs. The other compounds all have values of this axis close to zero (between −0.2 and +0.2), indicating that they have little effect on this component.

It is now possible to solve the preceding equation for each location (sample) and give a *score* for each. This value will be based on the concentration of each analyte in the sample and the loadings calculated previously. The corresponding scores plot for the PAH data shown earlier can be seen in Figure 9.5.

The scores plot includes an oval which represents the two dimensional 95% confidence limit, also known as the Hotelling $T^2$. Therefore, 95 out of 100 samples should fall inside this line. In the case shown in Figure 9.5, one (site 11) is close to this limit. The positions of the sample sites (numbered 1 to 15) within this figure are determined by their PAH content; those to the left of the zero line on PC1 are relatively enriched in the 5+ ring compounds, as shown by the inset contributions plot for site 7. The samples to the right

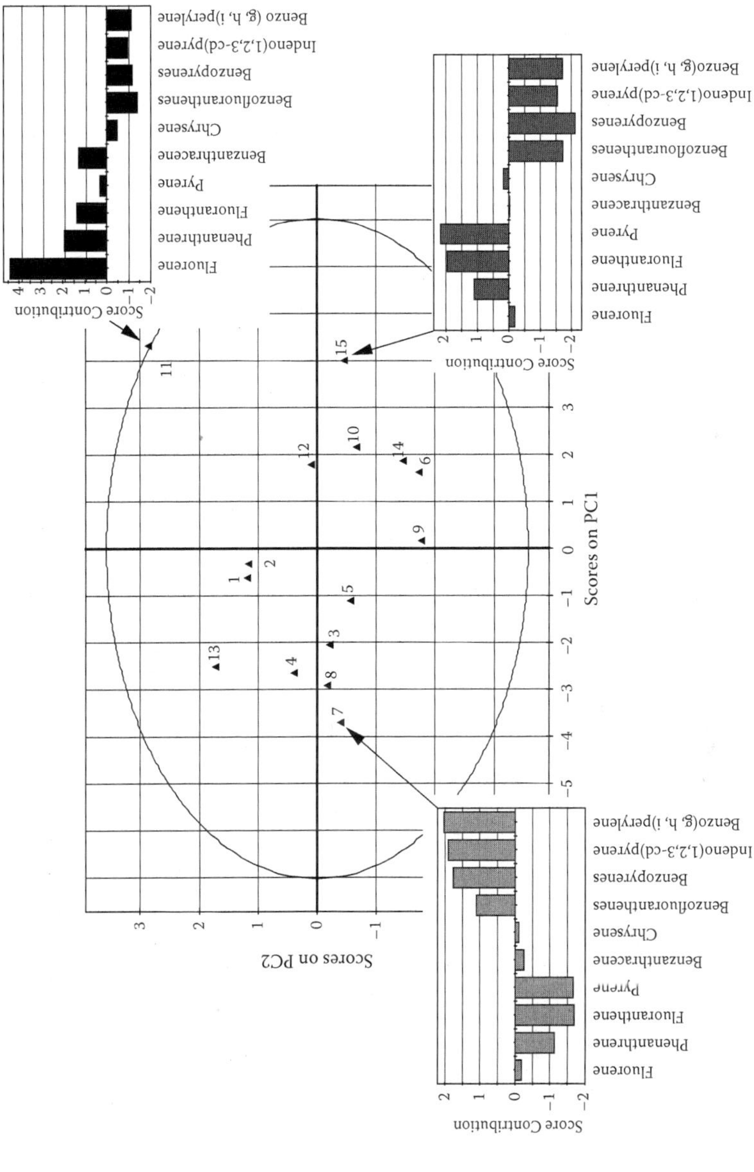


**Figure 9.5** The scores plot corresponding to the PAH data in the example from Guanabara Bay, Brazil. The bar diagrams show the relative contributions to each score from the PAHs analysed.

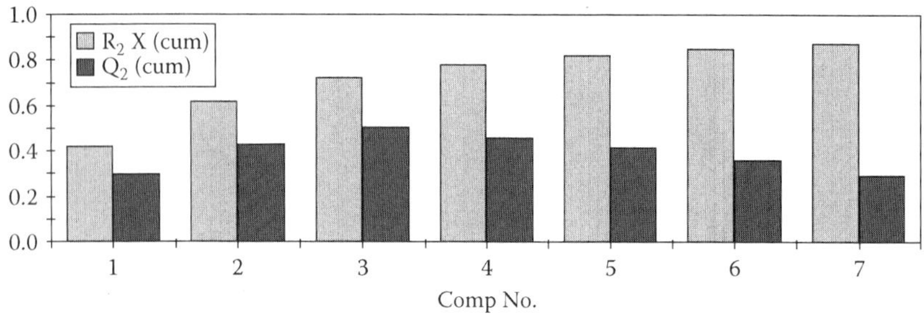


**Figure 9.6** The explained variance ($R^2$) and predictable variance ($Q^2$) for a set of principal components fitted to environmental data.

are relatively enriched in the smaller PAHs and depleted in the 5+ ring compounds. This can be seen in both the insets for site 11 and 15. However, site 11 loads towards the top of the figure (positively on PC2) and has a greater contribution from fluorene and benzanthracene and not chrysene.

It is usual to calculate two or more principal components as long as they are considered significant rather than just descriptors of random 'noise' in the data set. There are several criteria that may be used and a summary of six may be found in Johnson, Ehrlich, and Full (2002). A usual cut-off is around 5% of the variance explained by any one PC, although there are specific rules that commercial software employ based on the eigenvalues (usually greater than 1.0 or corresponding to sharp changes in explained fit; Jackson 1991). Principal components can be continually added, although the validity will reach a threshold and subsequent PCs are often noise. A typical plot that can be used to determine the number of valid PCs can be seen in Figure 9.6. In this case, after the addition of three PCs, the amount of predictable variance ($Q^2$) begins to decline. This can also be seen in the accompanying table of data (Table 9.1). The fourth PC is classed as 'not significant', so the maximum number of PCs that may be used in the interpretation of these data is three. The explained variance ($R^2$) continues to rise and will be asymptotic to 1.0.

Having determined the maximum number of valid PCs, by using a combination of the loadings and scores plot we are able to identify which compounds co-vary and which sites have similar attributes based on the chemical or biological analyses made. Often this is sufficient to demonstrate visually which sources are most closely related to which environmental samples. There are many examples of the use of PCA in determining the source of oil in spill situations (Burns et al. 1997; Lavine et al. 1995; Stout et al. 2001; Wang and Fingas 2003) and other environmental associations (D'Amore et al. 2005; Morrissey, Bendell-Young, and Elliott 2005; Qiang 2005).

In an environmental forensics context, PCA may be used to demonstrate which chemicals are behaving the same in a set of samples and are therefore

**Table 9.1 Values Associated with Principal Components Shown in Figure 9.6**

| | | | Type: PCA-X observations (N) = 25, variables (K) = 33 | | | | | |
|---|---|---|---|---|---|---|---|---|
| PC | $R^2X$ | $R^2X$(cum) | Eigenvalues | $Q^2$ | Limit | $Q^2$(cum) | Significance | Iterations |
| 1 | 0.42 | 0.42 | 10.5 | 0.299 | 0.0682 | 0.299 | R1 | 15 |
| 2 | 0.2 | 0.62 | 5 | 0.189 | 0.0707 | 0.431 | R1 | 16 |
| 3 | 0.101 | 0.721 | 2.53 | 0.128 | 0.0734 | 0.504 | R1 | 17 |
| 4 | 0.0549 | 0.776 | 1.37 | −0.0925 | 0.0762 | 0.458 | NS | 49 |
| 5 | 0.0427 | 0.819 | 1.07 | −0.0783 | 0.0794 | 0.416 | R2 | 38 |
| 6 | 0.0303 | 0.849 | 0.759 | −0.1 | 0.0828 | 0.358 | NS | 56 |
| 7 | 0.0256 | 0.875 | 0.641 | −0.115 | 0.0865 | 0.293 | NS | 48 |

*Notes:* $R^2$ is the explained variance and $Q^2$ is the variance predictable from the PCs. The significance can be seen according to whether they fit rule 1 (R1) or rule 2 (R2). NS means that the PCs are not significant. A fuller explanation of the rules may be found at the Umetrics Web site (www.umetrics.com).

from a similar source. It must be remembered, however, that this association does not on its own demonstrate guilt. There needs to be a mechanism by which the potential source can reach the affected environmental samples, and cause and effect will need to be demonstrated. This association needs to be demonstrated in all the significant PCs, although it may be possible to show that some chemicals are changing due to degradation or evaporation processes in the environment. In a similar manner, samples that co-vary may be shown; this is especially useful when showing a pure source material on a plot with a range of environmental samples. Classic examples of this include biomarkers within oils (Stout et al. 2001).

### *Normality and Transformations*

Many environmental chemical concentrations exhibit non-normal distributions (Limpert, Stahel, and Abbt 2001); this may not be the case for bulk analytes in sediments such as Al or Ca, but is often the case for trace contaminants such as Pb and Zn: Compare Fe as a bulk analyte with Cu as a trace analyte in Figure 9.7.

In Figure 9.7, the Fe component approximates to a normal distribution while the trace metal, Cu (potentially a contaminant from anthropogenic sources), is skewed toward low values. In the latter case the data are not normally distributed; however, if the data are transformed to their $\log_{10}$, they may become normal. This type of distribution is called log-normal and is common for such elements and compounds (Cabaniss et al. 2000). With biological data (species abundances, biomass, etc.), a square root transformation is often used (Arcos, Racotta, and Ibarra 2004; Hall et al. 1997; Koster, Racca, and Pienitz 2004).

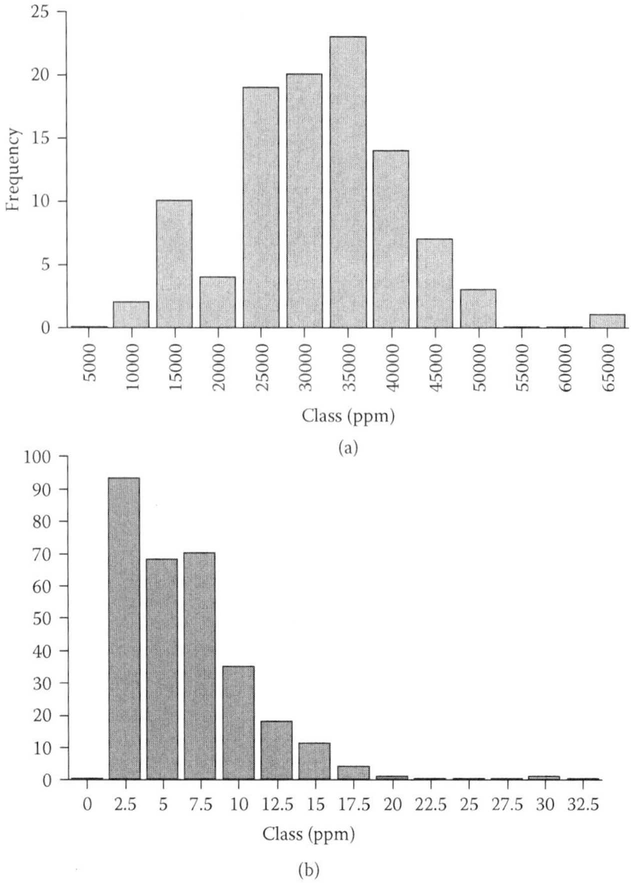


**Figure 9.7** Frequency histograms for (a) a bulk constituent of sediments, Fe, and (b) a trace constituent, Cu. (Data from Mudge, Birch, and Matthai. *Environmental Forensics,* 4(4): 305–312, 2003.)

## *Mean Centring to Unit Variance*

In PCA, what is important is the relative proportion of each compound, element or species, rather than the concentration per se. Wide differences in concentration and in the range of values between samples would not help in such an analysis. Therefore, in most PCA, the data are automatically transformed by mean centring to unit variance. The process and rationale can be seen in Figure 9.8. Here the length of the bars represents the concentration range and the horizontal line is the mean.

## *Using Proportions and Closure*

Despite mean centring to unit variance, if no further transformations are made to the data, the first PC may just represent a concentration gradient. This may be useful in some instances, but there are simpler ways to dem-

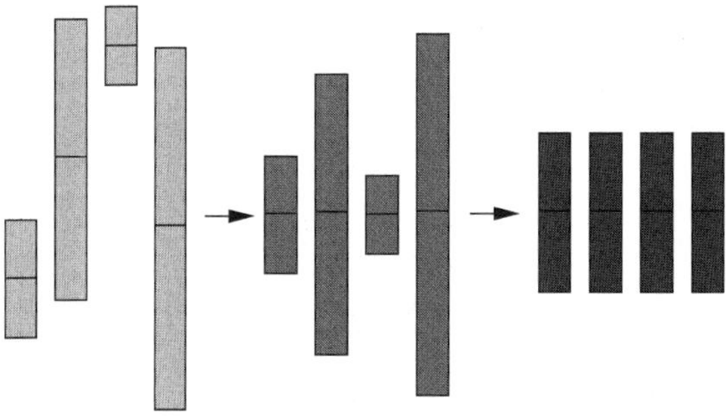


**Figure 9.8** Process of mean centring to unit variance to remove the effects of wide concentration ranges and different means.

onstrate these effects without the use of PCA. Therefore, the most common transformation made to data is to convert data to proportions or percentages of the total concentration or by class of compound (e.g., sterols and PAHs treated independently so that the sum of the proportions is 2.0; Mudge and Duce 2005). In this way, the internal relationship between each chemical compound or species in biological data is used independently of the total concentration or number of individuals in any sample. An example of the difference can be seen in Figure 9.9, which uses the same data but as raw concentrations in one case and as proportions in another.

If raw concentration data are used to generate the principal components, concentration itself has a major effect on the values up to a threshold (Figure 9.10a). The score for each site is also directly related to the value of PC1 (shown in Figure 9.10b; 5 has been added to the score to make all values positive and therefore able to be plotted on a log–log figure). The usefulness of PC1 is reduced as a signature or source discriminator and PC2 may take on that role.

However, in the case where proportion data are used to generate the PCs (Figure 9.9b), it is possible to see which compounds co-vary and have similar sources. For instance, most of the *iso* and *anteiso* fatty acids cluster together, indicating a bacterial source (Kaneda, 1967; Kolattukudy, Croteau, and Buckner 1976), while the long chain saturated fatty acids cluster in a separate location toward the top of the figure, indicating a terrestrial (vascular) plant source (Tulloch 1976; Wakeham et al. 1997). Distinct vectors can be seen within the two-dimensional distribution that are absent from the raw data loadings plot (Figure 9.9a). In this example, both the mean variable concentrations and the total sample concentrations are independent of the loadings or scores for PC1, respectively.

The potential problem with using proportion data is that as one value goes up, others must come down as the total is fixed (Johansson, Wold, and

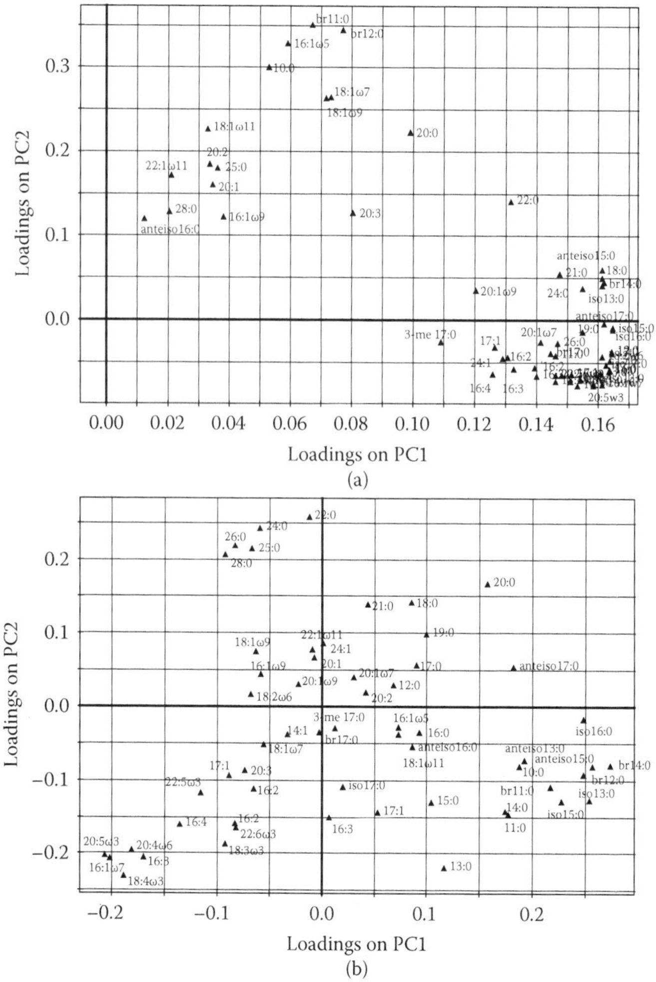


**Figure 9.9** The loadings for fatty acid data as (a) raw data and (b) proportion data. (From Mudge, S. M. et al. *Organic Geochemistry,* 29(4): 963–977, 1998.)

Sjodin 1984). This can lead to a condition of closure where changes in the signature are artificially created by the transformation method. For example, if two samples had the same chemical signature for most compounds, but a single chemical was present in a greater concentration in one of the samples due to the addition of another source, all the other values would decrease and the signature of the underlying chemicals would be lost.

The effects of closure can be reduced by using a log-ratio transformation (Yunker and Macdonald 2003). In this case, the data are normalised first to the concentration total (proportions) and then to the geometric mean to produce a data set that is not affected by negative bias or closure. With some data, this is not necessary (Mudge and Duce 2005), but it may be for others.

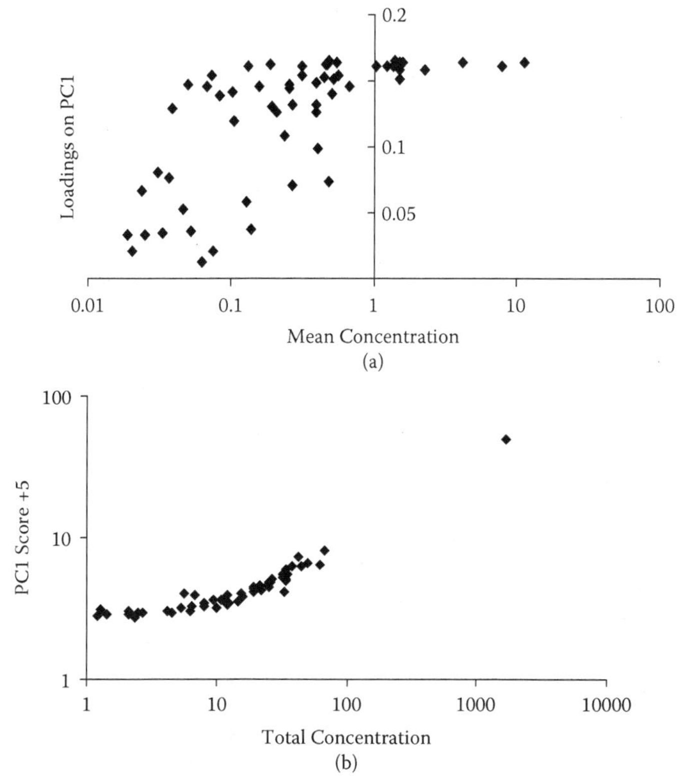


**Figure 9.10** (a) The mean concentration of each fatty acid has a major effect on the loading factors in PC1 using raw concentration data. (b) The total concentration of all compounds in each sample correlates to the score.

## *Zeros*

Analyses of data with zero values in the matrix can be a problem; transformations become more complex and there is the question of how many zeros can exist within a data set and it still be used meaningfully. Most software will handle missing values and replace them with the system default (e.g., –99), but there is the assumption that they are evenly distributed through the data set. If the zeros are systematically distributed (half of the samples were not analysed for this compound), bias may be built in and the results may not be as robust as they should be. Within environmental data there are usually three types of zeros. Depending on the nature of the zero, how it is treated in any subsequent multivariate analysis changes.

**Chemical data—less than the limit of detection.** In this case the zero that is sometimes used is not correct, as the compound may well be present in the sample but the analytical method does not have the necessary sensitivity or recovery to measure its presence. By implication, if we had a better analytical method, we would detect it. One of the accepted methods of treating such data is to enter a value that is half of the limit of detection. This will vary between compounds, so several possible values may be entered in any one data set. The advantage of doing this is twofold: (1) that analyte is available in the data set to have an influence on the signature, and (2) if a log transformation is used, the number can remain in the matrix.

**Chemical data—not analysed.** If the analytical result for a particular compound was not available, this needs to be coded appropriately in the data set. In this case it is not a zero and it is not less than the limit of detection; we just do not know the value because we did not measure it. The best solution to this situation is to leave the cell blank within the data matrix (Figure 9.2) and allow the software to enter a missing value notation.

**Biological data—species not present.** This is the only example here where a zero is the appropriate value to enter. Provided your methods are such that you would be able to observe the occurrence of that species, if it is not there, it is a zero in the data set. Problems may now ensue from any transformations that may reject that zero (e.g., log transform). However, transforms are possible if 1.0 is added prior to logging the data ($\log_{10}(x + 1)$). In this case, the log of 1.0 is 0, so observations in which the absence of a species can be as important as the presence of one can be preserved and used to characterise a sample.

### *Summary: Checklist for PCA*

1. Collect quality data in the first instance that you have confidence in.
2. Consider the validity of zeros in the data set and treat appropriately.
3. Use proportion data to remove the concentration effect and allow the signature to be seen.
4. Check for closure in such data by using the log-ratio transform.
5. Determine the number of valid principal components by sequential addition.
6. Plot the loadings and scores to explain the underlying rationale in each component.

## PLS

The partial least squares technique was developed by Wold et al. (1984) and has evolved into a powerful tool in environmental forensics (e.g., Mudge and Duce 2005; Yunker et al. 1995). In essence, PLS performs PCA on data that are defined as the signature (Geladi and Kowalski 1986). This data set, which can be chemical, physical, or biological in nature, is called the X-block and, ideally, will be a pure source sample. However, it could be made up of environmental samples that have a high proportion of a single source, such as coal deposits or oil from a spill. Although it is possible to use a single sample to define the signature, it is better to have several samples, which may cover a range of time and space to ensure that a representative profile is obtained.

Since the samples come from the same source, although the concentrations may vary, PCA will generate a principal component 1 that explains most of the variance in the analytical data, typically >90%. This projection or vector in $n$-dimensional space, where $n$ is the number of chemical compounds analysed, can be described by a series of loading factors on each compound; those compounds that have a major impact on PC1 will have high loadings (either positive or negative), whereas those that are relatively unimportant and therefore do not have a major influence on the data will have values close to zero. PC2 is fitted orthogonally to the first component so that no component of PC1 influences PC2. Once the first two PCs have been elucidated, their projection can be described in terms of the two sets of loadings in the same way as PCA. Each potential source can be characterised in this manner independently of each other and will have a unique set of loading factors that define that signature.

These projections, which represent the signatures defined in terms of the variables used, can now be applied to the environmental data called the Y-block. The amount of variance explained and predicted by each X-block signature can be quantified. Table 9.2 shows the model fit for one X-block signature with the associated $R^2$ and $Q^2$ values. The results can be shown graphically for each site, either through a scatter plot of the weightings on each sample or as the total variance explained and predicted. If the signature is similar to that of the environmental data, a high value for the explained variance is produced. Conversely, if a poor fit is produced, the explained variance is also small. Each signature can be fitted in turn and all are fitted independently of each other. If none of them explain the variation seen in the data, the fits will be small in every case. A fuller treatment of the PLS methodology, including the matrix manipulations used, can be found in Geladi and Kowalski (1986). The advantage of PLS over other methods (e.g., simple ratios) is the way in which it uses all compounds and develops a signature based on the internal relationships of each one.

An example of the amount of variance that can be explained ($R^2$) or predicted ($Q^2$) in surface sediment samples from the Ria Formosa Lagoon, Portugal, can be seen in Figure 9.11. In this case, sterol concentrations were measured in 59 sediments, and four of these adjacent to the sewage discharge (sites 33 to 36) were used to develop the X-block signature.

The amount of variance that can be explained by the signature (*goodness of fit*, $R^2$) is greater than the amount of variance that can be predicted (*goodness of prediction*, $Q^2$) (Eriksson et al. 1999). Therefore, the amount of variance that can be predicted from the X-block model is a better measure of the signature contribution to each individual sample. The values for the black bars ($Q^2$) vary between 0.0, where no amount of the signature is present in

**Table 9.2 Predicted and Explained Variance from a PLS Analysis of Metals in Southeast Australia**

| Type: PLS observations (N) = 8, variables (K) = 305 (X = 3, Y = 302) | | | | | | | | | | |
|---|---|---|---|---|---|---|---|---|---|---|
| A | $R^2X$ | $R^2X$(cum) | Eigenvalues | $R^2Y$ | $R^2Y$(cum) | $Q^2$ | Limit | $Q^2$(cum) | Significance | Iterations |
| 1 | 0.966 | 0.966 | 2.9 | 0.611 | 0.611 | 0.502 | 0.05 | 0.502 | R1 | 4 |
| 2 | 0.0304 | 0.997 | 0.0912 | 0.161 | 0.773 | −0.116 | 0.05 | 0.452 | R2 | 3 |

*Source:* Mudge, S. M. et al. *Environmental Forensics,* 4(4): 305–312, 2003.

*Notes:* The first PC in the X-block explains 96.6% of the variance in that signature data that, on average, explains 61.1% of the variance in the Y-block. The amount of variance in the Y-block that can be predicted from the X-block first PC is 50.2%.

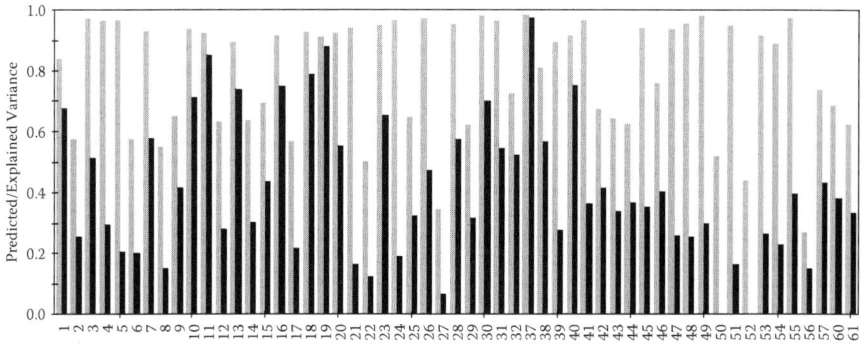


**Figure 9.11** The amount of variance that can be explained (gray bars) or predicted (black bars) in surface sediment samples from the Ria Formosa Lagoon, Portugal, using four samples collected adjacent to the sewage discharge. (Data from Mudge, S. M. et al. *Water Research,* 33(4): 1038–1048, 1999.)

the sample, to 0.975 (97.5% of the variance predictable) in the case of site 37. Since this sample was geographically close to those used to generate the X-block signature, this is not surprising.

### *Overlap between Signatures*

In general, the more compounds that are used, the better the specificity of the signature is. There are cases, however, where this is not the case. If the samples all contain several compounds or species common to all sites, including those used to develop the signature, the selectivity of the methods is less than it might be. In such cases, it would be better from a diagnostic viewpoint to remove those compounds or species from the PLS analysis. This can be achieved through simple inspection of the data and removal of the common component. The relative importance of each compound within the signature can be viewed through its PCA loadings; in a PLS context, this can be visualised with a plot of the PC1 loadings in the signature versus the PC1 loadings in the environmental data (e.g., Figure 9.12).

Those compounds that are relatively more important in the X-block—the samples used to generate the signature—will have either more positive or more negative values than the corresponding value in the Y-block. In the example shown in Figure 9.12, the Pb in the signature is more important than in the rest of the environmental samples. The converse is true of the Mn, which has a greater importance in the Y-block. In deciding which chemicals are most important in determining the signature and are therefore most diagnostic, this type of plot can be used.

PCA can be used independently of the PLS technique to determine the number of potential sources that may be present in the Y-block. The scores plot from such an analysis will group sites according to their chemical composition; those that co-vary are likely to have the same or a similar source.

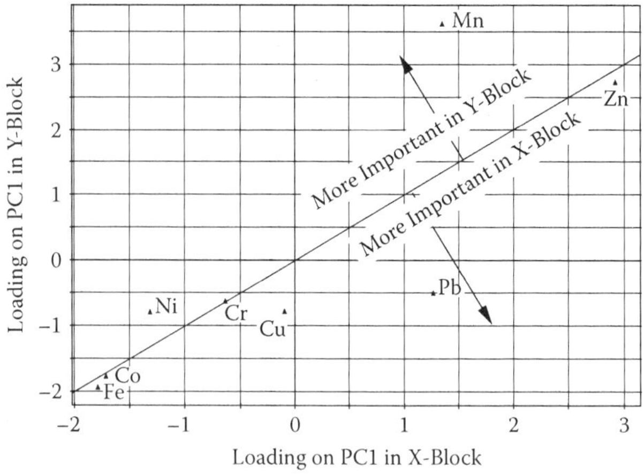


**Figure 9.12** The PC1 loading factors for each compound used to make the signature (X-block) plotted against the loadings in the Y-block. (Data from Mudge, S. M. et al. *Environmental Forensics,* 4(4): 305–312, 2003.)

Inspection of the groupings may provide an insight into the number of sources, although care must be exercised when dealing with mixtures of variable composition. This PCA technique may also be used to explore the source data and determine the groupings within the possible source materials.

Two cases can be seen in Figure 9.13. In the first figure, the four signatures fail to predict more than 80% of the variance in the samples. In one case SP(Int), none of the variance was predictable. The second figure is for samples taken from the same locations but several weeks later; in this case, the four signatures predicted more than 100% of the variance (1.0). In four cases, it is approximately 180% of the variance. This is due to overlap between the signatures and it may be that simply presenting the data normalised to 100% would be appropriate. However, in the first case, this would not be right and significant overimportance may be ascribed to some signatures. In both of these cases, it would be better to explore the commonality between variables and reduce the data set accordingly. In one example using metals (Mudge et al. 2003), complete separation between adjacent river catchments could be achieved by reducing the number of variables to four comprising the diagnostic metals in the system and leaving out the bulk mineral (background) elements.

## *Summary: Checklist for PLS*

1. Conduct PCA on the data set to get a measure of the number of likely sources within the environmental data.
2. Collect sufficient source samples or samples from sites with only one source in them (if possible) to characterise the signatures in both time and space.

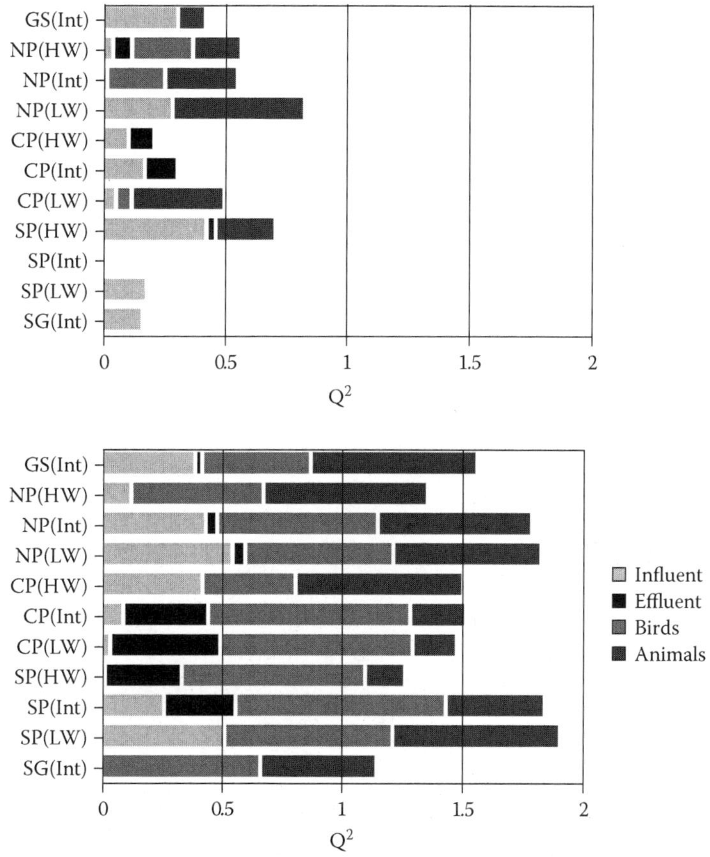


**Figure 9.13** The predictable variance in a series of samples from Blackpool Beach, United Kingdom, based on four lipid biomarker signatures for sewage influent and effluent, birds and animals. (Mudge, S. M., *The source of organic matter on Blackpool Beaches (2000)*. Bangor: University of Wales, 2001.)

3. When using multiple signatures to describe the composition of samples, check for overlap between signatures and treat accordingly (normalise or choose variables more selectively).

## Geostatistics

Geostatistics is the correlation and interpretation of data in a spatial context; the most common usage of geostatistics is the production of contour plots. In an environmental forensic context, data for several analytes may be collected over a wide area in order to determine the existence of gradients away from potential sources. These statistics may be used in court to demonstrate that a gradient exists along one axis but not along another, thereby identifying the source party (e.g., Casado et al. 1994). These can be powerful tools, but as with any method, they require quality data and a rigorous application of analysis to be valid. This starts with production of a variogram.

## Variogram/Semivariogram Analysis

The first empirical works using these methods were done by Krige (1951). The statistical theory was later developed by Matheron (1963), who described the variogram as a curve representing the degree of continuity of mineralisation (in a geological context). In general, the variogram is a function of increasing distance $h$, since on average the farther samples are from one another, the more their properties are dissimilar. Usually, the computation of the semi-variance is the first step in geostatistical analysis. The semivariance $\gamma(h)$ is expressed by the following equation:

$$\gamma(h) = \frac{1}{2n(h)} \sum_{i=1}^{n-h} [z(x_{i+h}) - z(x_i)]^2$$

where $n(h)$ is the number of sample pairs at each width $h$ and $z(x_i)$ and $z(x_{i+h})$ are the concentrations of the analyte $x$ at $i$ and $i + h$.

The graphical representation of the semivariance $\gamma(h)$ as a function of the step width is called a semivariogram and the representation with $2\gamma(h)$ is called a variogram (Figure 9.14). Thus, these functions describe the relations between spatially correlated data. The semivariogram has to be constructed on the basis of experimental data obtained by sampling over the area under investigation. An important advantage of geostatistical methods is the fact that the sampling points are not necessarily regularly distributed. For environmental investigation, this point is of particular importance because not all locations in the area can be sampled due to accessibility and safety considerations. An experimental semivariogram can be modelled by fitting a simple function to the data points; linear, spherical, or exponential models are often used.

According to the preceding equation, the theoretical semivariogram function has the value $\gamma = 0$ for $h = 0$. Semivariograms obtained from experimental data often have a positive value of intersection with the $\gamma(h)$-axis. This point of intersection is named the *nugget effect*. This term was coined by the mining industry and indicates an unexplained random variance that characterises the microheterogeneity at the sampling location. From this point, the semivariance rises up to the variance of the data set, named the *sill*. Up to this point the regionalised variables at the sampling locations are correlated. Subsequently, they must be considered spatially independent. This distance

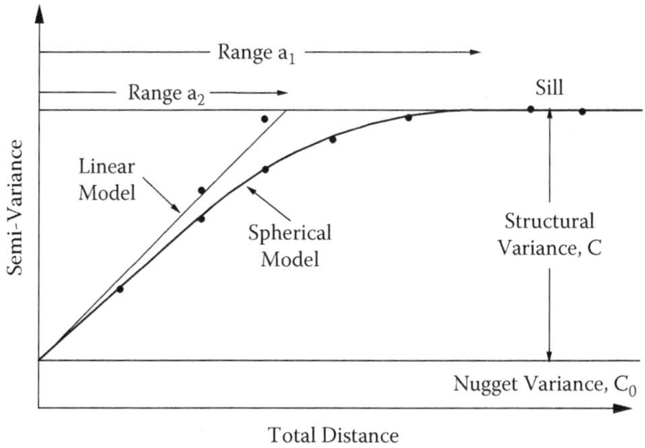


**Figure 9.14** An example of a semivariogram indicating the key elements of the plot. (After Trangmar et al. 1985.) The x-axis is the distance between pairs of points and the y-axis shows the semivariance for each pair. In the simple linear gradient, the data would describe a line (linear model) from the origin of the axes; in this example the model starts from a positive (nonzero) value due to the nugget effect. At some distance between points (range), there is no longer any change in the variance with increasing distance.

of correlation is called *range*. An example of an experimental semivariogram with nugget effect, sill, and range is demonstrated in Figure 9.14.

When $\gamma(h)$ equals the sill at all values of $h$, the experimental semivariograms are said to exhibit a pure nugget effect. This effect arises from very large point-to-point variation at short distances of separation and indicates a total absence of spatial correlation at the sampling scale used.

The robustness of the variogram is affected by the distribution of the population. Also, the larger the number of samples from which it is computed, the more precisely it is estimated. However, there are many examples from fewer than 50 points and some as few as 24. Some papers have attempted to justify the use of such small sample numbers on the grounds that each semivariance is calculated from many comparisons, perhaps following the misleading advice of Journel and Huijbregts (1978). Variograms computed from small sample numbers almost always appear erratic, and in many instances the principal reason is that there are too few data (Webster and Oliver 1992).

If the data used to generate the variogram describe a gradient, then it is possible to use an interpolation method to regularise the data onto a grid; the most widely used method for this is kriging. However, if the data are randomly distributed across the spatial coordinates, then kriging is inappropriate and it is better to use a classed posting of the data to show spot values without inferring a gradient between them.

A series of examples can be seen in Figure 9.15. In the first case, there is a large nugget effect superimposed on a gradient that does not reach a sill

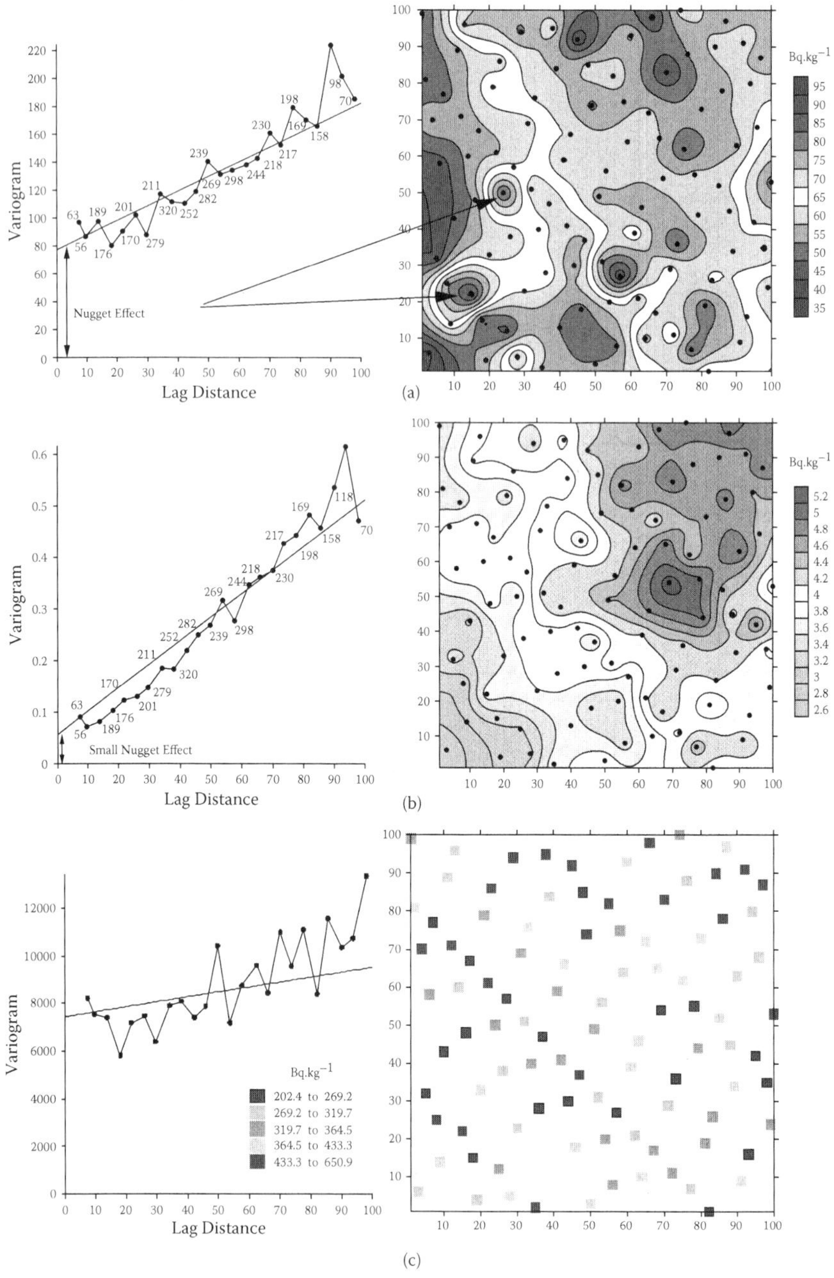

**Figure 9.15 (See colour insert following page 336.)** Three examples of variograms: (a) gradient across square but with a large nugget effect; (b) gradient across square but with a very small nugget effect; and (c) no gradient. The contour plots or classed postings show the spatial distribution of the data. The point labels on the variogram indicate the number of pairs of points used at that log distance.

within the region it was measured over (in this case, 100 m). Two examples of nuggets can be seen where high values exist outside the gradient and are surrounded by low values without a gradient. In the second example, there is only a small nugget effect and the data describe a trend across the sample region from lower left to upper right. In the third case, however, there is no significant gradient across the sample square with a random distribution of concentrations. A contour plot based on an interpolation between the points would not be valid in this case, so a classed posting showing each individual sample value as a colour is more appropriate. In this figure, red squares (for high values) and blue (for low values) can be seen adjacent to each other.

### *Kriging*

The variogram therefore has a central role in the spatial analysis. It summarises the variation of a compound or other property within a region. It is essential for kriging and even though estimates obtained are fairly stable, their estimation variances, and hence their confidence limits, are sensitive to the variogram. It is difficult to incorporate the error in the variogram into the final kriging variance, so it is important to estimate the variogram itself as well.

Kriging is a special regression method for interpolation of spatially or temporally correlated data under variance minimisation. The normal distribution of the data is an important condition. If the original data are not normally distributed, which is often the case for trace components in environmental samples (Limpert et al. 2001), the logarithm of the data or otherwise transformed data have to be applied to obtain a normal distribution of the data set. The function for simple point kriging is:

$$z^*(x_0) = \sum_{i=1}^{n} [\lambda_i z(x_i)]$$

where $z^*(x_0)$ is the estimated value of the random function $Z$ at the unsampled location $x_0$ and $\lambda_i$ are the weights of the $z(x_i)$. These weights $\lambda_i$ can be obtained by means of the kriging system:

$$\sum_{j=1}^{n} \lambda_i \gamma(x_i, x_j) + \mu = \gamma(x_i, x_0), \qquad i = 1, 2, \ldots, n$$

with $(x_i, x_j)$ = distance between the sampling points $x_i$ and $x_j$; $(x_i, x_0)$ = distance between the sampling points $x_i$; the unsampled point $x_0$, which shall be estimated; and $\mu$ = Lagrange multiplier.

The empirical weights have to be normalised with the following condition to obtain a distortion-free estimation value:

$$\sum_{i=1}^{n} \lambda_i = 1$$

The kriging variance is given by:

$$\sigma_K^2 = \sum_{i=1}^{n} \gamma(x_i, x_0) + \mu$$

As with many of these functions, they can be performed without in-depth knowledge by using computer packages. In this case, both the construction of variograms and contouring were conducted using the SURFER package from Golden Software (Colorado). An example of the variogram and the interpolated values shown as a contour map can be seen in Figure 9.16. The data used to make the contour map are the same in each case, although the first was created without using the variogram model and shows the nugget effect. The second contour map used the variogram model, and the overall trend underlying the data can be much more readily visualised.

### *Anisotropy*

The dispersion of contaminants occurs by physical processes that frequently have preferred orientations. For example, at the mouth of a river the coarse material settles out fastest, while the finer material takes longer to settle. Chemicals that associate with sediments will also follow this pattern as well with higher concentrations associated with the muds, which have a larger surface area. Thus, the closer one is to the shoreline, the coarser the sediments (and the lower the concentration) are, while further from the shoreline the sediments are finer (and the concentration is higher). When interpolating at a point, an observation 100 m away in a direction parallel to the shoreline is more likely to be similar to the value at the interpolation point than another observation 100 m away in a direction perpendicular to the shoreline. Anisotropy takes these trends in the data into account during the gridding process (SURFER Users Guide, Golden Software, Colorado).

In the kriging process, points closer to an interpolation location are given more weight than points farther away. If, as in the preceding example, the points in one direction have more similarity than points in another direction, it is advantageous to give points in a specific direction more weight in determining the value of the new interpolated point. The relative weighting is defined by the anisotropy ratio. In an environmental forensic context, this

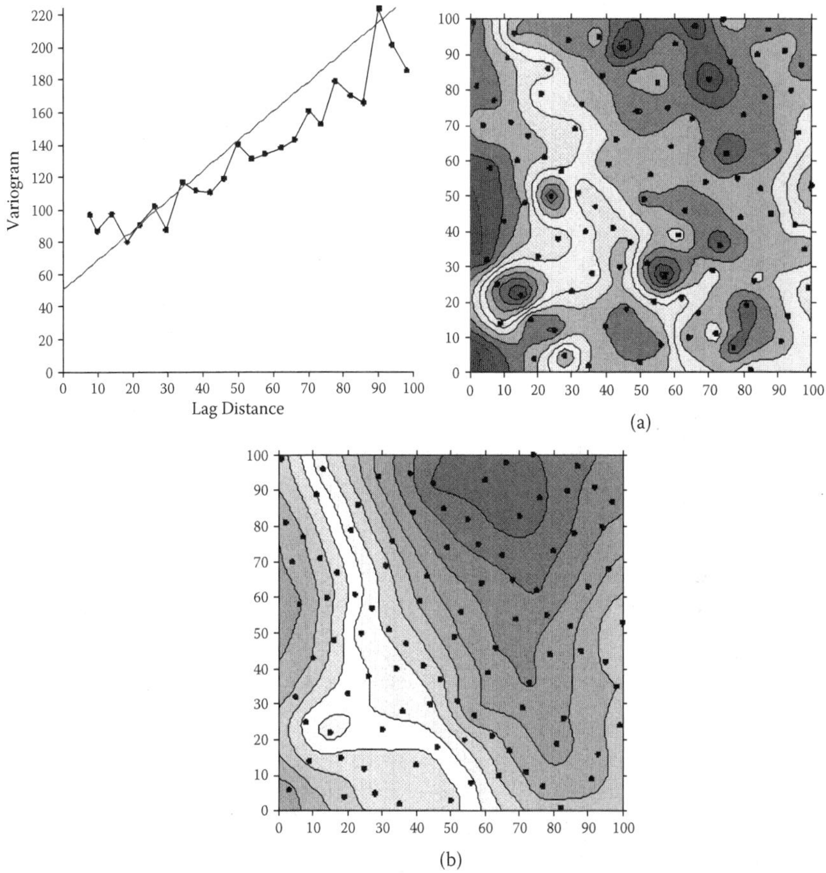


**Figure 9.16 (See colour insert following page 336.)** The variogram for data across a square with either (a) the contour produced without using the variogram model or (b) the same data interpolated using the model generated by the variogram.

directional aspect may be used to demonstrate that a gradient exists in one direction (away from a source) and not in another.

The anisotropy settings include a ratio and an angle. Anisotropy can usefully be visualised as an ellipse (Figure 9.17). The ellipse is specified by the lengths of its two orthogonal axes and by an orientation angle. The ratio is the maximum range divided by the minimum range. An anisotropy ratio less than two is considered mild, while an anisotropy ratio greater than four is considered severe. Typically, when the anisotropy ratio is greater than three, its effect is clearly visible on grid-based maps. The angle is the preferred orientation (direction) of the major axis in degrees.

The anisotropy function may be used to demonstrate preferred dispersion directions as well as the cross-plume spread; those plumes of contaminants that follow a narrow path (river channel, aquifer) will have large values

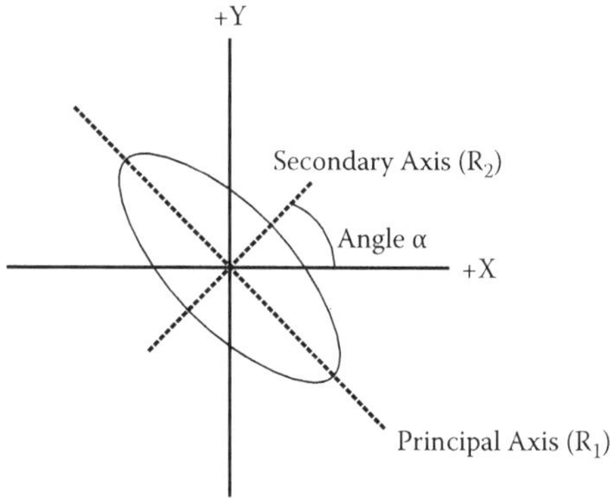


**Figure 9.17** A schematic visualisation of anisotropy in data. The ratio is defined by the $R_1/R_2$ and the angle is the angle of the principal axis relative to the positive X direction measured counterclockwise.

of the ratio, while radial dispersion will have a value closer to one. These functions together have been used to indicate the most likely source of a contaminant (Rouhani, personal communications).

### *Summary: Checklist for Geostatistics*

1. Design a sampling program such that sufficient data may be collected to make subsequent interpolations statistically valid.
2. Choose sample locations according to the physical processes that may lead to the dispersion of contaminants. In some cases there may be little spread away from a river channel but there will be considerable travel along it.
3. Conduct variogram analysis to determine if there are trends and gradients across the sample grid. At the same time, consider anisotropy to demonstrate direction travel and spreading.
4. If the variogram indicates a gradient across a sample grid, interpolation through kriging may be valid and so a contour plot can be produced. Conversely, if the variogram indicates a pure nugget effect (randomly distributed data), a classed posting may be the most appropriate spatial figure that can be produced.

## References

Arcos, F. G., I. S. Racotta, and A. M. Ibarra. (2004) Genetic parameter estimates for reproductive traits and egg composition in Pacific white shrimp *Penaeus* (Litopenaeus) *vannamei. Aquaculture,* 236(1–4): 151–165.

Burns, W. A., P. J. Mankiewicz, A. E. Bence, D. S. Page, and K. R. Parker. (1997) A principal-component and least-squares method for allocating polycyclic aromatic hydrocarbons in sediment to multiple sources. *Environmental Toxicology and Chemistry,* 16(6): 1119–1131.

Cabaniss, S. E., Q. H. Zhou, P. A. Maurice, Y. P. Chin, and G. R. Aiken. (2000) A log-normal distribution model for the molecular weight of aquatic fulvic acids. *Environmental Science & Technology,* 34(6): 1103–1109.

Casado, L. S., S. Rouhani, C. A. Cardelino, and A. J. Ferrier. (1994) Geostatistical analysis and visualisation of hourly ozone data. *Atmospheric Environment,* 28(12): 2105–2118.

D'Amore, J. J., S. R. Al-Abed, K. G. Scheckel, and J. A. Ryan. (2005) Methods for speciation of metals in soils: A review. *Journal of Environmental Quality,* 34(5): 1707–1745.

Eriksson, L., E. Johansson, N. Kettaneh-Wold, and S. Wold. (1999) *Introduction to multi- and megavariate data analysis using projection methods (PCA & PLS).* Umea, Sweden: Umetrics.

Geladi, P., and B. R. Kowalski. (1986) Partial least squares regression: A tutorial. *Analytica Chimica Acta,* 185: 1–17.

Hall, R. I., P. R. Leavitt, J. P. Smol, and N. Zirnhelt. (1997) Comparison of diatoms, fossil pigments and historical records as measures of lake eutrophication. *Freshwater Biology,* 38(2): 401–417.

Hewitt, E. J., and S. M. Mudge. (2004) Detecting anthropogenic stress in an ecosystem: 1. Meiofauna in a sewage gradient. *Environmental Forensics,* 5(3): 155–170.

Hopkins, F. E., and S. M. Mudge. (2004) Detecting anthropogenic stress in an ecosystem: 2. Macrofauna in a sewage gradient. *Environmental Forensics,* 5(4): 213–223.

Jackson, J. E. (1991) *A user's guide to principal components.* New York: John Wiley & Sons.

Johansson, E., S. Wold, and K. Sjodin. (1984) Minimising effects of closure on analytical data. *Analytical Chemistry,* 56(9): 1685–1688.

Johnson, G. W., R. Ehrlich, and W. Full. (2002) Principal components analysis and receptor models in environmental forensics. In *Introduction to environmental forensics,* ed. B. L. Murphy and R. D. Morrison, 461–515. New York: Academic Press.

Journel, A. G., and C. J. Huijbregts. (1978) *Mining geostatistics.* London: Academic Press.

Kaneda, T. (1967) Fatty acids of the genus *Bacillus. Journal of Bacteriology,* 93: 894–903.

Kolattukudy, P. E., R. Croteau, and J. S. Buckner. (1976) Biochemistry of plant waxes. In *Chemistry and biochemistry of natural waxes,* ed. P. E. Kolattukudy, 289–347. Amsterdam: Elsevier.

Koster, D., J. M. J. Racca, and R. Pienitz. (2004) Diatom-based inference models and reconstructions revisited: Methods and transformations. *Journal of Paleolimnology,* 32(3): 233–245.

Krige, D. G. (1951) A statistical approach to some basic mine valuation problems on the Witwatersrand. *Journal of Chemical, Metallurgical & Mining Society of South Africa,* 52: 119–139.

Lavine, B. K., H. Mayfield, P. R. Kromann, and A. Faruque. (1995) Source identification of underground fuel spills by pattern-recognition analysis of high-speed gas chromatograms. *Analytical Chemistry,* 67(21): 3846–3852.

Limpert, E., W. A. Stahel, and M. Abbt. (2001) Log-normal distributions across the sciences: Keys and clues. *Bioscience,* 51(5): 341–352.

Macii, D., P. Carbone, and D. Petri. (2003) Management of measurement uncertainty for effective statistical process control. *IEEE Transactions on Instrumentation and Measurement,* 52(5): 1611–1617.

Matheron, G. (1963) Principles of geostatistics. *Economic Geology,* 58: 1246–1266.

Morrissey, C. A., L. I. Bendell-Young, and J. E. Elliott. (2005) Identifying sources and biomagnification of persistent organic contaminants in biota from mountain streams of southwestern British Columbia, Canada. *Environmental Science & Technology,* 39(20): 8090–8098.

Mudge, S. M. (2001). *The source of organic matter on Blackpool Beaches (2000).* Bangor: University of Wales.

______. (2002a). Aspects of hydrocarbon fingerprinting using PLS—New data from Prince William Sound. *Environmental Forensics,* 3(3–4): 323–329.

______. (2002b). Reassessment of the hydrocarbons in Prince William Sound and the Gulf of Alaska: Identifying the source using partial least squares. *Environmental Science & Technology,* 36(11): 2354–2360.

Mudge, S. M., D. J. Assinder, and A. T. Russell. (2001) Micro-scale variability of contaminants in surface sediments: The implications for sampling. R&D Technical Report P3-057/TR, Environment Agency, UK, 88.

Mudge, S. M., M. Bebianno, J. A. East, and L. A. Barreira. (1999) Sterols in the Ria Formosa Lagoon, Portugal. *Water Research,* 33(4): 1038–1048.

Mudge, S. M., G. F. Birch, and C. Matthai. (2003) The effect of grain size and element concentration in identifying contaminant sources. *Environmental Forensics,* 4(4): 305–312.

Mudge, S. M., and C. E. Duce. (2005) Identifying the source, transport path and sinks of sewage derived organic matter. *Environmental Pollution,* 136(2): 209–220.

Mudge, S. M., J. A. East, M. J. Bebianno, and L. A. Barreira. (1998) Fatty acids in the Ria Formosa Lagoon, Portugal. *Organic Geochemistry,* 29(4): 963–977.

Qiang, S. (2005) Multivariate analysis, description, and ecological interpretation of weed vegetation in the summer crop fields of Anhui Province, China. *Journal of Integrative Plant Biology,* 47(10): 1193–1210.

Quesenberry, C. P. (1993) The effect of sample-size on estimated limits for (X)overbar and X-control charts. *Journal of Quality Technology,* 25(4): 237–247.

Stout, S. A., V. S. Magar, R. M. Uhler, J. Ickes, J. Abbott, and R. Brenner. (2001) Characterisation of naturally occurring and anthropogenic PAHs in urban sediments—Wycoff/Eagle Harbour Superfund Site. *Environmental Forensics,* 2(4): 287–300.

Tulloch, A. P. (1976) Chemistry of waxes of higher plants. In *Chemistry and biochemistry of natural waxes,* ed. P. E. Kolattukudy, 235–287. Amsterdam: Elsevier.

Wakeham, S. G., C. Lee, J. I. Hedges, P. J. Hernes, and M. L. Peterson. (1997) Molecular indicators of diagenetic status in marine organic matter. *Geochimica et Cosmochimica Acta,* 61(24): 5363–5369.

Wang, Z., and M. F. Fingas. (2003) Development of oil hydrocarbon fingerprinting and identification techniques. *Marine Pollution Bulletin,* 47(9–12): 423–452.

Webster, R., and M. A. Oliver. (1992) Sample adequately to estimate variograms of soil properties. *Journal of Soil Science,* 43: 177–192.

Wold, S., C. Albano, W. J. Dunn, U. Edlund, K. Esbensen, P. Geladi, S. Hellberg, E. Johansson, W. Lindberg, and M. Sjöström. (1984) Multivariate data analysis in chemistry. In *Chemometrics: Mathematics and statistics in chemistry,* ed. B. R. Kowalski. Dordrecht, Holland: D. Reidel Publishing Company.

Yunker, M. B., and R. W. Macdonald. (2003) Petroleum biomarker sources in suspended particulate matter and sediments from the Fraser River Basin and Strait of Georgia, Canada. *Organic Geochemistry,* 34(11): 1525–1541.

Yunker, M. B., R. W. Macdonald, D. J. Veltkamp, and W. J. Cretney. (1995) Terrestrial and marine biomarkers in a seasonally ice-covered arctic estuary—integration of multivariate and biomarker approaches. *Marine Chemistry,* 49(1): 1–50.

# 10 Identification of Air Pollution Sources via Modelling Techniques

IAN COLBECK

## Contents

Introduction .......... 310
Emission Sources .......... 311
Modelling Introduction .......... 312
    Modelling Guidelines .......... 313
    Types of Models .......... 314
    Types of Modelling Framework .......... 316
    Box Models .......... 317
Gaussian Models .......... 318
    Introduction .......... 318
    Basic Dispersion Equation .......... 318
    Plume Rise .......... 319
    Pasquill–Gifford Dispersion Parameters .......... 320
    Model Uncertainty .......... 325
    Model Validation .......... 325
    Types of Air Dispersion Models .......... 326
Dispersion Modelling Applications .......... 327
    September 1998 Air Pollution Episode .......... 327
    U.K. Foot-and-Mouth Disease Epidemic in 2001 .......... 330
Source Apportionment .......... 334
    Receptor Modelling .......... 334
        Chemical Mass Balance (CMB) .......... 335
        Principal Components Analysis .......... 336
        Positive Matrix Factorisation .......... 337
    Comparison with CMB .......... 338
    Receptor Modelling in Action .......... 339
        Chemical Mass Balance .......... 339
        Principal Component Analysis .......... 340
        Positive Matrix Factorisation .......... 342
Concluding Remarks .......... 344
References .......... 345

## Introduction

The quality of the air that we breathe can have important impacts on our health and quality of life. The World Health Organisation (WHO) has recently estimated that air pollution from particulate matter claims an average of 8.6 months from the life of every person in the European Union. Current policies to reduce emissions of air pollutants in the EU by 2010 are expected to save 2.3 months of life for the EU population: the equivalent of preventing 80,000 premature deaths and saving over 1 million years of life (WHO 2005). In the EU, the estimated annual monetary benefit from decreased population mortality attributed to particulate matter is €58–161 billion, and savings on the costs of diseases attributed to particulate matter account for €29 billion. In the United Kingdom, a report by the Committee on the Medical Effects of Air Pollutants (COMEAP 1998) concluded that the deaths of between 12,000 and 24,000 vulnerable people may be brought forward each year and that between 14,000 and 24,000 hospital admissions and readmissions may also result from poor air quality. These effects are attributed to just three pollutants: particulate matter ($PM_{10}$), which is estimated to bring forward 8,100 deaths annually; sulphur dioxide (3500 deaths); and ozone (from 700 to 12,500 deaths). In addition to these health concerns is the increasing evidence that air pollution may be strongly influenced by climate change.

Air quality is used by many governments as a key headline indicator of sustainable development. The U.K. National Air Quality Strategy, published in 1997 and revised in 2007, set up a strong framework for tackling air pollution. It established objectives for eight key pollutants, based on the best available medical and scientific understanding of their effects on health.

Air pollution is not stationary; it does not recognise national or international borders. Due to this mobility, it is essential to understand the relationship between the source identification and dispersion within the atmosphere. The distance that a pollutant will travel from its place of origin will be dictated by a range of factors, including its rate of emission, loss via deposition processes, emission height, and chemical reactivity. In order to improve air quality it is essential that the various sources are identified. In many cases it is useful and sometimes obligatory to know the fate of such pollutants. Many regional and national governments have air quality compliance regulations that require emitters of pollution to be able to demonstrate that their emissions present no danger to the environment as a whole and human health in particular. Quantifying the relative contributions of various pollutant sources in a given air shed becomes critical in developing control and mitigation strategies to meet regulatory standards. Frequently, adversarial groups or litigation are involved, requiring well-documented data. To achieve this, either the pollutants have to be physically tracked in real time, which is generally impractical, or some other method has to be employed.

Although measurements form an important aspect of monitoring, measurements alone are rarely sufficient to arrive at the best possible description of the desired concentration or deposition space or time fields. Models are often needed to establish larger scale average exposure and deposition fields, which cannot easily be derived from measurements. The reason is simply that observations are made at a few locations and may therefore not be very representative of larger areas. Substantial uncertainty can be introduced if measured data are extrapolated or interpolated into large domains and models are therefore used to generate best estimates in situations where measurements are lacking or cannot be made. Models are also necessary if the relative impact of various sources (source categories, emissions from different regions or countries) or emission scenarios have to be investigated.

## Emission Sources

Emission inventories are an important element of air quality management strategies. Determination of the character of pollutant emission sources can aid the quantification of the sources and level of potential impact. They provide best estimates for pollutant production rates for known sources and represent a reasonable basis for the developing and targeting of emission reduction air pollutant management strategies. This approach, however, is comparatively crude because it assumes that: (1) all pollution sources are known and fairly accurately characterised, and (2) there is a pro rata relationship between emissions at a pollution source and exposure at other sites. Despite this they are still widely used (Cadle et al. 2005; Dore et al. 2005; Lin et al. 2005; Taghavi, Cautenet, and Arteta 2005), although in recent years there has been a move away from emission-based strategy development towards receptor site impact-based strategy development. In this case, monitoring data and/or mathematical modelling are used to identify or predict the sources of pollutants at receptor sites of significant concern (e.g., urban population centres), to estimate the relative proportions of these impacting pollution sources, and, ultimately, to develop more effectively targeted air pollution management strategies.

Air pollution may have natural (volcanic eruptions, vegetal emissions, dust storms, etc.) or anthropogenic (i.e., linked to human activity, such as fixed or mobile combustion sources, industrial emissions, etc.) origins. As human activity disturbs natural systems, the distinction may become blurred. Pollutants can also be divided into primary pollutants, which are emitted directly (e.g., $SO_2$), and secondary pollutants, which are created by chemical transformations in the atmosphere. Pollutants originate from different sources, such as point sources (e.g., chimneys, liquid waste discharge pipes), line sources (e.g., highways, airline routes), or area sources (e.g., gas stations, urban or industrial complexes). The sources can also be stationary or mobile. The emissions may be categorised as 'planned', when it is economically or technically impossible to completely remove all the contaminants in a discharge and hence the process operation allows pollutants to be discharged to the environment at known and controlled rates; 'fugitive', when pollutants are released in an unplanned way, often without first passing through the entire process (e.g., losses from pipe work or wind-blown dust from stockpiles); or 'accidental', which may result from plant failure or from an accident involving either equipment or operator error. Accidental emissions can give rise to very high concentrations, but they occur infrequently. For many common air pollutants, road transport is the largest source of emissions. In the United Kingdom traffic contributes about 49% of the total emission of oxides of nitrogen, 60% of carbon monoxide emissions, and 27% of $PM_{10}$ emissions (AQEG 2004).

## Modelling Introduction

Computer simulations using mathematical models are highly effective at calculating the expected fate of pollutants. Simple models can be used for screening purposes to determine whether the pollutants under investigation do have the potential to cause harm. If a screening model suggests that the concentration of a given pollutant will be below harmful values (which may be set by a regulatory threshold), then no further action is necessary. However, if the model indicates concentrations may be harmful, then further investigation using more complex techniques will be required.

It should be noted that models, however complex, are simply decision-making tools; their results need to be interpreted in context and they are generally used in combination with measurements of concentration and meteorology. The model chosen must be fit for the purpose intended. For example, a simple screening model requires little understanding of the physical and chemical processes involved and interpretation of results is usually straightforward. More complex problems require some thought as to choice of model.

Air pollution models are commonly used for mitigation strategies, risk assessment, and planning applications (Zannetti 2001). In the United Kingdom the Department of Transport (2006) considered, in depth, the suitability of models to assess air quality around Heathrow Airport and reviewed the fitness for purpose of many models. Only specific models are now used.

## Modelling Guidelines

The Royal Meteorological Society (RMetS) produced a set of guidelines in 1995 that were intended to promote the use of best practice in the use of mathematical models of atmospheric dispersion, emphasising the principle of fitness for purpose in the selection of modelling procedures and the importance of effective communication in the documentation of reported results (http://www.rms.org). They did not try to give situation-specific technical advice (e.g., how to model a dense gas spill or which plume rise formula to use). Instead, they tried to identify and expound certain principles of good practice that apply to many modelling situations.

In 2004, the Atmospheric Dispersion Modelling Liaison Committee updated and extended the RMetS guidelines:

1. Statement of content and objectives—to explain the situation being modelled and the purpose of the dispersion calculations, giving a clear account of the relationship between the objectives and the modelling procedures adopted to achieve them. A site visit is essential.
2. Justification of choice of modelling procedure—to demonstrate the *fitness for purpose* of the modelling procedure.
3. Use of software implementations of modelling procedures—to provide a fully documented account of the details of the model and its conversion into valid software.
4. Input data—to show how the data requirements of the model have been met and to explore the implications on the assessment in cases where there are deficiencies in the available data.
5. Presentation of results and conclusions—to ensure that the findings of the exercise are successfully communicated, including any limitations.
6. Explicit quantification—to ensure that best use is made of the opportunity to express results in quantitative terms.
7. Sensitivity analysis—to expose how the results depend upon choices and assumptions made in respect of variables whose values may be debatable.
8. Uncertainty and variability—to ensure that these issues are addressed in respect of uncertainties in model parameters, the inherent variability of dispersion behaviour, and variations that are likely to be displayed between the results of one model and another.
9. Quality assurance of models—to demonstrate that the model used has been subjected to an evaluation procedure establishing its suitability for a specified range of tasks.
10. Auditability—to ensure that there is a clear and transparent account of the exercise for inspection by interested parties.

Careful consideration needs to be given to the selection of meteorological data. With further contractions in the network of weather stations, the use of representative meteorological data is of more concern as time goes on. It is particularly important that the data are representative of the area under study. The nearest meteorological observing station may not always be the most appropriate if there are, for example, coastal influences or major urban effects. However, the principle of assessing how representative input data are is common to all parameters. At this point it should be noted that any model, however complex, is only as good as the data upon which it is fed. If any of the data used, from emission rates to meteorology, contains large errors or is inappropriately chosen, then the results of the model will need to be treated with caution. Most common errors are associated with the input data. If, for example, the predicted concentrations appear to be unusually high or low, this can usually be traced to an error in the source emission rate. Dispersion models can only directly predict concentrations arising from the sources that have been input to them. In all cases, it will be necessary to take account of background pollutant levels, which may come from other sources outside the immediate study area. Hence, information on regional sources as well as local sources may be required.

## Types of Models

To determine the sources of pollutants in an air shed, two types of models are currently available. These are *dispersion* models and *receptor* models. Dispersion models are predictive in nature (i.e., they predict the impact of a specific source at a specific location from an emission inventory, other source parameters, meteorological conditions, and topography). Receptor models, on the other hand, are demonstrative in nature; that is, they provide an estimate of the impact of various sources at a given location at a given time based on the chemical characteristics of the various source emissions and the chemical character of the ambient samples collected at any receptor. Receptor modelling has been primarily used with particles, while dispersion modelling has been routinely used with both particulate and gaseous pollutants. Receptor modelling is independent of meteorological conditions and emission rate variability and can permit the simultaneous calculation of numerous source contributions to an air shed.

Models can be either physical (e.g., scale models in wind tunnels) or mathematical. Mathematical models can be either analytical, using formulae which can yield an exact analytical solution, or numerical, where approximate numerical solutions are found using numerical integration techniques. An analytical model is a very simplified equation that has an exact solution. A numerical model approximates the partial differential equations describing parameters such as diffusion or chemical reactions. Numerical models, though still simplifications of the actual situation, are typically much more complex than analytical models. Every model, whether it is a simple analytical model or a complex numerical one, will have particular applicability and usefulness in various atmospheric investigations.

The equation solved by an analytical model is a simplification of the more complex three-dimensional flow or transport equations. Initially, prior to the development and widespread use of computers, there was a need to simplify the three-dimensional equations because it was not possible to solve them easily. The simplifications involve reducing the flow to one dimension and the transport equation to one or two dimensions. Analytical models are typically steady state and one dimensional, although some are two or even three dimensional. Because of the simplifications inherent with analytical models, it is not possible to account for field conditions that change with time or space. Analytical models are best used:

- for initial assessments where a high degree of accuracy is not needed;
- prior to beginning field activities to aid in designing data collection;
- to check results of numerical model simulations; or
- where field conditions support the simplifying assumptions found in the analytical models.

Numerical models are capable of solving the more complex equations that generally describe multidimensional flow, transport, and chemical reactions, although there are one-dimensional numerical models. Numerical models use approximations (e.g., finite differences, or finite elements) to solve the differential equations describing flow or transport. The approximations require that the model domain and time be divided into discrete packages, sometimes referred to as discretisation. In this process, a network of grid cells or elements represents the model domain and time is represented as a series of steps. The accuracy of numerical models depends upon the accuracy of the input data, the size of the space and time discretisation (the greater the discretisation steps are, the greater is the possible error), and the numerical method used to solve the model equations. In addition to complex multidimensional problems, numerical models may be used to simulate very simple flow and transport conditions that may be just as easily simulated using analytical models. However, numerical models are generally used to simulate problems that cannot be accurately described using analytical models.

For these reasons, numerical models are much more useful in environmental studies, so we shall focus on these. These numerical models can be further divided into diagnostic and prognostic models:

- Diagnostic models are based on actual meteorological measurements; they contain no time-dependency terms and therefore cannot be used for forecasting.
- Prognostic models contain full time-dependent equations.

Diagnostic models have the advantage of using measurement data and being able to adapt to actual conditions. They are, however, dependent on the accuracy and density of the data and can only be used post hoc. While prognostic models incorporate meteorological physics, they are incapable of using available data to modify their forecasts.

These numerical models can be yet further divided into deterministic and statistical models:

- Deterministic models are based on mathematical descriptions of atmospheric processes where causes (emissions) lead to effects (pollution). They typically require two to three classes of model input data:
  - emission data: pollutant of interest, source locations, and emission rates;
  - dispersion data: local meteorological measurements such as wind direction and wind speed, and possibly local geographic or architectural data; and
  - reactivity data: sometimes estimates of the rate of reaction of airborne species (e.g., rates of generation and degradation) are also included to account for airborne transformation.
- Statistical models are semi-empirical, based on actual measurements.

It should be noted, however, that the chaotic behaviour of the Earth–atmosphere system makes a truly deterministic model impossible. The more complex the model is (i.e., the more parameters it has), the more sensitive it is to errors in the input data. However, such models can generate approximate descriptions of local air quality at multiple receptor sites and provide useful tools for the prediction of general local air quality in the absence of actual receptor site monitoring data and investigation of 'what if?' questions when developing air quality management strategies.

## Types of Modelling Framework

There are two types of coordinate framework in common use:

- Eulerian: the whole system under discussion is described relative to a fixed coordinate system (e.g., relative to the Earth); and
- Lagrangian: the coordinates move with the system, assuming a constant wind speed.

These two reference frameworks are shown in Figure 10.1.

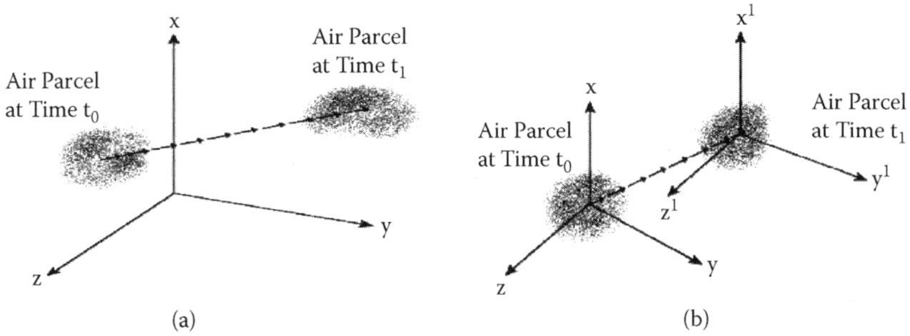


**Figure 10.1** (a) Eulerian framework; (b) Lagrangian framework.

## Box Models

The simplest air pollution model is the single box model. This consists of a Eulerian box inside which mass conservation of pollutant applies. The box is usually of the order of tens of kilometres on a side (i.e., large enough to encompass a city). The model can be applied to both inert and reactive pollutants. For reactive species, the mass balance equations have to include the necessary chemistry. The single box model is effectively one dimensional and cannot respond to rapid changes in initial conditions. A variation called the slug model allows concentrations to vary in both the x and z (vertical) dimensions.

The basic equation for this model is given by:

$$q = \frac{St}{V} \tag{10.1}$$

where $V$ = volume of box, $S$ = source or sink rate, $t$ = residence time, and $q$ = steady-state concentration of pollutant in the box (Figure 10.2).

This concept can be extended to multiple box models where the output of one box feeds the input of the next. It can also be used in a Lagrangian form, where the box moves with the (constant) wind vector to provide time-varying concentrations. The advantage of this is that it requires less computing power because integration takes place only over the volume of the box rather than over the entire coordinate system. The disadvantages are the assump-

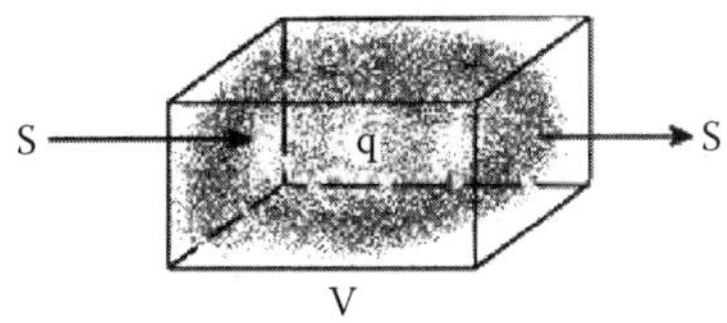


**Figure 10.2** The standard box model.

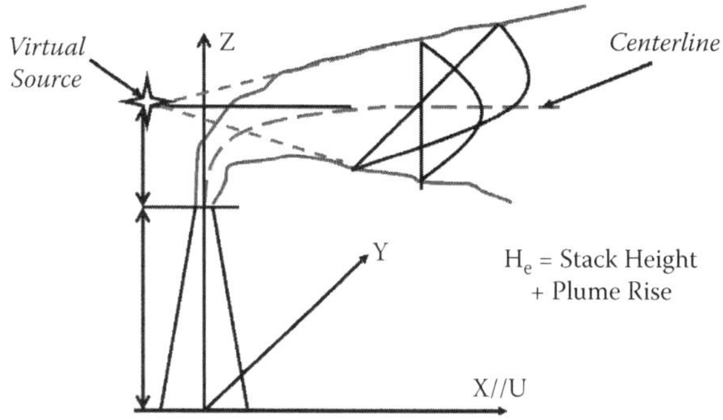


**Figure 10.3** Schematic diagram of Gaussian plume model distribution.

tion of the constant wind vector and the difficulty of comparing the outputs with real measurement data, which is Eulerian (fixed).

## Gaussian Models

### Introduction

Atmospheric dispersion models often assume a Gaussian distribution of gasses and particles emitted from a stack under steady-state conditions. Gaussian models can be either Eulerian or Lagrangian or sometimes a combination of the two. They give a 'snapshot' fixed in time from a given set of input parameters.

### Basic Dispersion Equation

In the Gaussian plume approach, the spread of the plume in the vertical and horizontal directions is assumed to occur by simple diffusion perpendicular to the direction of the mean wind (Figure 10.3). The concentration $C$ at any point $(x, y, z)$ is given by:

$$C(x,y,z)= \frac{Q}{2\pi\sigma_y\sigma_z U}\exp\left(-\frac{y^2}{2\sigma_y^2}\right)\left\{\exp\left[\frac{(z-H_e)^2}{2\sigma_z^2}\right]+\exp\left[-\frac{(z+H_e)^2}{2\sigma_x^2}\right]\right\} \tag{10.2}$$

where

$Q$ is the pollutant mass emission rate in μg s$^{-1}$
$U$ is the wind speed (m s$^{-1}$)

$H_e$ is the effective stack height (m) given by the height of the stack plus the plume rise

$x$, $y$, and $z$ are the lateral, transverse, and vertical directions (m) downwind with the base of the stack as the coordinate origin

The parameters $\sigma_y$ and $\sigma_z$ measure the extent of plume growth and in the Gaussian formalism are the standard deviations of the horizontal and vertical concentrations in the plume, respectively. Note that the concentration ($C$) is directly proportional to the emission rate and that, by increasing the effective stack height, the distance downwind that the maximum concentration, $C_{max}$, will occur increases and its value will decrease with increasing $H_e$. $C_{max}$ is roughly proportional to $H_e^{-2}$. This 'tall stack' policy underpinned air pollution control in the United Kingdom in the 1970s.

Implicit assumptions used in Gaussian modelling include:

- Continuous emission or emission times are equal to or greater than travel times to the downwind position of interest so that diffusion in the direction of travel can be ignored.
- The material diffused is a stable gas or aerosol (<20 μm diameter) that remains suspended in the air over long periods of time.
- Mass is conserved through reflection at surfaces.
- There are steady-state conditions during the time interval for which the model is used, usually 1 hour.
- Constant wind speed $U$ with height is assumed.
- Constant wind direction with height is assumed.
- The wind shear effect on horizontal diffusion is not considered (effect becomes large after ~10 km).
- The dispersion parameters are assumed to be independent of $z$ and functions of $x$ (and hence $U$ alone).
- The averaging time of all quantities ($U$, $\sigma_x$, $\sigma_y$, $\sigma_z$) is assumed to be the same.

## Plume Rise

An important consideration in air dispersion modelling studies is to estimate the effective height of the emission source at which the plume becomes essentially level. It is rarely equal to the physical height of the stack due to its buoyancy and momentum. In 1969, Briggs published his now classical critical review of the entire plume rise literature, in which he proposed a set of plume rise equations that have since become widely known as 'the Briggs equations'. Briggs divided air pollution plumes into these four general categories:

- cold jet plumes in calm ambient air conditions;
- cold jet plumes in windy ambient air conditions;
- hot, buoyant plumes in calm ambient air conditions; and
- hot, buoyant plumes in windy ambient air conditions.

Briggs considered the trajectory of cold jet plumes to be dominated by their initial velocity momentum and the trajectory of hot, buoyant plumes to be dominated by their buoyant momentum to the extent that their initial velocity momentum was relatively unimportant. Although Briggs proposed plume rise equations for each of the preceding plume categories, it is important to emphasise that the Briggs equations that have become widely known and widely used are those that he proposed for bent-over, hot buoyant plumes. In the Briggs model the buoyancy of an emission is determined via the buoyancy flux parameter, $F$:

$$F = gvr^2\left(\frac{T_s - T}{T_s}\right) \tag{10.3}$$

where

$v$ is the efflux velocity from the stack
$T_s$ is the stack exit temperature
$g$ is the acceleration due to gravity
$r$ is the inside radius of the stack
$T$ is the ambient temperature (°C)

This formula can then be used to determine plume rise under certain meteorological conditions (Turner 1994). It is generally accepted that the Briggs equations could over- or underpredict actual plume rises by 20%.

## Pasquill–Gifford Dispersion Parameters

Critical input parameters into equation 10.2 are the standard deviations ($\sigma$) of the plume along the axes. They are determined by the prevailing atmospheric turbulence in the boundary layer. Small-scale turbulent motions tend to dominate plume growth close to the point of emission, while large-scale eddies dominate at greater distances. Additionally, small-scale eddies are associated with short time scales and larger eddies with longer time scales. Hence, $\sigma_y$ and $\sigma_z$ increase in value with distance from the source and with time or sampling period.

The values may be obtained by either theoretical or empirical approaches. They have been measured in numerous field studies as a function of distance from the source, with sampling periods in the range of 3–60 minutes. The

**Table 10.1 Meteorological Conditions Defining Pasquill Turbulence Types**

| Surface wind speed ($m\ s^{-1}$) | Day—incoming solar radiation | | | Night—cloud cover | |
|---|---|---|---|---|---|
| | Strong | Moderate | Slight | Thinly overcast | Mostly cloudy |
| <2 | A | A–B | B | | |
| 2–3 | A–B | B | C | E | F |
| 3–5 | B | B–C | C | D | E |
| 5–6 | C | C–D | D | D | D |
| >6 | C | D | D | D | D |

most common tabulated data are the Pasquill–Gifford sigmas, formulated for low-level releases over relatively smooth terrain at distances of a few thousand metres from a source (Pasquill 1961; Gifford 1962).

In order to relate the state of atmospheric convection to simply observable parameters, Pasquill developed a simple quantitative rating scheme consisting of six stability classes: highly unstable (A), moderately unstable (B), slightly unstable (C), neutral (D), slightly stable (E), and moderately stable (F). These classes are summarised in Table 10.1. The Pasquill stability classes are presented as they are defined by the prevailing meteorological conditions of (1) surface wind speed measured at 10 m above ground level, and (2) daytime incoming solar radiation or the nighttime percentage of cloud cover.

The resulting Pasquill–Gifford (P–G) $\sigma_y$ and $\sigma_z$ curves under varying conditions of stability are shown in Figures 10.4 and 10.5. For use in dispersion models, it is convenient to have analytical expressions for $\sigma_y$ and $\sigma_z$ (Turner 1994). In the rural mode the equations used to calculate $\sigma_y$ (in metres) are of the form:

$$\sigma_y = 465.117(x)\tan(\theta) \tag{10.4}$$

where

$$\theta = 0.0175[c - d\ln(x)] \tag{10.5}$$

In equations 10.4 and 10.5, the downwind distance $x$ is in kilometres and the stability class-dependent coefficients, $c$ and $d$, are listed in Table 10.2. The equation used to calculate $\sigma_z$ is of the form:

$$\sigma_z = ax^b \tag{10.6}$$

where the downwind distance $x$ is in kilometres and $\sigma_z$ is in metres. The coefficients $a$ and $b$ are given in Table 10.3.

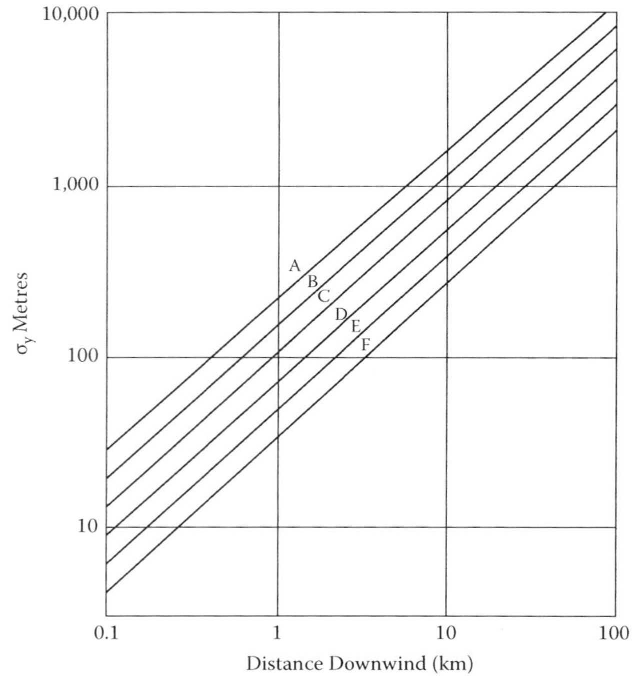


**Figure 10.4** Pasquill–Gifford horizontal dispersion parameter, $\sigma_y$, as a function of Pasquill stability class and distance downwind from the source.

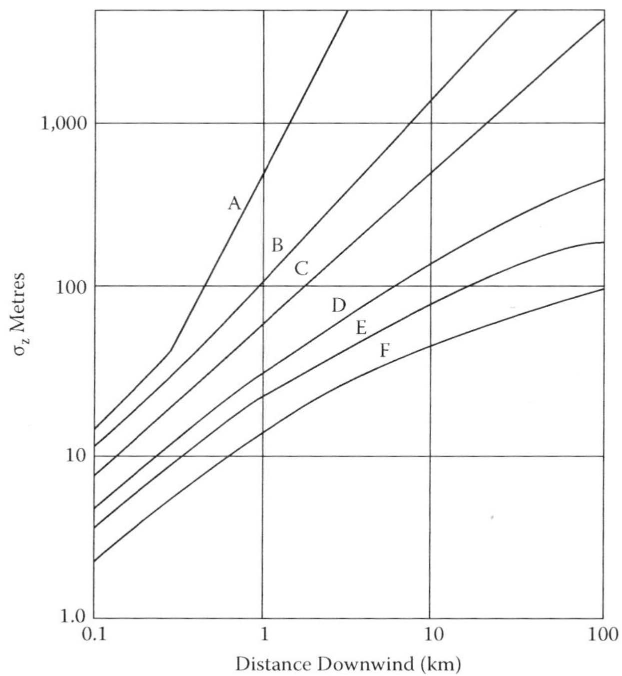


**Figure 10.5** Pasquill–Gifford vertical dispersion parameter, $\sigma_z$, as a function of Pasquill stability class and distance downwind from the source.

**Table 10.2 Coefficients *c* and *d* for Use in Equation 10.5 to Calculate the Pasquill–Gifford Horizontal Dispersion Parameter[a]**

| Pasquill stability class | *c* | *d* |
|---|---|---|
| A | 24.167 | 2.5334 |
| B | 18.333 | 1.8096 |
| C | 12.5 | 1.0857 |
| D | 8.333 | 0.72382 |
| E | 6.25 | 0.54287 |
| F | 4.1667 | 0.36191 |

[a] $\sigma_y$ is in metres and $x$ is in kilometres.

**Table 10.3 Coefficients *a* and *b* for Use in Equation 10.6 to Calculate the Pasquill-Gifford Vertical Dispersion Parameter**

| Pasquill stability class | *x* (km) | *a* | *b* |
|---|---|---|---|
| A | <0.10 | 122.8 | 0.9447 |
| | 0.10–0.15 | 158.08 | 1.0542 |
| | 0.16–0.20 | 170.22 | 1.0932 |
| | 0.21–0.25 | 179.52 | 1.1262 |
| | 0.26–0.30 | 217.41 | 1.2644 |
| | 0.31–0.40 | 258.89 | 1.4094 |
| | 0.41–0.50 | 346.75 | 1.7283 |
| | 0.51–3.11 | 453.85 | 2.1166 |
| | >3.11 | 5000 | 5000 |
| B | <0.20 | 90.673 | 0.93198 |
| | 0.21–0.40 | 98.483 | 0.98332 |
| | >0.40 | 109.3 | 1.0971 |
| C | All | 61.141 | 0.91465 |
| D | 0.3 | 34.459 | 0.86974 |
| | 0.31–1.00 | 32.093 | 0.81066 |
| | 1.01–3.00 | 32.093 | 0.64403 |
| | 3.01–10.00 | 33.504 | 0.60486 |
| | 10.01–30.00 | 36.65 | 0.56589 |
| | >30.00 | 44.053 | 0.51179 |
| E | <0.10 | 24.26 | 0.8366 |
| | 0.10–0.30 | 23.331 | 0.81956 |
| | 0.31–1.00 | 21.628 | 0.7566 |

Continued

**Table 10.3 Coefficients *a* and *b* for Use in Equation 10.6 to Calculate the Pasquill-Gifford Vertical Dispersion Parameter (Continued)**

| Pasquill stability class | $x$ (km) | $a$ | $b$ |
|---|---|---|---|
| | 1.01–2.00 | 21.628 | 0.63077 |
| | 2.01–4.00 | 22.534 | 0.57154 |
| | 4.01–10.00 | 24.703 | 0.50527 |
| | 10.01–20.00 | 26.97 | 0.46713 |
| | 20.01–40.00 | 35.42 | 0.37615 |
| | >40.00 | 47.618 | 0.29592 |
| F | <0.20 | 15.209 | 0.81558 |
| | 0.21–0.70 | 14.457 | 0.78407 |
| | 0.71–1.00 | 13.953 | 0.68465 |
| | 1.01–2.00 | 13.953 | 0.63227 |
| | 2.01–3.00 | 14.823 | 0.54503 |
| | 3.01–7.00 | 16.187 | 0.4649 |
| | 7.01–15.00 | 17.836 | 0.41507 |
| | 15.01–30.00 | 22.651 | 0.32681 |
| | 30.01–60.00 | 27.074 | 0.27436 |
| | >60.00 | 34.219 | 0.21716 |

These sigma data can be applied for releases over flat, rural terrain. However, dispersion in the urban environment usually produces greater rates of spread than these field data expressions. For urban dispersion, a second set of curves is often used. Table 10.4 shows the equations used to determine $\sigma_y$ and $\sigma_z$ for the urban option. These expressions were determined by Briggs as reported by Gifford (1976) and represent a best fit to urban vertical diffusion data reported by McElroy and Pooler (1968). Although the Briggs functions are assumed to be valid for downwind distances less than 100 m, the user is cautioned that concentrations at receptors less than 100 m from a source

**Table 10.4 Formulae to Calculate the Pasquill–Gifford Horizontal ($\sigma_y$) and Vertical ($\sigma_z$) Dispersion Parameter in an Urban Environment[a]**

| Pasquill stability category | $\sigma_y$ (Metres) | $\sigma_z$ (Metres) |
|---|---|---|
| A | $0.32 \times (1.0 + 0.0004\ x)^{-1/2}$ | $0.24 \times (1.0 + 0.001\ x)^{1/2}$ |
| B | $0.32 \times (1.0 + 0.0004\ x)^{-1/2}$ | $0.24 \times (1.0 + 0.001\ x)^{1/2}$ |
| C | $0.22 \times (1.0 + 0.0004\ x)^{-1/2}$ | $0.20 \times$ |
| D | $0.16 \times (1.0 + 0.0004\ x)^{-1/2}$ | $0.14 \times (1.0 + 0.0003\ x)^{-1/2}$ |
| E | $0.11 \times (1.0 + 0.0004\ x)^{-1/2}$ | $0.08 \times (1.0 + 0.0015\ x)^{-1/2}$ |
| F | $0.11 \times (1.0 + 0.0004\ x)^{-1/2}$ | $0.08 \times (1.0 + 0.0015\ x)^{-1/2}$ |

[a] $x$ in metres.

may be suspect. Likewise, Pasquill's coefficients could be in error by ±25%, especially when used for nonlevel, complex terrain and for large distances ranging up to 50 km or more.

## Model Uncertainty

Any atmospheric dispersion model will always have a degree of error due to a variety of factors. These include uncertainties in the emissions data and in the recorded meteorological data and simplifications made in the model algorithms that describe the atmospheric dispersion and chemical processes.

There are essentially two main types of errors or uncertainty in dispersion models. Systematic errors occur when the model shows the same trend at all times (e.g., the model consistently underpredicts concentrations when compared against the true value for a given application). This introduces a bias to the modelling predictions. The systematic error may be considered synonymous with the accuracy of the model predictions (i.e., how close the predicted value is to the true value).

There are also likely to be random errors. These random errors may be considered synonymous with the precision of the model (i.e., how wide the scatter is or residual variability of the predicted values compared with the true value) once the bias has been allowed for. Despite their limitations Gaussian models have enjoyed a wide degree of popularity, mainly due to their simplicity and the fact that they give reasonable results. They can also be formulated for line sources, area sources, mobile sources, and complex sources.

## Model Validation

In 1991, a European initiative was launched for increased cooperation and standardisation of atmospheric dispersion models for regulatory purposes. A new generation of models was emerging with physically more justifiable parameterisations of dispersion processes. A need was felt for these new models to be developed in a well-organised manner and turned into practical, generally accepted tools fit for the various needs of decision makers. On this background it was decided to organise a series of workshops, 'Harmonisation within Atmospheric Dispersion Modelling for Regulatory Purposes', to promote the use of new-generation models within atmospheric dispersion modelling, and in general improve modelling culture. These workshops are held on a regular basis (see http://www.harmo.org/default.asp).

Various attempts have been made to establish databases and evaluation methodologies for respective models. One such 'toolbox' not only covers the classic single-stack problem but also has the potential to be extended to other dispersion problems: the 'ASTM Standard Guide for Statistical Evaluation of Atmospheric Dispersion Model Performance' (ASTM 2000).

The fundamental premise of the ASTM procedure is that observations and model predictions should *not* be compared directly and that observations should be properly averaged before comparison. The comparison takes place within *regimes,* which can be defined according to atmospheric stability and distance to the source. The ASTM procedure then calculates performance measures based on *regime averages* (i.e., averaging over all experiments within a regime), rather than the values for individual experiments.

An alternative is the 'model validation kit' developed as part of the European harmonisation initiative (see http://www.harmo.org/Kit/Download/Kit_UsersGuide.pdf). It is a collection of four field data sets as well as software for model evaluation and addresses the classic problem of dispersion from a single point source with the emphasis on operational short-range models (Olesen 2001). Modelled pollutant concentration should not be expected to match observed concentrations exactly, even for a matching averaging time. The mismatch between the observed and modelled concentrations generally increases when modelling short-term pollutant concentrations (such as over 10 minutes).

The methods used to evaluate the performance of air quality models have been reviewed by Chang and Hanna (2004). The focus of their paper was on the statistical evaluation. They concluded that there was not a single best performance measure or best evaluation methodology and recommended that a suite of different performance measures be applied. Suggestions were given concerning the magnitudes of the performance measures expected of 'good' models. For example, a good model should have a relative mean bias less than about 30% and a relative scatter less than a factor of about two.

## Types of Air Dispersion Models

There are numerous types of air dispersion models available. They have been developed for a variety of pollutants, time scales, and length scales. Atmospheric phenomena at any specific scale are influenced by the ensemble of interacting atmospheric processes occurring at various scales. The phenomena at the local and urban scales have a horizontal extension of tens of metres to 500 km and a time scale of several minutes to several days. Scale separation has proved to be a successful approach for atmospheric modelling because different approximations and parameterisations can be applied for different phenomena occurring in the different scales. This often results in a nesting approach where, for example, an urban area may be modelled at the local scale (up to 10 km × 10 km) to deal with street canyons and then at an urban scale (up to 500 km × 500 km).

Various catalogues of dispersion models are available on the Internet. The U.S. Environmental Protection Agency (EPA) provides descriptions and documentation for three types of air quality models: dispersion, photochemical, and receptor models (http://www.epa.gov/ttn/scram/). Also provided are source code and associated users' guides and documentation for preferred/recommended models, screening models, and alternative models. In 1996, the European Topic Centre on Air and Climate Change documented around 30 models, with emphasis on models that may be readily applied in support of environmental decision making (Moussiopoulos et al. 1966). This work has evolved into the model documentation system (http://air-climate.eionet.eu.int/databases/MDS/index_html), which aims to provide guidance to any model user in the selection of the most appropriate model for his or her application. The searchable database contains over 100 models.

## Dispersion Modelling Applications

Following the Chernobyl disaster in 1986, the meteorological office began work to develop a model to predict the impact of such incidents. The nuclear accident model (NAME) allows such predictions. NAME is a mesoscale and long-range, multiple-particle dispersion model that is now applied to a wide range of atmospheric pollution problems, ranging from emergency responses to daily air-quality forecasts.

Over the years, NAME has been applied to a number of atmospheric releases, including radioactive releases, the Kuwaiti oil fires, major industrial fires and chemical spills, and two major volcanic eruptions that occurred in Iceland. Both of these eruptions resulted in aircraft having to be rerouted to avoid potentially dangerous ash clouds.

### September 1998 Air Pollution Episode

An air pollution episode occurred across much of the Midlands and South Yorkshire (United Kingdom) on September 2, 1998. Although this particular episode was small in comparison with the smogs of the 1950s, it occurred across a widespread area of the Midlands and South Yorkshire and generated significant public concern. Monitoring stations recorded peak sulphur dioxide concentrations up to six and a half times higher than the air quality standard; concentrations were high enough for asthmatic individuals to experience adverse effects.

The Environment Agency investigated this episode and was able to link the meteorology with the pattern of public complaint and results from air quality monitoring stations to establish the sequence of events. The Meteorological Office utilised NAME to predict the observed air pollution concentrations. For the first time in the United Kingdom, this allowed an attribution of measured ground-level concentrations to emissions from specific power stations and other industrial sources at large upwind dis-

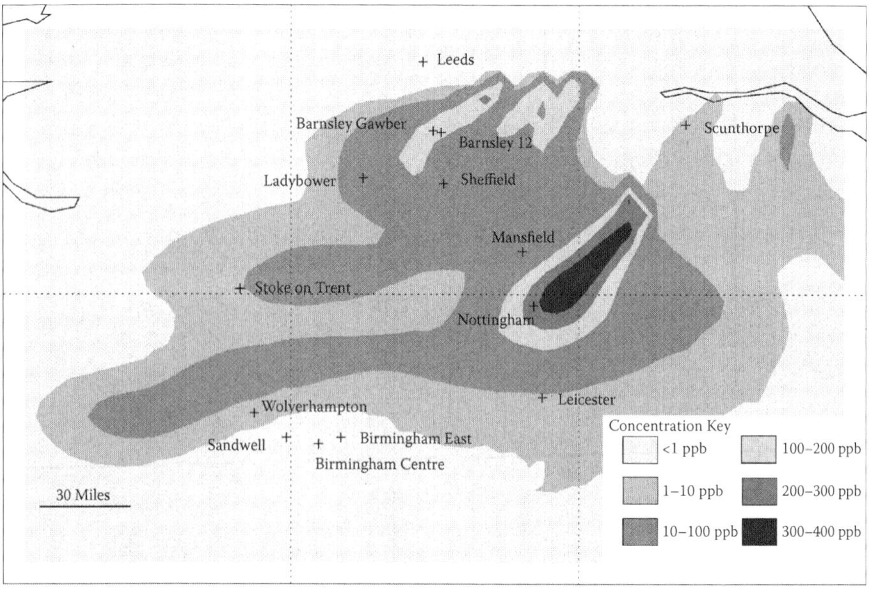


**Figure 10.6 (See colour insert following page 336.)** Sulphur dioxide concentration plot for 16:00 GMT.

tances. In order to numerically model the pollution episode, release data for the period leading up to and through the episode were required from those sources identified as being potential contributors. Sources included those regulated by the environment agency or by a local authority under Local Authority Air Pollution Control or the Clean Air Act 1993 or that do not come under direct regulation (such as traffic or residential). Power generation, refineries, iron and steel and other major combustion processes are significant sources accounting for nearly 90% of the annual sulphur dioxide emitted in the United Kingdom.

Input data included stack height and diameter and the average release rate of $SO_2$, $NO_2$, NO and particulates; temperature and release were used together with relevant meteorological data to predict atmospheric concentrations. Typical output, in this case the $SO_2$ concentration at 16:00, is shown in Figure 10.6. The match between monitoring results and NAME prediction was considered exceptionally good, taking into account that only the 16 most significant sources were included in the input data for NAME (Table 10.5). As each particle released into the NAME simulated atmosphere was labelled with its time of release and its source, at any given point and time it was possible to calculate the air concentration of pollutants and also the source of the particles.

Table 10.6 shows the NAME-predicted aggregated sulphur dioxide mass (as a percentage of the total) at four monitoring stations for the period 11:00 to 23:00 GMT for the lowest 50 m of the boundary layer. The incident was

**Table 10.5 Processes Selected for the NAME Modelling Exercise**

| Location | Process |
|---|---|
| Bolsover | Fuels |
| Buxton | Cement |
| Cottam | Power generation |
| Drax | Power generation |
| Eggborough | Power generation |
| Ferrybridge | Power generation |
| High Marnham | Power generation |
| Hope | Cement |
| North Killingholme | Refinery |
| Ratcliffe-on-Soar | Power generation |
| Rugeley | Power generation |
| South Ferriby | Cement |
| South Killingholme | Refinery |
| Scunthorpe | Iron and steel |
| West Burton | Power generation |
| Willington | Power generation |

**Table 10.6 Aggregated Sulphur Dioxide Mass for the Period 11:00 to 23:00 for the Lowest 50 m of the Boundary Layer**

| Source | Birmingham Centre (%) | Ladybower (%) | Nottingham Centre (%) | Stoke-on-Trent Centre (%) |
|---|---|---|---|---|
| Bolsover | | | 0.05 | 40.27 |
| Buxton | | | | 0.98 |
| Cottam | 38.50 | | 32.90 | |
| Drax | 11.27 | | 6.49 | |
| Eggborough | 0.78 | 0.84 | 0.52 | 4.51 |
| Ferrybridge | 0.97 | 99.16 | 1.46 | 53.29 |
| High Marnham | 10.69 | | 7.86 | |
| Ratcliffe | 4.76 | | | |
| West Burton | 33.03 | | 50.61 | |

driven by the weather occurring on September 1 and 2, 1998. A period of low wind speed allowed pollutants from a range of industrial sources to accumulate in the atmosphere to a high concentration before dispersal. Overall, the NAME model performed well in modelling the episode. The highest sulphur dioxide peak was recorded at a monitoring station in Nottingham, where NAME modelled the peak to within a factor of two. It also predicted that 95% of the sulphur dioxide detected at Birmingham, Ladybower, Nottingham, and Stoke-on-Trent originated from coal-fired power stations in the Aire and Trent Valleys.

## U.K. Foot-and-Mouth Disease Epidemic in 2001

Between February and September 2001 the United Kingdom experienced the worst epidemic of foot-and-mouth disease (FMD) since records began; this involved 2030 cases spread across the country. Six million animals were killed and losses to agriculture and the food chain were £3.1 billion. Some £2.5 billion was paid by the government in compensation for slaughtered animals and payments for disposal and cleanup costs. In such an incident it is vital to identify not only the initial source but also how it was transmitted in order to prevent and control any future outbreaks of infectious diseases in livestock.

The FMD virus is highly contagious and spreads in four main ways:

- Animals in direct or indirect contact with diseased animals can catch the disease. Infected animals begin to release the virus a few days before signs of the disease develop. Pigs are especially contagious because they produce large numbers of virus particles.
- The disease spreads physically, by the movement of contaminated animals or people and vehicles unwittingly moving the virus around. This is why infected farms have disinfectant as part of their biosecurity at their gates.
- FMD spreads through the air. Depending on the weather and the local landscape, the disease can travel quite far. For example, airborne spread of FMD from Brittany probably caused an outbreak on the Isle of Wight in 1981.
- In the past, pigs fed on swill that had not been properly treated were at risk of contracting FMD and other diseases. The improper use of these waste food products has been implicated in several previous outbreaks and the use of pig swill has now been banned.

The most common mechanism by which FMD is spread among ruminants and from pigs to ruminants is by the movement of infected animals and airborne transmission to susceptible animals of infectious droplets and droplet nuclei. These particles originate mainly from the respiratory tract and are emitted in the exhaled breath of infected animals. The spread of the FMD virus by the wind is possible, but requires the simultaneous occurrence of particular epidemiological and climatic conditions. However, when these conditions are united, spread of the disease can be both rapid and extensive and involve areas well beyond disease-control areas. The pathogen can then travel a considerable distance (several kilometres) in the air on a gently moving wind with low turbulence (Donaldson 1972; Sorensen et al. 2000; Donaldson and Alexandersen 2001; Alexandersen, Brotherhood, and Donaldson 2002). Because turbulence is generally less marked over water than over land, airborne spreading up to 250 km over water can occur (Quinn and Markey 2001), leading to suggestions that the 1981 outbreak of FMD on the Channel Islands and the Isle of Wight could have been caused by windborne viruses from France (Donaldson et al. 1982; Sorensen et al. 2000). The fact that the viruses involved in the contemporaneous French and British outbreaks were virtually identical reinforced this possibility. Airborne spread was also implicated in the 1967 outbreak (Sellers and Gloster 1980; Gloster, Freshwater et al. 2005).

Although the first FMD outbreak was confirmed in pigs in an abattoir in Essex on February 20, the origin for that outbreak and for the whole epidemic is considered to have been a pig finishing unit at Burnside Farm, Heddon on the Wall, Northumberland, United Kingdom (Defra 2002). The epidemiological inquiry indicated that there were two routes of spread from the Burnside Farm. First, the movement of diseased pigs or pigs recovering from FMD on February 8 and 15 resulted in infection being transferred to the abattoir in Essex. Second, there was airborne spread of disease from Burnside Farm to sheep on nearby premises (Prestwick Hall Farm, Ponteland) and the subsequent sale of 16 infected sheep at Hexham market on February 13. These sheep were further sold at other markets, where they infected other sheep, people, or vehicles, thereby spreading the FMD virus widely in England and Wales and the bordering counties of southern Scotland.

The spread of FMD by the wind is uncontrollable, but simulation models have been developed that can be used to predict the risk of virus dissemination and the probable direction and distance of spread. The objective of these model simulations is to calculate FMD virus concentrations at ground level and at distances in general up to few hundred kilometres from the sources. Those values can be directly compared with experimentally determined threshold values of minimum infection doses. This information can assist control procedures during emergencies, enabling the manpower for surveillance activities to be directed in the most efficient manner.

The basic quantities involved in mathematical modelling of airborne spread of FMD include minimum infectious doses for the different cloven-hoofed animal species, their inhalation rates, and the virus concentration in the inhaled air. The latter depends on the rate of virus excretion and the meteorological conditions.

Cattle and pigs affected by FMD emit the airborne virus for a period of 4–5 days, while sheep can excrete the virus for up to 7 days. Cattle and pigs excrete maximally during the early acute stages of the disease; however, sheep behave differently, with maximum airborne virus emission occurring for 1–2 days before the onset of clinical disease. Pigs are by far the most

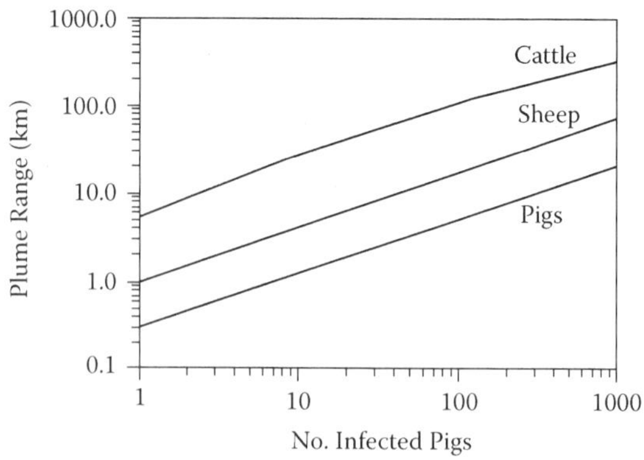


**Figure 10.7** Range of an infectious FMD virus plume as a function of the number of pigs at source and recipient animals of different species downwind. Reproduced with permission. (Sorensen, J. H. et al., *Physics and Chemistry of the Earth Part B—Hydrology Oceans and Atmosphere,* 26(2): 93–97, 2001.)

potent emitters of airborne virus (Sellers 1969; Donaldson et al. 1982; Donaldson 1987; Sanson 1994; Knowles et al. 2001). Cattle and sheep excrete similar amounts of airborne virus but in significantly less quantities than pigs. An important finding that influenced predictions of airborne disease spread was made shortly after the start of the epidemic. It was established that pigs infected with the 2001 U.K. virus emitted 300 times less amounts of virus than had been found with previous strains experienced in the United Kingdom (Alexandersen and Donaldson 2002). Consequently, the probability of airborne disease infection was much less than in previous outbreaks.

The biological decay of airborne FMD is mainly influenced by the relative humidity, and it has been shown that all strains retain their maximum infectivity at relative humidities above 60% (Donaldson 1972). Hence, in the United Kingdom the potential for the survival of airborne FMD virus is high for most of the year. For a given strain of FMD virus and a given holding of infected animals of different species, it is possible to estimate the potentially largest range of an infectious FMD virus plume. The range depends on the virus excretion rate, the minimum infectious dose, and the atmospheric conditions. Typical length scales are shown in Figure 10.7.

Bearing in mind that first outbreaks were recorded in the eastern part of the United Kingdom, initial investigations considered the long-range spread of the disease to the near continent. Two models were used to evaluate the risk: the NAME model, described earlier, and the Danish emergency response model of the atmosphere (DERMA), developed at the Danish Meteorological Institute for similar purposes (Sorensen 1998). The two models showed close agreement and indicated that the risk of airborne transmission over intercontinental ranges was extremely low (Gloster et al. 2003; Mikkelsen et al. 2003).

There have been several papers describing the airborne spread of FMD during the early stages of the 2001 epidemic (Alexandersen et al. 2003; Gloster et al. 2003; Mikkelsen et al. 2003). In addition to the NAME and DERMA, two other models were used: a 10-km Gaussian plume model and the real-time emergency preparedness and decision support model RIMPUFF. The former model was developed in 1980 for the specific purpose of predicting the local area at risk from an FMD virus source (Gloster et al. 1981; Gloster, Sellers, and Donaldson 1982). The model calculates the daily and total inhaled virus dosage at 1-km intervals from the source. This model takes no account of topography. The RIMPUFF model releases virus particles in puffs; a single puff contains the equivalent of a 10-minute release. The puffs grow in size due to local turbulence; they rise due to the growth and follow the local winds.

Simulation of the virus plumes from Burnside Farm was performed with the four different models all run with actual weather data during the period of airborne virus emission, estimated to have commenced around February 1 and to have ended on February 24, 2001, when the animals were culled. The daily virus excretion rates from Burnside Farm were calculated initially from estimations of the number of infected pigs on the premises, the stage of their clinical disease, and historical data from a virus production model and then with details specific to the U.K. 2001 strain. The results indicated that the infection at Prestwick Hall was consistent with airborne disease spread from Burnside Farm. During the first half of February 2001, southwesterly winds often combined with stable stratified atmospheric conditions over Northumberland. These conditions would have produced narrow plumes containing high concentrations of virus. Detailed studies (Gloster et al. 2003; Mikkelsen et al. 2003) concluded that a minimum of 7 out of the 12 outbreaks within 9 km of Burnside Farm were consistent with airborne disease spread.

More recently, Gloster, Champion et al. (2005) extended the investigation to three more typical cases: Longtown and Penrith, where cattle were the source of the airborne virus, and Witton-le-Wear, which involved a small number of pigs. The NAME model was again used in the simulation studies. The Longtown study demonstrated that sites up to 16 km away from the source may be infected by airborne virus. The Penrith study indicated that the disease may be spread by airborne viruses over a distance of 4–5 km, despite there being only 20 infected cattle emitting the virus. Finally, the Witton-le-Wear outbreak showed that pigs have the potential to produce very large quantities of airborne virus and that local topography may affect the pattern of disease spread.

However, overall in 2001, airborne spread seems not to have been a common cause of secondary outbreaks; only 1% of cases have been attributed to this route. This was primarily due to the low quantities of virus emitted by animals infected with the 2001 U.K. virus strain, combined with the lack of disease in the prolific virus emitters: pigs. In Cumbria, spread along valley floors in the direction of the prevailing wind has led to comments that wind-borne spread was more significant than the published figures suggest.

## Source Apportionment

In source apportionment, regardless of procedure, mathematical and sometimes intuitive methods are used to estimate the relative contributions of different sources of a range of *species of interest* at a given *receptor site.* In atmospheric pollutant source apportionment, the receptor sites are most often areas of significant concern (e.g., population centres) and the species of interest are atmospheric pollutants. An improved understanding of relative contribution of various pollution sources to air pollutants at such sites is fundamental to the future development of more successful targeted air pollution management strategies and continued improvement of ambient air quality. Source identification and apportionment will increase in importance as the need to seek 'damages' from polluters grows.

### Receptor Modelling

Receptor modelling is the term applied to the use of measurements of a pollutant at a single receptor site. Source contributions are then determined as the best-fit combination of pollutant source emission profiles needed to rationalise these receptor site observations. Receptor modelling has been primarily used with particles. At the simplest level this method can involve discrimination between primary and secondary components of particulate matter on the basis of their differing chemical compositions. At a more sophisticated level it can involve detailed statistical analysis of the temporal patterns of individual components of particulate matter with a view to identifying those associated in time that thus have common sources. While the techniques of receptor modelling are insufficiently precise to allow clear discrimination of all of the primary sources considered in the emissions inventories or even to provide an exact separation of primary and secondary components, they are capable of identifying and quantifying the contributions of major sources.

The fundamental principle of receptor modelling is that mass conservation can be assumed and a mass balance analysis can be used to identify and apportion sources of contaminants in the atmosphere. This relationship can be expressed as:

$$x_{ij} = \sum_{k=1}^{p} c_{ik} s_{kj} \tag{10.7}$$

where $x_{ij}$ is the measurement of *i*th species in the *j*th sample, $c_{ik}$ is the gravimetric concentration of the *i*th element in material from the *k*th source, and $s_{kj}$ is the airborne mass concentration of material from the *k*th source contributing to the *j*th sample.

The following fundamental, natural, and physical constraints must be obeyed when developing any model for identifying and apportioning the sources (Hopke 1985):

- The model must explain the observations; the model must reproduce the original data.
- Predicted source compositions must be positive.
- A source cannot emit a negative mass, so the predicted source contributions to an aerosol must be positive.
- The total sum of the predicted elemental mass concentrations for each source must be less than or equal to the total measured mass for each element.

Various receptor modelling approaches have been used to rationalise this expression. These methods include chemical mass balance for source apportionment, empirical orthogonal functions for identification of the locale and strengths of emission sources, and principal components analysis for source identification (e.g., Hopke 1999; Brown and Hafner 2005; Paatero et al. 2005; Thurston et al. 2005).

### *Chemical Mass Balance (CMB)*

For the CMB approach, the source profiles (i.e., the fractional amount of the species in the emissions from each source type) and the receptor concentrations, with appropriate uncertainty estimates, are required (Figure 10.8). It is assumed that all of the species contributing to the measured concentrations at the receptor have been identified and that the measurement errors are random, uncorrelated, and normally distributed about a mean value of zero. The effective variance least squares fitting procedure is used to address the problem iteratively (Hopke 1997, 1999). Several problems can arise from the CMB method. If applied to an area, the question of which sources and source profiles should be included in the model must be addressed. Profiles used in studies in other areas may be applicable to only that specific source. Additionally, emission profiles, such as those from motor vehicles, change with time. New fuels, engines, and control technologies have resulted in dramatic changes in vehicle emissions over the past 20 years. One final problem is the possible existence of two sources with similar fingerprints or a source whose profile is a linear combination of other source profiles. Despite these uncertainties, the U.S. EPA has approved a CMB model as part of the implementation planning process (http://www.epa.gov/scram001/receptor_cmb.htm).

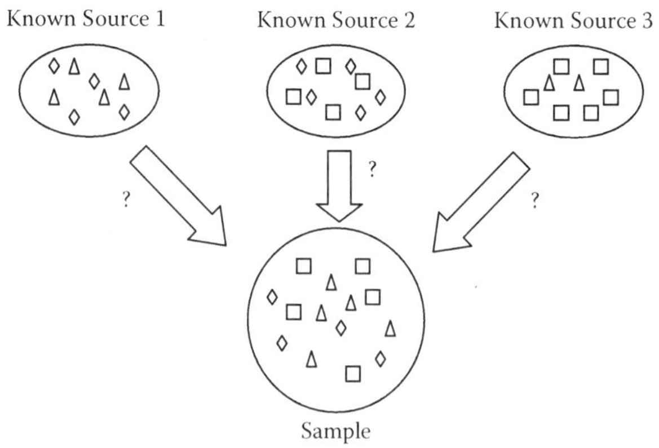


**Figure 10.8** Basic principles of CMB source apportionment.

For many locations, the sources are either unknown or the compositions of the local particulate emissions have not been measured. Thus, it is desirable to estimate the number and compositions of the sources as well as their contributions to the measured particulate matter. The multivariate data analysis methods that are used to solve this problem are generally referred to as *factor analysis.*

### *Principal Components Analysis*

The most common form of factor analysis is principal components analysis (PCA). The PCA results are generally calculated using an eigenvector analysis of a correlation matrix (Hopke 1985; Henry 1991). The major assumptions of PCA modelling are (Figure 10.9):

- There is constant composition of emission sources.
- Chemical species with the PCA analysis do not interact with each other and their concentrations are linearly additive.
- The variability of the concentrations is dominated by changes in source contributions and not by changes in source composition or by measurement uncertainty.
- Measurement errors are random and uncorrelated.
- There are many more samples than source types for statistically meaningful calculations.
- The impact of processes that affect all sources equally, such as atmospheric dispersion, is smaller than the impact of processes that influence individual sources, such as wind direction.

There are numerous examples of PCA applications in the literature (e.g., APEG 1999; Bruno et al. 2001; Graney, Dvonch, and Keeler 2004; Brown and Hafner 2005; Gomez et al. 2005; Toscano et al. 2005).

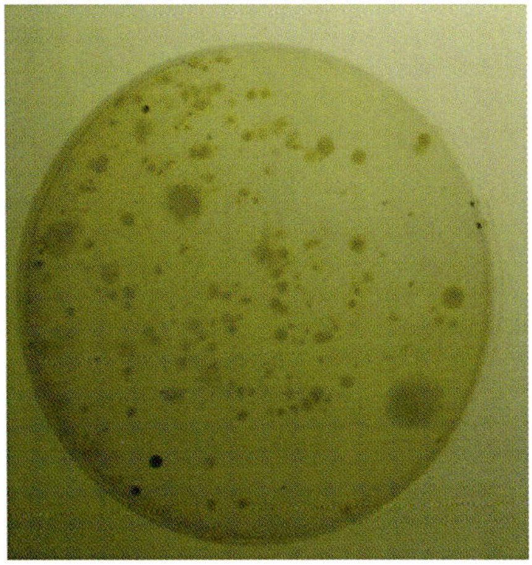

**Figure 6.1** Selective isolation plating of an environmental sample.

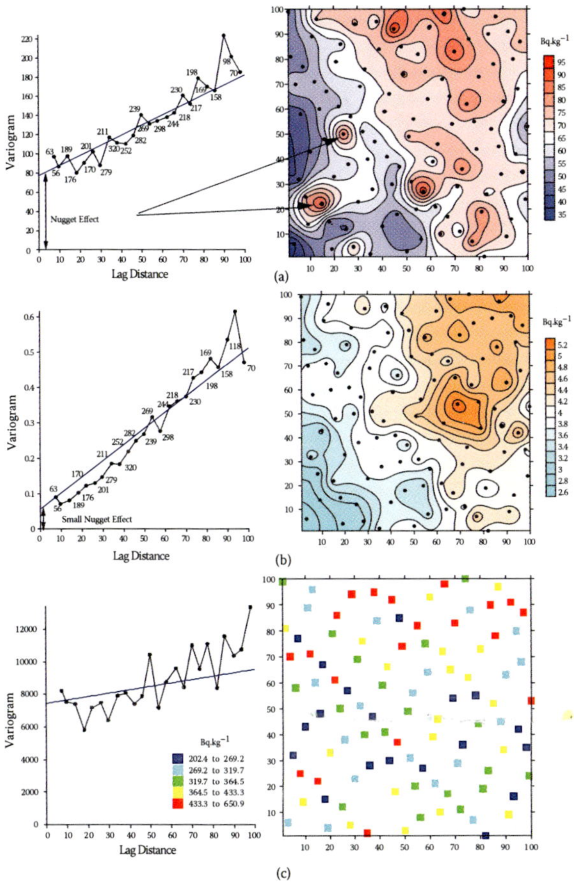

**Figure 9.15** Three examples of variograms: (a) gradient across square but with a large nugget effect; (b) gradient across square but with a very small nugget effect; and (c) no gradient. The contour plots or classed postings show the spatial distribution of the data. The point labels on the variogram indicate the number of pairs of points used at that log distance.

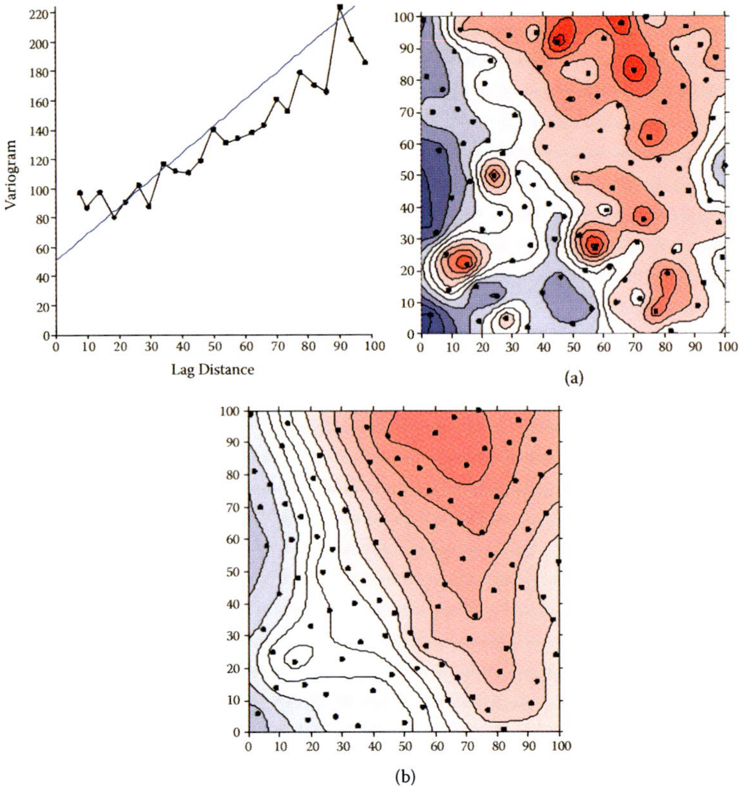


**Figure 9.16** The variogram for data across a square with either (a) the contour produced without using the variogram model or (b) the same data interpolated using the model generated by the variogram.

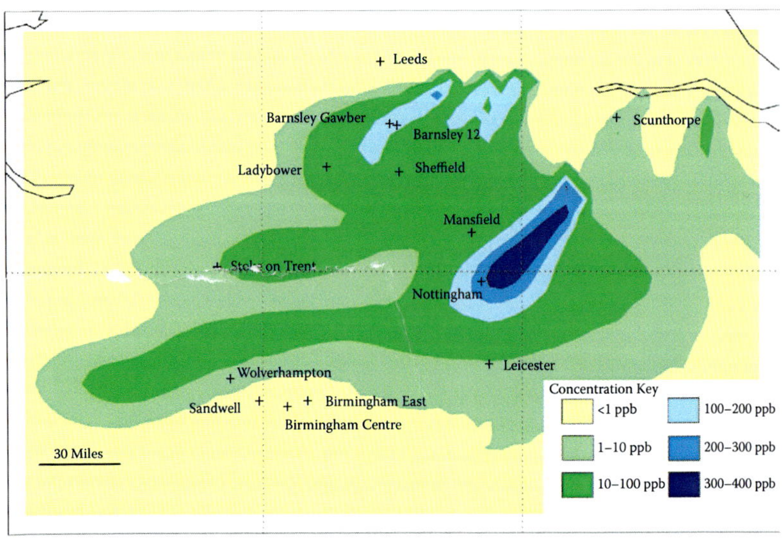


**Figure 10.6** Sulphur dioxide concentration plot for 16:00 GMT.

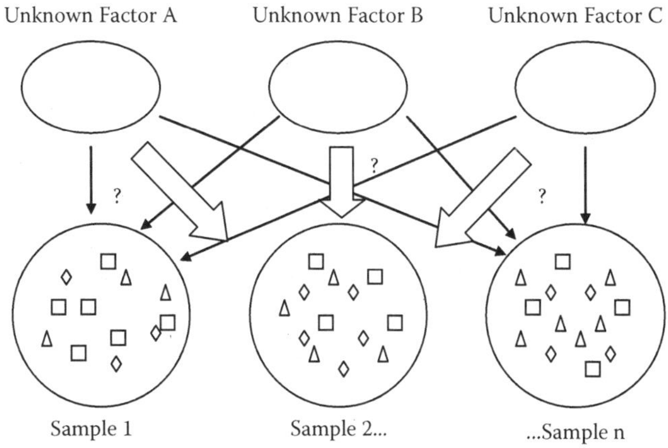


**Figure 10.9** Basic principles of PCA source apportionment.

The main limitations of PCA include:

- the requirement of a large number of receptor samples;
- judgment to determine how many 'factors' to retain (see chapter 9);
- the need to judge what sources are represented by each factor; and
- the possibility of obtaining negative values for chemical components or factors that cannot clearly be connected to sources.

### *Positive Matrix Factorisation*

On the positive side, PCA can be used without needing source emission compositions as inputs, and it can help identify important missing sources. However, Paatero and Tapper (1993) and Paatero (1997) have shown that PCA could result in scaling of the data and that this could lead to distortions in the analysis; thus, an alternative formulation of the factor analysis problem is required. One such approach is positive matrix factorisation (PMF), where the problem of nonoptimal scaling has been explicitly addressed. The main differences between PCA and PMF are that PMF constrains solutions to be >0 (so 'factors' do not have negative chemical components); PMF can account for uncertainties in the input measurements and can handle missing or below-detection-limit input data.

PMF solves the equation:

$$X = GF + E \tag{10.8}$$

where $X$ is the matrix of measured values, $G$ and $F$ are the factor matrices to be determined, and $E$ is the matrix of residuals, the unexplained part of $X$. In the PMF model, the solution is a weighted least squares fit, where the known standard deviations for each value of $X$ are used for determining the weights of the residuals in matrix E. The objective of PMF is to minimise the sum of the weighted residuals. PMF uses information from all samples by weighting the squares of the residuals with the reciprocals of the squares of the standard deviations of the data values.

In environmental pollution problems, one row of *X* would consist of the concentrations of all chemical species in one sample, and one column of *X* would be the concentration of one species for each of the samples. One row of the computed *F* matrix would be the source profile for one source, and the corresponding column of *G* would be the amount of this source in each individual sample. Required input matrices for PMF are *X*, the measured values, and $X_{\text{std-dev}}$, the standard deviations (uncertainties) of the measured values. PMF requires that all values and uncertainties be positive values; therefore, missing data and zero values must be omitted or replaced with appropriate substitute values.

PMF was initially applied to data sets of major ion compositions of daily precipitation samples collected over a number of sites in Finland (Juntto and Paatero 1994) and samples of bulk precipitation (Anttila et al. 1995) in which it was possible to obtain considerable information on the sources of these ions. More recently, there has been a series of applications of PMF to various source/receptor modelling problems (Henry 1997; Hopke 2003; Brown and Hafner 2005; Hien, Bac, and Thinh 2005; Wang and Shooter 2005; Yuan et al. 2006). PMF is somewhat more complex and harder to use than PCA, but it appears to provide improved resolution of sources and better quantification of impacts of those sources than PCA (Huang, Rahn, and Arimoto 1999).

## Comparison with CMB

Both CMB and PMF provide quantitative estimates of the source contributions. In CMB analysis, source profiles are provided; in PMF, the source profiles are estimated. If some of the source profiles are known, they can be used in PMF to constrain the extracted source profiles and thereby reduce the rotational indeterminacy. With PMF, it is not possible to precisely assign errors to the source profiles and contributions. In a CMB analysis, it is possible to define error estimates to each source contribution value. Also, because the CMB analysis is done on a sample-by-sample basis, there can be errors in the estimated source contributions because of the variations that can occur in the source profiles. PMF uses all of the data and thus estimates the average source profile over the time interval during which samples were acquired. Hence, there are some similarities in the process and the outcome, but there are also some important differences in what is being estimated, the input data required, and the estimates of the uncertainties in the calculated values.

## Receptor Modelling in Action

### *Chemical Mass Balance*

There are a considerable number of papers on applications of CMB for source apportionment (e.g., Hopke 1997 and references therein; Watson et al. 2002). It has been widely used to develop pollution control strategies. Recent CMB source apportionment studies for $PM_{10}$ and $PM_{2.5}$ have been conducted in urban areas such as Las Vegas, Nevada (Chow et al. 1999); Birmingham, United Kingdom (Hopkins, Lewis, and Seakins 2005); Seoul, South Korea (Park and Kim 2005); and California (Chow et al. 1995, 1996, 2000; Schauer et al. 1996; Schauer and Cass 2000; Hannigan, Busby, and Cass 2005; Sattler and Lijestrand 2005). Recent rural CMB studies have been conducted near the Grand Canyon (Malm and Gebhart 1997; Eatough, Farber, and Watson 2000) and near the Mt. Zirkel wilderness in northwestern Colorado (Watson et al. 2002). Detailed descriptions of the validation of CMB for hydrocarbon source apportionment have been documented (Fujita et al. 1994; Kenski et al. 1995; Thijsse, van Oss, and Lenschow 1999; Vega et al. 2000; Hellen, Hakola, and Laurila 2003; Srivastava 2004).

Hundreds of source profiles have been compiled for use in the CMB modelling, representing emissions from fugitive dust (e.g., paved and unpaved road dust, soil dust, storage pile dust), motor vehicle exhaust, vegetative burning, marine aerosol, industrial emissions, and other aerosol sources (Watson, Chow, Lowenthal, et al. 1994; Watson, Chow, Lu, et al. 1994; Chen, Lin, and Chou 2001; Vega et al. 2001; Watson and Chow 2001; Samara et al. 2003; Chow et al. 2004). Major and trace elements, ionic components (sulphate, ammonium, nitrate salts), and carbon species (organic/elemental) are sufficient to account for most of the particle mass in both ambient PM and source emissions. Elements, in particular, are good fitting species in the CMB modelling due to their chemical stability (i.e., they are not subjected to transformations during transport from source to receptor). Although the use of widely accessible library profiles is inevitable, using profiles from other study areas and times may substantially bias source contribution estimates (Chow and Watson 2002).

In recent work (Ward and Smith 2005), a CMB receptor model was used to apportion the sources of the fine aerosol fraction in the Missoula Valley, Montana. The results revealed that the largest source throughout the year at both the urban and rural background sites was wood combustion, contributing an average of 41% to the fine fraction. The second largest source of $PM_{2.5}$ throughout the year was diesel exhaust (19%), followed by secondary ammonium nitrate (17%), a kraft recovery boiler from a local pulp mill (14%), hog fuel boilers (6%) from the local wood products industry, and street sand (5%). During the forest fire season, CMB results of samples collected during smoky conditions revealed that 81% of the ambient $PM_{2.5}$ came from wood combustion (forest fires).

Samara (2005) used a similar CMB model to apportion total suspended particulate in Macedonia, Greece—an area with intensive lignite burning for power generation. He found that diesel burning in vehicular traffic and in the power plants for generator start-up was the major contributor to ambient total suspended particulate (TSP) levels. Other sources with significant contributions were domestic coal burning, vegetative burning (wood combustion and agricultural burns), and refuse open-air burning. Fly ash escaping the electrostatic precipitators of the power plants was a minor contributor to ambient TSP.

Beijing, the capital of China, is facing very serious air pollution problems, including extremely high concentrations of suspended particles in the atmosphere. The annual average concentration of $PM_{2.5}$ in Beijing is approximately 110 μg $m^{-3}$ (He et al. 2001; Zheng et al. 2005), whereas sites in Europe or the United States are of the order of 30 μg $m^{-3}$. Hence, there is great need to control particulate matter concentrations in order to improve visibility, protect human health, and reduce ecological damage; thus, a clear understanding of the composition, concentration, and sources of these airborne fine particles is required. Zheng et al. (2005) collected aerosol samples simultaneously at five sites in Beijing including rural and urban areas. They were analysed to obtain detailed information of particle-phase and solvent-extractable organic compounds in $PM_{2.5}$, which were used as tracers for source apportionment. More than 100 organic compounds, including *n*-alkanes, PAHs, *n*-alkanoic acids, *n*-alkenoic acids, resin acids, aliphatic and aromatic dicarboxylic acids, and key tracer organic compounds such as levo-glucosan were identified and quantified in each sample.

The major sources of $PM_{2.5}$ mass in Beijing averaged over five sites on an annual basis were determined as dust (20%), secondary sulphate (17%), secondary nitrate (10%), coal combustion (7%), diesel and gasoline exhaust (7%), secondary ammonium (6%), biomass aerosol (6%), cigarette smoke (1%), and vegetative detritus (1%). The lowest $PM_{2.5}$ mass concentration was found in January, but the contribution of carbonaceous aerosol to $PM_{2.5}$ mass was maximal during this season, accounting for 57% of the mass (Figure 10.10). During the cold season, the contributions from coal combustion and biomass aerosol to $PM_{2.5}$ mass increased due to their use as a fuel for heating. In April 2000, the impact of dust storms was so significant that dust alone constituted 36% of $PM_{2.5}$ mass. On average, the model resolved 88% of the sources of the $PM_{2.5}$ mass concentrations in Beijing.

### *Principal Component Analysis*

In order to enhance the accuracy of emission source identification, PCA has been used in many studies (Ho, Lee, and Chiu 2002; Park, Kim, and Kang 2002; Fang et al. 2004; Poupard et al. 2005; Wenzel et al. 2006). An example of recent use is given by Almeida et al. (2005), who used it to identify possible sources of particulate matter in Lisbon and to determine their mass

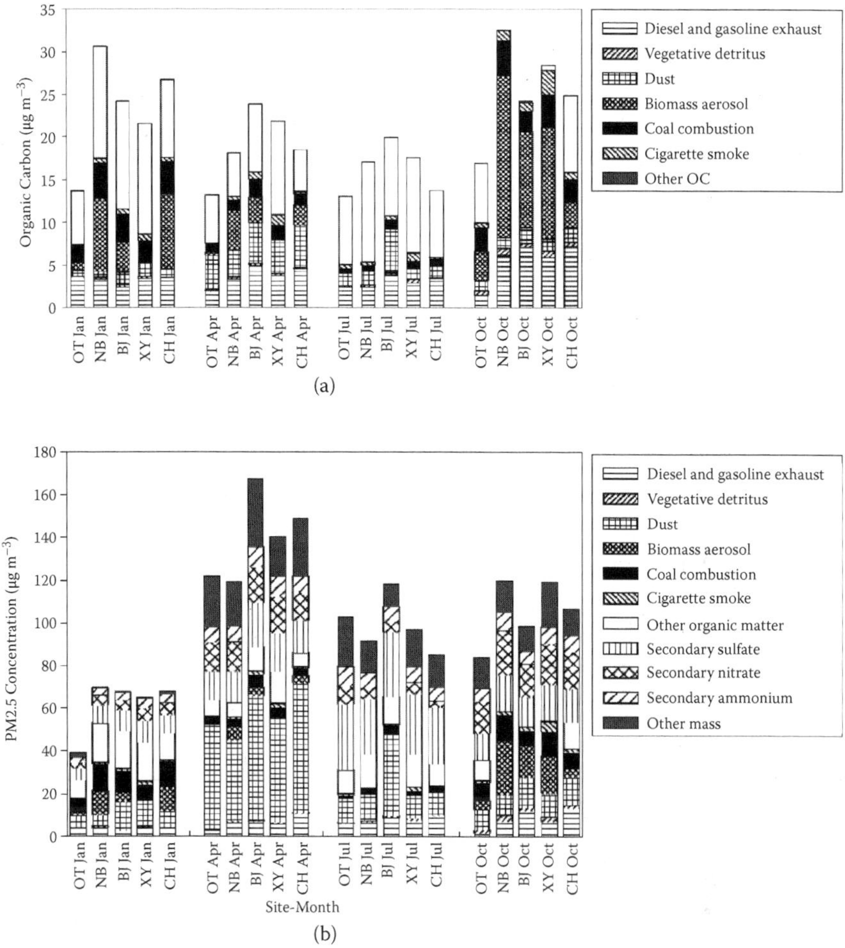


**Figure 10.10** (a) Source apportionment of fine organic carbon in Beijing and (b) primary and secondary sources of $PM_{2.5}$ in Beijing. 'Other mass' refers to the difference between the sum of identified sources and the measured $PM_{2.5}$ concentration. OT, NB, BJ, XY, and CH refer to sampling sites. Reproduced with permission. (Zheng, M. et al., *Atmospheric Environment*, 39(22): 3967–3976, 2005.)

contribution. Seven main groups of sources were identified: soil, sea, secondary aerosols, road traffic, fuel-oil combustion, coal combustion, and a Se/Hg emission source. In $PM_{2.5}$, secondary aerosol and vehicle exhaust contributed, on average, 25 and 22% to total mass, respectively, while sea spray and soil represented, respectively, 47 and 20% of the coarse fraction mass loading. After identifying the contribution of PM sources, they concluded that abatement strategies to improve local air quality should focus on traffic and on nonmobile combustion processes emitting sulphur and $NO_X$, which are conducive to the formation of secondary aerosols.

Godoy, Godoy, and Artaxo (2005) analysed air samples in and around a coal-burning thermoelectric complex in southern Brazil. PCA was applied in order to identify and quantify the aerosol particle sources. These profiles were compared with those observed on the U.S. EPA speciate data bank and a good similarity was observed. The power station contribution to the fine and coarse particle concentration was identified and quantified as 25 and 15%, respectively. This is a significant impact based on a single source and has important effects on the regional air quality. In addition, the emissions from the power station were the main source of elements such as As, Bi, Cd, Pb, Sb, and Se in both aerosol fractions, ranging from 34 up to 83% in mass. They concluded that suitable analysis of trace elements in aerosols coupled with receptor models provides an excellent method for quantitative aerosol source apportionment in industrial complexes.

Ravindra et al. (2006) applied PCA to find the possible emission sources of PAHs in Flanders, Belgium. These results show that the vehicular emission is a major source of PAHs in Flanders, although other anthropogenic sources such as incinerators, petroleum/oil burning, coke production, and wood/coal combustion also have an impact on the total PAH levels.

A two-step procedure that combines an air dispersion model with a receptor model was used to identify the key sources that contribute to air levels of suspended particulate matter at four sites in San Nicolas, Argentina (Gomez et al. 2005). The dispersion model was used to estimate the contribution of the thermal power plant. In San Nicolas, 70% of the $PM_{10}$ sampled at three of the four monitoring sites could be attributed to the power plant in those scenarios where winds connected the facility's point sources with the sampling locations. The contribution to the measured $PM_{10}$ levels from the remaining sources that were present was confirmed by way of receptor models. Principal component analysis was performed on the standard matrix of composition profiles. The diagonalisation of the covariance matrix was used as a screening procedure to differentiate the most likely contributing sources from those that were not significant. By means of this technique, it was possible to generate more reliable source information for the chemical mass balance problem, and the original 16 candidate profiles were reduced to four distinct sources.

### *Positive Matrix Factorisation*

Positive matrix factorisation has been used in numerous studies to infer pollution sources and to investigate such parameters as urban–rural differences, regional–local contrasts, and seasonal variations (Thailand: Chueinta, Hopke, and Paatero 2000; Hong Kong: Lee, Chan, and Paatero 1999; Rhode Island: Huang et al. 1999; Alaska: Polissar et al. 1999; Vermont: Polissar, Hopke, and Poirot 2001; Chile: Jorquera and Rappengluck, 2004; Bangladesh: Begum et al. 2004). Some of the recent investigations using PMF for different locations across the globe are briefly summarised next.

Liu et al. (2005) used PMF to infer the sources of $PM_{2.5}$ observed at four sites in Georgia and Alabama in the United States. One pair of urban and rural sites in each state was used to examine the regional and urban influence on $PM_{2.5}$ concentrations. Eight factors were resolved for the two urban sites and seven factors were resolved for the two rural sites. Sulphate, nitrate, and soil factors showed regional characteristics with similar seasonal variation patterns and low-frequency variations. The soil factor had high source contribution peaks during April 2001, July 2001, and July 2002. The April event reflected the intercontinental dust transport from Asia and the two July events likely reflected dust transport from Sahara. The sulphate and soil factors had the highest correlations among the sites. The regional factors contributed to about 40–50% of the total $PM_{2.5}$ masses. The seasonal patterns of the wood smoke factors were different between urban and rural sites. The dominant wood smoke source for the urban areas was residential wood burning characterised by high concentration in winter and that for the rural areas was local agricultural burning with high contributions during spring time. The strongest local contributing factor to the primary fine particle masses for the urban sites was traffic, which on average contributed 17% to the $PM_{2.5}$ mass. The strongest local contributing factor to the fine particle masses for the rural sites was wood smoke, which on average contributed 20 and 29% to the $PM_{2.5}$.

Kim et al. (2005) successfully applied PMF to hourly speciated nonmethane volatile organic compound (NMVOC) data from three sites in the Houston area. Seven sources were extracted from three monitoring sites. The measured total NMVOC was well reconstructed, suggesting that the identified sources accurately represented the ambient data at the three monitoring sites. Similar source types were identified at three sites showing their spatial distribution in the entire Houston Ship Channel area: flare emissions characterised by $C_2$–$C_3$ alkanes, butanes, and pentanes; industrial sources characterised by butane and pentane; natural gas/propane characterised by ethane and propane; refineries characterised by benzene, toluene, and xylenes; and motor vehicles characterised by benzene, ethylbenzene, toluene, and xylene. Flare emissions contributed the most, accounting for 25–38% of NMVOC at three monitoring sites.

Two source–receptor models—the chemical mass balance model (CMB) and positive matrix factorisation (PMF)—were used by Jimenez et al. (2006) to estimate smoke intrusion from regional agricultural burning in Pullman, Washington. The CMB results showed major contributions of $PM_{2.5}$ from soil (38%), vegetative burning (35%), and sulphate aerosol (20%) and much less from vehicles (2%) and cooking (1%). The three-source profiles generated by PMF were consistent with those selected for CMB modelling.

Table 10.7 shows the average source contributions to fine aerosol mass concentrations in the study of Jimenez et al. (2006) and that of Kim et al. (2003), who undertook similar investigations in Spokane, Washington, a larger city approximately 120 km north of Pullman, Washington. Allowing

**Table 10.7 Comparison of PMF and CMB Methods**

| Average source contribution in Spokane[a] | | | Average source contribution in Pullman[b] | | | | |
|---|---|---|---|---|---|---|---|
| Mass contribution | PMF (mg $m^{-3}$) | % | Mass contribution | CMB (mg $m^{-3}$) | % | PMF (mg $m^{-3}$) | % |
| Vegetative burning | 5.28 ± 0.14 | 44 | Vegetative burning | 3.96 ± 0.13 | 35 | 1.81 ± 1.57 | 17 |
| Airborne soil | 1.01 ± 0.04 | 8 | Airborne soil | 4.55 ± 0.03 | 38 | 6.20 ± 5.22 | 57 |
| Sulphate aerosol | 2.30 ± 0.04 | 19 | Sulphate aerosol | 2.22 ± 0.06 | 20 | 2.82 ± 2.14 | 26 |
| Motor vehicle | 1.29 ± 0.04 | 11 | Motor vehicle | 0.19 ± 0.01 | 2 | — | — |
| Nitrate aerosol | 1.04 ± 0.05 | 9 | Cooking | 0.12 ± 0.04 | 1 | — | — |
| Chlorine rich | 0.68 ± 0.03 | 6 | Unexplained | 0.42 ± 0.24 | 4 | — | — |
| Metal processing | 0.29 ± 0.01 | 3 | | | | | |

[a] Kim, E. et al., *Atmospheric Research,* 66(4): 291–305, 2003.

[b] Jimenez, J. et al., *Atmospheric Environment,* 40: 639–650, 2006.

for differences in vehicle density and the number of paved roads, the results show good agreement.

Song et al. (2006) successfully used the PMF model to apportion $PM_{2.5}$ from different sources in Beijing. The interpretation of the PMF analytical results was improved by comparison to earlier CMB results. They found that the contribution from coal consumption found in this study was 19%, which is not surprising given that coal combustion provides Beijing with over 50% of its energy. Road dust was shown to contribute 9%, whereas a figure of 20% was obtained from the CMB results (Zheng et al. 2005). The contributions from secondary sulphates and nitrates and motor vehicles were comparable. However, contributions from other organic matter (13%), cigarette smoke (1%), and vegetative debris (1%) were recognised in the CMB results but not apportioned in the PMF analysis. The other organic matter was considered to have been derived from fossil fuel combustion. The total percentage of mass concentrations determined by the CMB model was 6% higher than by PMF. The main reason for this can be attributed to the different principles on which the receptor models operate. CMB uses chemical source profiles in a non-negatively constrained regression model in order to obtain the contributions, whereas PMF obtains both the source profiles and the contributions from the observed concentrations using bilinear assumptions.

## Concluding Remarks

The quality of the air that we breathe can have important impacts on our health and quality of life; it is therefore important to us all. Understanding air pollution and identifying its sources provide a sound scientific basis for its management and control. Measurements alone cannot be used directly by policymakers to establish an effective strategy for solving air quality problems. A combination of state-of-the-art measurements with state-of-the-art models is the best approach for making real progress toward understanding the atmosphere. Air quality models are valuable air quality management tools. Models are mathematical descriptions of pollution transport, dispersion, and related processes in the atmosphere. They estimate the air pollutant concentration at many locations and provide a cost-effective way to analyse impacts over a wide spatial area, where factors such as meteorology, topography, and emissions from nearby sources could be important. A number of factors often limit the application of these models, including the need for spatially resolved, time-dependent emission inventories and meteorological data. Receptor models attack the source contribution identification problem in reverse order, proceeding from concentrations at a receptor backwards to responsible emission sources, without reconstructing the dispersion patterns. Overall, this is an area of continuing rapid growth and development.

## References

Alexandersen, S., I. Brotherhood, and A. I. Donaldson. (2002) Natural aerosol transmission of foot-and-mouth disease virus to pigs: Minimal infectious dose for strain O-1 Lausanne. *Epidemiology and Infection,* 128(2): 301–312.

Alexandersen, S., and A. I. Donaldson. (2002) Further studies to quantify the dose of natural aerosols of foot-and-mouth disease virus for pigs. *Epidemiology and Infection,* 128(2): 313–323.

Alexandersen, S., R. P. Kitching, L. M. Mansley, and A. I. Donaldson. (2003) Clinical and laboratory investigations of five outbreaks of foot-and-mouth disease during the 2001 epidemic in the United Kingdom. *Veterinary Record,* 152(16): 489–496.

Almeida, S. M., C. A. Pio, M. C. Freitas, M. A. Reis, and M. A. Trancoso. (2005) Source apportionment of fine and coarse particulate matter in a suburban area at the Western European Coast. *Atmospheric Environment,* 39(17): 3127–3138.

Anttila, P., P. Paatero, U. Tapper, and O. Jarvinen. (1995) Source identification of bulk wet deposition in Finland by positive matrix factorisation. *Atmospheric Environment,* 29(14): 1705–1718.

APEG. (1999) Source apportionment of airborne particulate matter in the United Kingdom. Report of the Airborne Particle Expert Group. London, DETR.

AQEG. (2004) Nitrogen dioxide in the United Kingdom. Report of the UK Air Quality Expert Group, AQEG. Prepared for the Department of the Environment Food and Rural Affairs, the Scottish Executive, the Welsh Assembly and the Department of the Environment in Northern Ireland. Defra Publications, London.

ASTM. (2000) *Standard guide for statistical evaluation of atmospheric dispersion model performance.* West Conshohocken, PA: American Society for Testing and Materials.

Begum, B. A., E. Kim, S. K. Biswas, and P. K. Hopke. (2004) Investigation of sources of atmospheric aerosol at urban and semi-urban areas in Bangladesh. *Atmospheric Environment,* 38(19): 3025–3038.

Brown, S. G., and H. R. Hafner. (2005) Multivariate receptor modelling workbook, prepared for the U.S. Environmental Protection Agency, Office of Research and Development, Research Triangle Park, NC, by Sonoma Technology, Inc., Petaluma, CA.

Bruno, P., M. Caselli, G. de Gennaro, and A. Traini. (2001) Source apportionment of gaseous atmospheric pollutants by means of an absolute principal component scores (APCS) receptor model. *Fresenius Journal of Analytical Chemistry,* 371(8): 1119–1123.

Cadle, S. H., T. C. Belian, K. N. Black, F. Minassian, M. Natarajan, E. J. Tierney, and D. R. Lawson. (2005) Real-world vehicle emissions: A summary of the 14th Coordinating Research Council On-Road Vehicle Emissions Workshop. *Journal of the Air & Waste Management Association,* 55(2): 130–146.

Chang, J. C., and S. R. Hanna. (2004) Air quality model performance evaluation. *Meteorology and Atmospheric Physics,* 87(1–3): 167–196.

Chen, K. S., C. F. Lin, and Y. M. Chou. (2001) Determination of source contributions to ambient $PM_{2.5}$ in Kaohsiung, Taiwan, using a receptor model. *Journal of the Air & Waste Management Association,* 51(4): 489–498.

Chow, J. C., D. Fairley, J. G. Watson, R. Demandel, E. M. Fujita, D. H. Lowenthal, Z. Q. Lu, C. A. Frazier, G. Long, and J. Cordova. (1995) Source apportionment of wintertime Pm(10) at San Jose, Calif. *Journal of Environmental Engineering—ASCE,* 121(5): 378–387.

Chow, J. C., and J. G. Watson. (2002) Review of $PM_{2.5}$ and $PM_{10}$ apportionment for fossil fuel combustion and other sources by the chemical mass balance receptor model. *Energy & Fuels,* 16(2): 222–260.

Chow, J. C., J. G. Watson, M. C. Green, D. H. Lowenthal, B. Bates, W. Oslund, and G. Torres. (2000) Cross-border transport and spatial variability of suspended particles in Mexicali and California's Imperial Valley. *Atmospheric Environment,* 34(11): 1833–1843.

Chow, J. C., J. G. Watson, M. C. Green, D. H. Lowenthal, D. W. DuBois, S. D. Kohl, R. T. Egami, et al. (1999) Middle- and neighbourhood-scale variations of $PM_{10}$ source contributions in Las Vegas, Nevada. *Journal of the Air & Waste Management Association,* 49(6): 641–654.

Chow, J. C., J. G. Watson, H. Kuhns, V. Etyemezian, D. H. Lowenthal, D. Crow, S. D. Kohl, J. P. Engelbrecht, and M. C. Green. (2004) Source profiles for industrial, mobile, and area sources in the Big Bend Regional Aerosol Visibility and Observational Study. *Chemosphere,* 54(2): 185–208.

Chow, J. C., J. G. Watson, D. H. Lowenthal, and R. J. Countess. (1996) Sources and chemistry of PM(10) aerosol in Santa Barbara County, CA. *Atmospheric Environment,* 30(9): 1489–1499.

Chueinta, W., P. K. Hopke, and P. Paatero. (2000) Investigation of sources of atmospheric aerosol at urban and suburban residential areas in Thailand by positive matrix factorisation. *Atmospheric Environment,* 34(20): 3319–3329.

COMEAP. (1998) *The quantification of the effects of air pollution on health in the United Kingdom.* London: HMSO.

Defra. (2002) Origin of the U.K. foot and mouth disease epidemic in 2001. Department for Environment, Food and Rural Affairs.

Department of Transport. (2006) Project for the sustainable development of Heathrow—Air quality technical report.

Donaldson, A. I. (1972) Influence of relative humidity on aerosol stability of different strains of foot-and-mouth disease virus suspended in saliva. *Journal of General Virology,* 15(Apr): 25.

______. (1987). Foot-and-mouth-disease—the principal features. *Irish Veterinary Journal,* 41(9): 325–327.

Donaldson, A. I., and S. Alexandersen. (2001) Relative resistance of pigs to infection by natural aerosols of FMD virus. *Veterinary Record,* 148(19): 600–602.

Donaldson, A. I., J. Gloster, L. D. J. Harvey, and D. H. Deans. (1982) Use of prediction models to forecast and analyse airborne spread during the foot-and-mouth-disease outbreaks in Brittany, Jersey and the Isle of Wight in 1981. *Veterinary Record,* 110(3): 53–57.

Dore, C. J., J. D. Watterson, T. P. Murrells, P. N.R., M. M. Hobson, S. L. Baggott, G. Thistlethwaite, J. W. L. Goodwin, et al. (2005) U.K. emissions of air pollutants 1970 to 2003, AEA Technology.

Eatough, D. J., R. J. Farber, and J. G. Watson. (2000) Second generation chemical mass balance source apportionment of sulphur oxides and sulphate at the Grand Canyon during the Project MOHAVE summer intensive. *Journal of the Air & Waste Management Association,* 50(5): 759–774.

Fang, G. C., Y. S. Wu, M. H. Chen, T. T. Ho, S. H. Huang, and J. Y. Rau. (2004). Polycyclic aromatic hydrocarbons study in Taichung, Taiwan, during 2002–2003. *Atmospheric Environment,* 38(21): 3385–3391.

Fujita, E. M., J. G. Watson, J. C. Chow, and Z. Q. Lu. (1994) Validation of the chemical mass-balance receptor model applied to hydrocarbon source apportionment in the Southern California Air-Quality Study. *Environmental Science & Technology,* 28(9): 1633–1649.

Gifford, F. A. (1962) Uses of routine meteorological observations for estimating atmospheric dispersion. *Nuclear Safety,* 2: 47–51.

______. (1976) Turbulent diffusion-typing schemes—Review. *Nuclear Safety,* 17(1): 68–86.

Gloster, J., R. M. Blackall, R. F. Sellers, and A. I. Donaldson. (1981) Forecasting the airborne spread of foot-and-mouth-disease. *Veterinary Record,* 108(17): 370–374.

Gloster, J., H. J. Champion, L. M. Mansley, P. Romero, T. Brough, and A. Ramirez. (2005) The 2001 epidemic of foot-and-mouth disease in the United Kingdom: Epidemiological and meteorological case studies. *Veterinary Record,* 156(25): 793–803.

Gloster, J., H. J. Champion, J. H. Sorensen, T. Mikkelsen, D. B. Ryall, P. Astrup, S. Alexandersen, and A. I. Donaldson. (2003) Airborne transmission of foot-and-mouth disease virus from Burnside Farm, Heddon-on-the-Wall, Northumberland, during the 2001 epidemic in the United Kingdom. *Veterinary Record,* 152(17): 525–533.

Gloster, J., A. Freshwater, R. F. Sellers, and S. Alexandersen. (2005) Reassessing the likelihood of airborne spread of foot-and-mouth disease at the start of the 1967–1968 U.K. foot-and-mouth disease epidemic. *Epidemiology and Infection,* 133(5): 767–783.

Gloster, J., R. F. Sellers, and A. I. Donaldson. (1982) Long-distance transport of foot-and-mouth-disease virus over the sea. *Veterinary Record,* 110(3): 47–52.

Godoy, M., J. M. Godoy, and P. Artaxo. (2005) Aerosol source apportionment around a large coal fired power plant—Thermoelectric complex Jorge Lacerda, Santa Catarina, Brazil. *Atmospheric Environment,* 39(29): 5307–5324.

Gomez, D. R., S. L. Reich, L. E. Dawidowski, and C. Vazquez. (2005) A combined analysis to identify airborne $PM_{10}$ sources. *Journal of Environmental Monitoring,* 7(1): 52–59.

Graney, J. R., J. T. Dvonch, and G. J. Keeler. (2004) Use of multi-element tracers to source apportion mercury in South Florida aerosols. *Atmospheric Environment,* 38(12): 1715–1726.

Hannigan, M. P., W. F. Busby, and G. R. Cass. (2005) Source contributions to the mutagenicity of urban particulate air pollution. *Journal of the Air & Waste Management Association,* 55(4): 399–410.

He, K. B., F. M. Yang, Y. L. Ma, Q. Zhang, X. H. Yao, C. K. Chan, S. Cadle, T. Chan, and P. Mulawa. (2001) The characteristics of $PM_{2.5}$ in Beijing, China. *Atmospheric Environment,* 35(29): 4959–4970.

Hellen, H., H. Hakola, and T. Laurila. (2003) Determination of source contributions of NMHCs in Helsinki (60 degrees N, 25 degrees E) using chemical mass balance and the Unmix multivariate receptor models. *Atmospheric Environment,* 37(11): 1413–1424.

Henry, R. C. (1991) Multivariate receptor models. In *Receptor modelling for air quality management,* ed. P. K. Hopke, 117–147. Amsterdam: Elsevier Science Publishers.

______. (1997) History and fundamentals of multivariate air duality receptor models. *Chemometrics and Intelligent Laboratory Systems,* 37(1): 37–42.

Hien, P. D., V. T. Bac, and N. T. H. Thinh. (2005) Investigation of sulphate and nitrate formation on mineral dust particles by receptor modelling. *Atmospheric Environment,* 39(38): 7231–7239.

Ho, K. F., S. C. Lee, and G. M. Y. Chiu. (2002) Characterisation of selected volatile organic compounds, polycyclic aromatic hydrocarbons and carbonyl compounds at a roadside monitoring station. *Atmospheric Environment,* 36(1): 57–65.

Hopke, P. K. (1985) *Receptor modelling in environmental chemistry.* New York: Wiley–Interscience.

______. (1997) Receptor modelling for air quality management. In *Air quality management,* ed. R. E. Hester and R. M. Harrison, 95–117. Cambridge: Royal Society of Chemistry.

______. (1999) An introduction to receptor modelling. In *Elemental analysis of airborne particles,* ed. S. Landsberger and M. Creatchman, 273–315. London: Gordon & Breach.

______. (2003) Recent developments in receptor modelling. *Journal of Chemometrics,* 17(5): 255–265.

Hopkins, J. R., A. C. Lewis, and P. W. Seakins. (2005) Analysis and applications of measurements of source dominated hydrocarbon concentrations from the PUMA campaigns in June/July 1999 and January/February 2000 at an urban background site in Birmingham, U.K. *Atmospheric Environment,* 39(3): 535–548.

Huang, S. L., K. A. Rahn, and R. Arimoto. (1999) Testing and optimising two factor-analysis techniques on aerosol at Narragansett, Rhode Island. *Atmospheric Environment,* 33(14): 2169–2185.

Jimenez, J., C.-F. Wu, C. Claiborn, T. Gould, C. D. Simpson, T. Larson, and L. J. S. Liu (2006). Agricultural burning smoke in eastern Washington—Part I: Atmospheric characterisation. *Atmospheric Environment,* 40: 639–650.

Jorquera, H., and B. Rappengluck. (2004) Receptor modelling of ambient VOC at Santiago, Chile. *Atmospheric Environment,* 38(25): 4243–4263.

Juntto, S., and P. Paatero. (1994) Analysis of daily precipitation data by positive matrix factorisation. *Environmetrics,* 5(2): 127–144.

Kenski, D. M., R. A. Wadden, P. A. Scheff, and W. A. Lonneman. (1995) Receptor modelling approach to VOC emission inventory validation. *Journal of Environmental Engineering—ASCE,* 121(7): 483–491.

Kim, E., S. G. Brown, H. R. Hafner, and P. K. Hopke. (2005) Characterisation of non-methane volatile organic compounds sources in Houston during 2001 using positive matrix factorisation. *Atmospheric Environment,* 39(32): 5934–5946.

Kim, E., T. V. Larson, P. K. Hopke, C. Slaughter, L. E. Sheppard, and C. Claiborn (2003). Source identification of $PM_{2.5}$ in an arid northwest U.S. city by positive matrix factorisation. *Atmospheric Research,* 66(4): 291–305.

Knowles, N. J., A. R. Samuel, P. R. Davies, R. P. Kitching, and A. I. Donaldson. (2001) Outbreak of foot-and-mouth disease virus serotype O in the U.K. caused by a pandemic strain. *Veterinary Record,* 148(9): 258–259.

Lee, E., C. K. Chan, and P. Paatero. (1999) Application of positive matrix factorisation in source apportionment of particulate pollutants in Hong Kong. *Atmospheric Environment,* 33(19): 3201–3212.

Lin, C. J., T. C. Ho, H. W. Chu, H. Yang, M. J. Mojica, N. Krishnarajanagar, P. Chiou, and J. R. Hopper. (2005) A comparative study of U.S. EPA 1996 and 1999 emission inventories in the west Gulf of Mexico coast region, USA. *Journal of Environmental Management,* 75(4): 303–313.

Liu, W., Y. H. Wang, A. Russell, and E. S. Edgerton. (2005) Atmospheric aerosol over two urban–rural pairs in the southeastern United States: Chemical composition and possible sources. *Atmospheric Environment,* 39(25): 4453–4470.

Malm, W. C., and K. A. Gebhart. (1997) Source apportionment of sulphur and light extinction using receptor modelling techniques. *Journal of the Air & Waste Management Association,* 47(3): 250–268.

McElroy, J. L., and F. Pooler. (1968) St. Louis dispersion study. U.S. Public Heath Service.

Mikkelsen, T., S. Alexandersen, P. Astrup, H. J. Champion, A. I. Donaldson, F. N. Dunkerley, J. Gloster, J. H. Sorensen, and S. Thykier-Nielsen. (2003) Investigation of airborne foot-and-mouth disease virus transmission during low-wind conditions in the early phase of the U.K. 2001 epidemic. *Atmospheric Chemistry and Physics,* 3: 2101–2110.

Moussiopoulos, N., E. Berge, T. Bohler, F. de Leeuw, K. E. Gronskei, and S. Mylona. (1966) Ambient air quality, pollutant dispersion and transport models. Copenhagen, European Environment Agency.

Olesen, H. R. (2001) Model validation kit—Recent developments. *International Journal of Environment and Pollution,* 16(1–6): 129–136.

Paatero, P. (1997) A weighted non-negative least squares algorithm for three-way 'PARAFAC' factor analysis. *Chemometrics and Intelligent Laboratory Systems,* 38(2): 223–242.

Paatero, P., P. K. Hopke, B. A. Begum, and S. K. Biswas. (2005) A graphical diagnostic method for assessing the rotation in factor analytical models of atmospheric pollution. *Atmospheric Environment,* 39(1): 193–201.

Paatero, P., and U. Tapper. (1993) Analysis of different modes of factor-analysis as least-squares fit problems. *Chemometrics and Intelligent Laboratory Systems,* 18(2): 183–194.

Park, S. S., and Y. J. Kim. (2005) Source contributions to fine particulate matter in an urban atmosphere. *Chemosphere,* 59(2): 217–226.

Park, S. S., Y. J. Kim, and C. H. Kang. (2002) Atmospheric polycyclic aromatic hydrocarbons in Seoul, Korea. *Atmospheric Environment,* 36(17): 2917–2924.

Pasquill, F. (1961) The estimation of the dispersion of windborne material. *Meteorological Magazine,* 90: 33–99.

Polissar, A. V., P. K. Hopke, P. Paatero, Y. J. Kaufmann, D. K. Hall, B. A. Bodhaine, E. G. Dutton, and J. M. Harris. (1999) The aerosol at Barrow, Alaska: Long-term trends and source locations. *Atmospheric Environment,* 33(16): 2441–2458.

Polissar, A. V., P. K. Hopke, and R. L. Poirot. (2001) Atmospheric aerosol over Vermont: Chemical composition and sources. *Environmental Science & Technology,* 35(23): 4604–4621.

Poupard, O., P. Blondeau, V. Iordache, and F. Allard. (2005) Statistical analysis of parameters influencing the relationship between outdoor and indoor air quality in schools. *Atmospheric Environment,* 39(11): 2071–2080.

Quinn, P. J., and B. K. Markey. (2001) A review of foot-and-mouth disease. *Irish Veterinary Journal,* 54(4): 183.

Ravindra, K., L. Bencs, E. Wauters, J. de Hoog, F. Deutsch, E. Roekens, N. Bleux, P. Berghmans, and R. Van Grieken. (2006) Seasonal and site-specific variation in vapour and aerosol phase PAHs over Flanders (Belgium) and their relation with anthropogenic activities. *Atmospheric Environment,* 40(4): 771–785.

Samara, C. (2005) Chemical mass balance source apportionment of TSP in a lignite-burning area of Western Macedonia, Greece. *Atmospheric Environment,* 39(34): 6430–6443.

Samara, C., T. Kouimtzis, R. Tsitouridou, G. Kanias, and V. Simeonov. (2003) Chemical mass balance source apportionment of $PM_{10}$ in an industrialised urban area of northern Greece. *Atmospheric Environment,* 37(1): 41–54.

Sanson, R. L. (1994) The epidemiology of foot-and-mouth-disease—Implications for New Zealand. *New Zealand Veterinary Journal,* 42(2): 41–53.

Sattler, M. L., and H. M. Lijestrand. (2005) Chemical mass balance model with fractionation for apportioning $PM_{2.5}$: A test case for Los Angeles traffic sources. *Journal of the Air & Waste Management Association,* 55(9): 1335–1344.

Schauer, J. J., and G. R. Cass. (2000) Source apportionment of wintertime gas-phase and particle-phase air pollutants using organic compounds as tracers. *Environmental Science & Technology,* 34(9): 1821–1832.

Schauer, J. J., W. F. Rogge, L. M. Hildemann, M. A. Mazurek, and G. R. Cass (1996). Source apportionment of airborne particulate matter using organic compounds as tracers. *Atmospheric Environment,* 30(22): 3837–3855.

Sellers, R. F. (1969) Airborne excretion of foot-and-mouth disease virus. *Journal of Hygiene—Cambridge,* 67: 671.

Sellers, R. F., and J. Gloster. (1980) The Northumberland epidemic of foot-and-mouth-disease, 1966. *Journal of Hygiene,* 85(1): 129–140.

Song, Y., Y. Zhang, S. Xie, L. Zeng, M. Zheng, L. G. Salmon, M. Shao, and S. Slanina. (2006) Source apportionment of $PM_{2.5}$ in Beijing by positive matrix factorisation. *Atmospheric Environment,* 40: 1526–1537.

Sorensen, J. H. (1998) Sensitivity of the DERMA long-range Gaussian dispersion model to meteorological input and diffusion parameters. *Atmospheric Environment,* 32(24): 4195–4206.

Sorensen, J. H., C. O. Jensen, T. Mikkelsen, D. K. J. Mackay, and A. I. Donaldson (2001). Modelling the atmospheric dispersion of foot-and-mouth disease virus for emergency preparedness. *Physics and Chemistry of the Earth Part B—Hydrology Oceans and Atmosphere,* 26(2): 93–97.

Sorensen, J. H., D. K. J. Mackay, C. O. Jensen, and A. I. Donaldson. (2000) An integrated model to predict the atmospheric spread of foot-and-mouth disease virus. *Epidemiology and Infection,* 124(3): 577–590.

Srivastava, A. (2004) Source apportionment of ambient VOCS in Mumbai city. *Atmospheric Environment,* 38(39): 6829–6843.

Taghavi, M., S. Cautenet, and J. Arteta. (2005) Impact of a highly detailed emission inventory on modelling accuracy. *Atmospheric Research,* 74(1–4): 65–88.

Thijsse, T. R., R. F. van Oss, and P. Lenschow. (1999) Determination of source contributions to ambient volatile organic compound concentrations in Berlin. *Journal of the Air & Waste Management Association,* 49(12): 1394–1404.

Thurston, G. D., K. Ito, T. Mar, W. F. Christensen, D. J. Eatough, R. C. Henry, E. Kim, et al. (2005) Workgroup report: Workshop on source apportionment of particulate matter health effects—Intercomparison of results and implications. *Environmental Health Perspectives,* 113(12): 1768–1774.

Toscano, G., A. Gambaro, I. Moret, G. Capodaglio, C. Turetta, and P. Cescon. (2005) Trace metals in aerosol at Terra Nova Bay, Antarctica. *Journal of Environmental Monitoring,* 7(12): 1275–1280.

Turner, D. B. (1994) *Workbook of atmospheric dispersion estimates: An introduction to dispersion modelling.* Boca Raton, FL: Lewis Publishers.

Vega, E., V. Mugica, R. Carmona, and E. Valencia. (2000) Hydrocarbon source apportionment in Mexico City using the chemical mass balance receptor model. *Atmospheric Environment,* 34(24): 4121–4129.

Vega, E., V. Mugica, E. Reyes, G. Sanchez, J. C. Chow, and J. G. Watson. (2001) Chemical composition of fugitive dust emitters in Mexico City. *Atmospheric Environment,* 35(23): 4033–4039.

Wang, H. B., and D. Shooter. (2005) Source apportionment of fine and coarse atmospheric particles in Auckland, New Zealand. *Science of the Total Environment,* 340(1–3): 189–198.

Ward, T. J., and G. C. Smith. (2005) The 2000/2001 Missoula Valley $PM_{2.5}$ chemical mass balance study, including the 2000 wildfire season—Seasonal source apportionment. *Atmospheric Environment,* 39(4): 709–717.

Watson, J. G., and J. C. Chow. (2001) Source characterisation of major emission sources in the Imperial and Mexicali valleys along the U.S./Mexico border. *Science of the Total Environment,* 276(1–3): 33–47.

Watson, J. G., J. C. Chow, D. H. Lowenthal, L. C. Pritchett, C. A. Frazier, G. R. Neuroth, and R. Robbins. (1994) Differences in the carbon composition of source profiles for diesel-powered and gasoline-powered vehicles. *Atmospheric Environment,* 28(15): 2493–2505.

Watson, J. G., J. C. Chow, Z. Q. Lu, E. M. Fujita, D. H. Lowenthal, D. R. Lawson, and L. L. Ashbaugh. (1994) Chemical mass-balance source apportionment of Pm(10) during the Southern California air-quality study. *Aerosol Science and Technology,* 21(1): 1–36.

Watson, J. G., T. Zhu, J. C. Chow, J. Engelbrecht, E. M. Fujita, and W. E. Wilson. (2002) Receptor modelling application framework for particle source apportionment. *Chemosphere,* 49(9): 1093–1136.

Wenzel, K. D., A. Huber, L. Weissflog, R. Kuhne, P. Popp, A. Kindler, and G. Schuurmann. (2006) Influence of different emission sources on atmospheric organochlorine patterns in Germany. *Atmospheric Environment,* 40: 943–957.

WHO. (2005) Particulate matter air pollution: How it harms health. Fact sheet EURO/04/05, World Health Organisation.

Yuan, Z. B., J. Z. Yu, A. K. H. Lau, P. K. K. Louie, and J. C. H. Fung. (2006) Application of positive matrix factorisation in estimating aerosol secondary organic carbon in Hong Kong and its relationship with secondary sulphate. *Atmospheric Chemistry and Physics,* 6: 25–34.

Zannetti, P. (2001) Environmental litigation—Air pollution models and modellers in court. *Air Pollution,* 9: 3–12.

Zheng, M., L. G. Salmon, J. J. Schauer, L. M. Zeng, C. S. Kiang, Y. H. Zhang, and G. R. Cass (2005). Seasonal trends in $PM_{2.5}$ source contributions in Beijing, China. *Atmospheric Environment,* 39(22): 3967–3976.

# Evidence and Expert Witnesses in Environmental Forensics Cases

# 11

ALLAN KANNER

## Contents

The Expert Process ... 354
*Daubert* and the Goal of Liberalising Proof ... 354
The Underlying Problem ... 360
Judges as Junk Scientists ... 360
Judges as Managers ... 362
Understanding the Cases ... 364
Legal Background ... 364
Overview ... 364
Private Environmental Law ... 364
Toxic Torts ... 364
Strict Liability ... 364
Nuisance and Trespass ... 365
Emotional Distress ... 367
Medical Monitoring ... 368
Public Environmental Law ... 368
Common Law Public Trust Doctrine ... 368
Common Law Public Nuisance Doctrine ... 369
Citizen Suits ... 370
Natural Resources ... 370
Future Trends in Natural Resource Damages ... 371
Environmental Justice ... 376
Conclusions ... 376
References ... 377

## The Expert Process

In order to fully appreciate the complexity of issues associated with the judicial recognition of an environmental forensic expert, a bit of background is needed.

### *Daubert* and the Goal of Liberalising Proof

In 1975, U.S. federal courts adopted the Federal Rules of Evidence (FRE) to govern the admissibility of facts and opinions that parties to a lawsuit may use to prove their case.[*] Article VII of the Rules (Rules 701–706) regulates the admissibility of expert or opinion testimony so crucial in cases involving questions of science.[†] In environmental cases, for example, the case often turns on the admission or exclusion of a party's expert, who is needed to establish the causal link between the defendant's activity and the plaintiff's injury.

Prior to an expert's testimony in a case, the opposing side has the opportunity to challenge the admissibility of his or her testimony. Prior to 1993, federal courts used the '*Frye* standard' to determine whether such testimony was admissible.[‡] In *Frye v. United States,* the D.C. Circuit held that an expert's underlying theory must 'have gained general acceptance in the particular field in which it belongs'.[§]

Since 1993, '*Daubert*' has been the single most important term in discussions of expert proof. In *Daubert v. Merrell Dow Pharmaceuticals, Inc.,*[¶] the Supreme Court held that the Federal Rules of Evidence—specifically Rule 702—had superseded *Frye's* 'general acceptance' test because it was deemed at odds with the 'liberal thrust' of the FRE.

Rule 702 provides that '[i]f scientific, technical, or other specialised knowledge will assist the trier of fact to understand the evidence or to determine a fact in issue', an expert 'may testify thereto'.[**] After reviewing Rule 702, the Court noted that '[n]othing in the text of this Rule establishes "general acceptance" as an absolute prerequisite to admissibility'.[††] When *Frye* pro-

[*] See Federal Rules of Evidence, available at http:// www.law.cornell.edu/rules/fre/index.html.

[†] Other legal systems have adopted similar codified rules governing admissibility of expert evidence. See, for example, U.K. Civil Procedure Rules, Part 35, 'Experts and Assessors', available at http://www.dca.gov.uk/civil/procrules_fin/contents/parts/part35.htm.

[‡] *Frye v. United States,* 293 F. 1013 (D.C. Cir. 1923).

[§] *Id.* at 1014.

[¶] *Daubert v. Merrell Dow Pharmaceuticals, Inc.,* 509 U.S. 579, 113 S.Ct. 2786, 125 L.Ed.2d 469 (1993).

[**] Federal Rules of Evidence 702.

[††] *Daubert v. Merrell Dow Pharmaceuticals, Inc.,* 509 U.S. 579, 588, 113 S.Ct. 2786, 125 L.Ed.2d 469 (1993); *Frye v. United States,* 293 F. 1013, 1014 (D.C. Cir. 1923). See also Daubert–Joiner–Kumho: *The Brave New World of Expert Evidence,* 15 Tox. L. Rep. (BNA) 1213 (Nov. 23, 2000); Leslie Lunney, *Protecting Juries from Themselves: Restricting the Admission of Expert Testimony in Toxic Tort Cases,* 48 SMU L. Rev. 103 (1994).

vided the governing test, the standard for the admission of expert testimony focused upon the question of scientific consensus rather than the quality of the scientific method.

That the Supreme Court characterised the *Frye* test as 'austere', specifically, Justice Blackmun explained that '[t]he drafting history [of the FRE] makes no mention of *Frye*, and a rigid "general acceptance" requirement would be at odds with the "liberal thrust" of the Federal Rules and their "general approach" of relaxing the traditional barriers to "opinion" testimony'.*

All of the justices agreed that nothing in the text of Rule 702 established 'general acceptance' as an absolute prerequisite to admissibility.† They found, given the 'Rules' permissive backdrop', that the assertion that they somehow 'assimilated *Frye*...unconvincing'.‡ Consequently, the majority sought to articulate an alternative, and ostensibly more liberal, standard in accordance with the FRE.

The *Daubert* court interpreted Rule 702 to impose two distinct requirements in the case of scientific expert evidence:

> (1) The evidence must be reliable, that is, the underlying methodology and procedure from which evidence is derived (not the conclusion drawn) must be based on scientific knowledge. The district court acts as a gate keeper under Federal Rule of Evidence 104(a) and makes a preliminary assessment of whether the reasoning or methodology underlying the testimony is scientifically valid. In deciding if the testimony is scientifically valid, the court looks to many factors, including whether the theory or technique can and has been tested, whether it has been subjected to peer review, the known or potential rate of error, and whether it has been generally accepted.
>
> (2) The evidence must be relevant, that is, it must assist the trier of fact either in understanding the evidence or in determining a fact in issue.

In *Daubert*, a product liability case involving a morning-sickness drug called Bendectin (Debendox in the United Kingdom), the Court considered the admissibility of the testimony of the plaintiff's expert interpreting epidemiological studies by others. The expert's testimony had been rejected by the trial court and the U.S. Court of Appeals for the Ninth Circuit under the *Frye* standard. The Supreme Court held that the adoption of Rule 703 had effectively liberalised and overruled the *Frye* test, 'given the Rules' permissive backdrop and their inclusion of a specific rule on expert testimony that does not mention "general acceptance"'.§ It substituted a case-specific inquiry

* *Daubert v. Merrell Dow Pharmaceuticals, Inc.*, at 588.

† *Daubert* at 588–89.

‡ *Id.* at 589.

§ 509 U.S. at 589.

by the trial judge, applicable not only to 'unconventional evidence' but also to other scientific testimony.

Under *Daubert*, a trial court judge faced with a proffer of expert scientific testimony must determine whether the expert is proposing to testify to (1) scientific knowledge that (2) will assist the trier of fact to understand or determine a fact at issue.* The *Daubert* court noted that many factors will bear on the inquiry and stated that it was not setting out a definitive checklist or test. Because Rule 702 clearly implies some degree of regulation, the *Daubert* court imposed conditions: 'The subject of an expert's testimony must be "scientific...knowledge"'—that is, 'an inference or assertion must be derived by the scientific method'.† In order to determine whether proffered evidence is 'scientific knowledge' that 'will assist the trier of fact [usually the jury] to understand or determine a fact in issue', the majority provided a list of four factors (hereafter the *Daubert* criteria) to assist the trial judge's assessment:

> Ordinarily, a key question to be answered in determining whether a theory or technique is scientific knowledge that will assist the trier of fact will be whether it can be (and has been) tested.‡
>
> Another pertinent consideration is whether the theory or technique has been subjected to peer review and publication. Publication (which is but one element of peer review) is not a sine qua non of admissibility; it does not necessarily correlate with reliability.§
>
> Additionally, in the case of particular scientific techniques, the court ordinarily should consider the known or potential rate of error...and the existence or maintenance of standards controlling the technique's operation.
>
> Finally, 'general acceptance' can yet have a bearing on the inquiry, in the sense that widespread acceptance can be an indicator of reliability, and 'a known technique which has been able to attract only minimal support within the community...may properly be viewed with scepticism'.¶

Notwithstanding the provision of specific criteria, the majority explained that any inquiry under Rule 702 should be 'flexible'.** In consequence, these factors were characterized as indicative rather than a 'definitive checklist or test'.††

* 509 U.S. at 592–593 and n.11.

† *Daubert* at 589–90.

‡ *Id.*, citing K. Popper, *Conjectures and Refutations: The Growth of Scientific Knowledge* 37 (5th ed. 1989) ('[T]he criterion of the scientific status of a theory is its falsifiability, or refutability, or testability').

§ *Id.*, citing see S. Jasanoff, *The Fifth Branch: Science Advisors as Policymakers* 61–76 (1990).

¶ *Daubert* at 593–4.

** *Daubert* at 593.

†† *Id.*

The Court in *Daubert* added that federal courts would also evaluate admissibility under other applicable federal rules. It specifically noted Rule 703, which limits facts or data upon which experts may rely to those 'reasonably relied upon by experts in the particular field',* and Rule 104, which permits the court to exclude relevant evidence if 'its probative value is substantially outweighed by the danger of unfair prejudice, confusion of the issues, or misleading the jury'.†

In closing, the *Daubert* court addressed defendants' concern that abandonment of the general acceptance test would result in a 'free-for-all' before the jury. Defendants would continue to have available the traditional means of attacking admissible evidence—'[v]igorous cross-examination, presentation of contrary evidence, and careful instruction on the burden of proof'.‡

In *General Electric Co. v. Joiner*,§ the Supreme Court held that the standard of review of a district court's decision with regard to the admissibility of expert opinion after *Daubert*¶ is abuse of discretion. The Court clarified two potential errors in applying the standard of review: that there is a difference in standard of review between decisions to admit expert testimony and decisions finding such testimony inadmissible and that a different standard of review should apply if the decision on admissibility is 'outcome determinative'. In both cases, no different standard of review applies. The Court concurred in the district court's evaluation that the proffered expert opinion did not rise above 'subjective belief or unsupported speculation'.** What this means in practice is that the trial court has a great deal of discretion to avoid trying a case on the merits.

In *Kumho Tire v. Carmichael*, the Court addressed *Daubert*'s applicability to expert testimony that is not classically 'scientific', such as engineering or other 'specialised knowledge'.†† The Court ruled that the *Daubert*

* *Daubert v. Merrell Dow Pharmaceuticals, Inc.*, 509 U.S. 579 (1993).

† *Daubert v. Merrell Dow Pharmaceuticals, Inc.*, 509 U.S. 579, 595 (1993).

‡ *Daubert v. Merrell Dow Pharmaceuticals*, 509 U.S. 579, 591 (1993). Since *Daubert* was decided, several circuits have applied it in affirming district court exclusion of expert testimony under the *Frye* standard. *In re Paoli R.R. Yard PCB Litig.*, 35 F.3d 717 (3d Cir. 1994), not overruled, but called into question by *Amorgianos v. AMTRAK*, 137 F. Supp. 2d 147, 164 (E.D.N.Y. 2001); *United States v. Jones*, 24F.3d 1177 (9th Cir. 1994) (affirming exclusion of expert testimony concerning voice identification); *O'Conner v. Commonwealth Edison Co.*, 13 F.3d 1090 (7th Cir. 1994) (affirming exclusion of expert testimony that plaintiff's cataracts were caused by a radiation dose thousands of times less than that commonly believed by experts to be required to cause this condition). Some courts not following the Federal Rules of Evidence have declined to apply a *Daubert*-type analysis, instead opting for the more restrictive *Kelly/Frye* general acceptance test. See, for example, *People v. Leahy*, 882 P.2d 321 (Cal. 1994); *State v. Coon*, 974 P.2d 386, 395 (Alas. 1999).

§ *GE v. Joiner*, 520 U.S. 114 (1997).

¶ *Daubert v. Merrell Dow Pharmaceuticals., Inc.*, 509 U.S. 579 (1993).

** *GE v. Joiner*, 520 U.S. 1114 (1997).

†† *Kumho Tire Co., Ltd. v. Carmichael*, 526 U.S. 137 (1999).

criteria, while originally established for scientific evidence, could be used to evaluate other types of expertise, and that *Daubert*'s fundamental holding—that an expert's testimony must be both relevant and reliable—applies to all experts.*

The drafters of the federal rules amended Federal Rule of Evidence 702 in 2000 to conform to or enrich the Supreme Court's definition of the district court's inquiry in *Daubert*:

> If scientific, technical or other specialized knowledge will assist the trier of fact to understand the evidence or to determine a fact in issue, a witness qualified as an expert by knowledge, skill, experience, training, or education may testify thereto in the form of an opinion or otherwise, if (1) the testimony is based upon sufficient facts or data, (2) the testimony is the product of reliable principles and methods, and (3) the witness has applied the principles and methods reliably to the facts of the case.†

The trial court will then attempt to understand and critically evaluate the expert's scientific or technical methodology, even when that evaluation could determine the outcome of a case. The federal trial courts were thus assigned a substantial task, well beyond the parameters of the general acceptance test of *Frye*: validation of the scientific technique that the expert employs in its broader application and its case-specific use.

Since the *Daubert–Joiner–Kuhmo* trilogy and the modification of Rule 702, the Supreme Court has been conservative in upholding the admission of expert witness testimony. *Weisgram v. Marley Company*‡ resolved a circuit split regarding Rule 50. The Supreme Court held that an appellate court has the power to direct a judgement as a matter of law when, after determining that evidence was erroneously admitted at trial, it finds the evidence insufficient for a submissible case.§ A court of appeals may, therefore, throw out expert testimony, find remaining evidence insufficient, and proceed to grant a judgement notwithstanding verdict.¶

The Court rejected the plaintiff's argument for an 'automatic remand'.** It focused on the fact that *Daubert* has put litigants on notice. 'It is implau-

* See, Peter J. Goss, Debra L. Worthington, Merrie Jo Stallard, & Joseph M. Price, *Clearing Away the Junk: Court-Appointed Experts, Scientifically Marginal Evidence, and the Silicone Gel Breast Implant Litigation*, 56 Drug L.J. 227, 231–234 (2001).

† Federal Rules of Evidence 702.

‡ *Weisgram v. Marley Company*, 528 U.S. 440 (2000).

§ *Weisgram v. Marley Company*, 528 U.S. 440 (2000).

¶ A judgement notwithstanding verdict, or 'j.n.o.v.', is a practice in American civil law whereby a judge may overrule a jury's decision shortly after their judgement. Appellate courts uphold a j.n.o.v. when they find no reasonable juror could have reached the verdict the jury reached in the case.

** *Weisgram v. Marley Company*, 528 U.S. 440 (2000).

sible to suggest...that parties will initially present less than their best expert evidence in the expectation of a second chance should their first try fail'.[*] Since *Daubert*, plaintiffs have had notice of the 'exacting standards of reliability'.[†]

Later appellate court reviews have not exercised this power, however. The court in *Toole v. Baxter Healthcare Corp.*[‡] noted, 'We review a trial court's evidentiary rulings on the admission of expert witness testimony for abuse of discretion'.[§]

Some appellate courts have been lenient as to when *Daubert* applies and how closely it should be followed. One court found that there is no requirement that a district court always hold a *Daubert* hearing prior to qualifying an expert witness under FRE 702.[¶] Federal Rule of Evidence 702 permits a district court to allow the testimony of a witness whose knowledge, skill, training, experience, or education will assist a trier of fact in understanding an area involving specialised subject matter.[**] Courts conform to the *Daubert* factors[††] to various extents. The court in *Fillebrown v. Steelcase, Inc.*[‡‡] noted that the '*Daubert* test for reliability is flexible and *Daubert*'s list of specific factors neither necessarily nor exclusively applies to all experts or in every case'.

It is worth noting that *Frye* remains important and that *Daubert* is still the minority view in the United States. About 98% of all U.S. civil and criminal cases are litigated in state courts. Only 16 states have expressly adopted the *Daubert* standard, while 19 states still adhere to the *Frye* standard. Among those 19 states, which encompass 55% of the U.S. population, are populated

[*] *Weisgram v. Marley Company*, 528 U.S. 440 (2000).

[†] *Weisgram v. Marley Company*, 528 U.S. 440 (2000).

[‡] *Toole v. Baxter Healthcare Corp.*, 235 F.3d 1307 (11th Cir. 2000).

[§] *Toole v. Baxter Healthcare Corp.*, 235 F.3d 1307, 1312 (11th Cir. 2000). Cf *Diefenbach v. Sheridan Transp.*, 229 F.3d 27, 30 (1st Cir. 2000).

[¶] *United States v. Solorio-Tafolla*, 324 F.3d 964 (8th Cir. 2003).

[**] *United States v. Solorio-Tafolla*, 324 F.3d 964, 965 (8th Cir. 2003). Cf. *United States v. Alatorre*, 222 F.3d 1098 (9th Cir. 2000).

[††] The *Daubert* factors are: (1) whether the scientific theory or technique can be tested, (2) whether it has been subject to peer review and publication, (3) the known or potential rate of error associated with the technique, and (4) whether the theory or technique has gained 'general acceptance'. *Daubert v. Merrell Dow Pharms.*, 509 U.S. 579, 593–95, 113 S. Ct. 2786, 125 L.Ed.2d 469 (1993).

[‡‡] *Fiflebrown v. Steelcase, Inc.*, No. 02-1080, 2003 U.S. App. Lexis 4393 (3d Cir. February 24, 2003) at *5.

states like California, New York, Florida, Illinois, Pennsylvania, Michigan, and New Jersey.*

## The Underlying Problem

*Daubert* came as a reaction to the proliferation of bad experts in the federal courts. It is true that some putative experts are not objective and neutral. Scientists, of course, work harder at being objective because of the limits and goals of their scientific disciplines, but this does not mean personal preference, greed, or ideologies never get in the way of their research. The scientific community has its share of ambition, suppression of truth, prejudice, plagiarism, manipulation of data, etc.† Against this backdrop, *Daubert* makes sense.

## Judges as Junk Scientists

*Daubert* only makes sense when judges themselves do not engage in junk science. This charge has been made in numerous cases. An interesting example is Judge Blake's analysis in *Newman v. Motorola*,‡ a cell phone cancer case. The argument is made by Gary Edmond and David Mercer§:

> While limits to peer review have been widely acknowledged, [Judge] Blake's scepticism and forensic investigation of the correspondence between [plaintiff's expert] Hardell and various journal editors places her judgment on what might appear to be a scientifically unaccountable basis. In a discussion of problems with the 'use and abuse' of research subpoenas and judicial misconceptions of the role of scientific peer review Sheila Jasanoff explained that:

* These jurisdictions include Alabama, Arizona, California, Washington, D.C., Florida, Illinois, Kansas, New Jersey (except for toxic torts), New York, North Dakota, Maryland, Michigan, Minnesota, Mississippi, Missouri, Pennsylvania, and Washington. *See Courtland Fibers v. Long,* 779 So.2d 198 (Ala. 2000) (*Frye* followed except for DNA); *Logerquisz v. McVay,* 1 P.3d 113 (Ariz. 2000); *People v. Leahy,* 8 Cal.4th 589 (1994); *Jones v. U.S.*, 548 A.2d 35 (D.C. 1988); *Florida Power & Light Co. v. Tursi,* 729 So.2d 995 (Fla. Dist. App. 1999); *Donaldson v. Illinois Public Service Co.*, 767 N.E.2d. 314 (Ill. 2002); *Kuhn v. Sandoz,* 14 P.3d 1170 (Kansas 2000); *Hutton v. Store,* 663 A.2d 1289 (MD 1995); *People v. Davis,* 72 N.W.2d 269 (Mi. 1955); *Goeb v. Thoraldson,* 615 N.W2d 800 (Minn. 2000); *Kansas City Southern Railway v. Johnson,* 798 So.2d 374 (Miss. 2001); *M.C. v. Yeargin,* 11 S.W.3d 604 (Mo. App.1999); *Store v. Doriguzzi,* 760 A.2d 336 (N.J. Sup. A.D. 2000) (*Daubert* for toxic torts); *City of Fargo v. McLaughlin,* 512 N.W2d 700 (N.D.1994); *Blum ex. rel. Blum v. Merrell Dow,* 764A.2d 1 (Pa. 2000); *Store v. Copeland,* 922 P.2d 1304 (Wash. 1996). (See also Bureau of National Affairs, Product Safety & Liability Reporter, 30(15): 328—341.)

† This is illustrated by Tel Aviv Medical School's professor of urology Alexander Kohn in his *False Prophets: Fraud and Error in Science and Medicine* (1986), and by Broad and Wade's *Betrayers of the Truth: Fraud and Deceit in the Halls of Science* (1982).

‡ 28 F. Supp.2d 769 (D.Md. 2002).

§ Daubert *and the Exclusionary Ethos: The Convergence of Corporate and Judicial Attitudes towards the Admissibility of Expert Evidence in Tort Litigation,* Law & Policy, 26(2) (April 2004): 231, 241–243.

> [s]cientific peer review is likely to differ markedly in its objectives and impact from review carried out by an expert in a litigation context. In legal review, the goal is neither to make good work better nor to retrieve what might be of value from work of lesser significance. It is instead, to seek to aggressively as possible discredit the proffered evidence and to deploy in the process all the sceptical resources that experts specifically engage for this purpose can muster (Jasanoff 1996: 113–14).

Having challenged the circumstances surrounding the publication of Hardell's work, Blake continued the critical assessment of Hardell's claims: 'The fact of publication, of course, does not eliminate the need to examine the results and methodology of the study, keeping the inquiry focused on relevance and validity as it relates to the causation opinions offered in this case' (*Newman* 2002 at 12). Blake criticised Hardell's 'methodology', highlighting problems of recall bias, lack of a demonstrated dose–response relationship, the relationship of ipsilateral causation to general causation; problems with subgroup comparisons; and, lastly, the reliance of a methodology for testing laterality that 'has not been used by any other scientist proffered to the court…nor…been replicated' (*Newman* 2002 at 14).

Furthermore, Blake's references to testing and replication provide a good example of the flexible ways ideal images of the scientific method can be used in legal settings to help deconstruct or marginalise particular forms of expertise. For example, Blake suggested that Hardell's work had not been replicated because '[t]he Inskip and Muscat studies [two alternative epidemiological studies] which tested laterality by other means and admittedly with a smaller number of people do not show increased risk' (*Newman* 2002 at 14).

Sociologists of science, most notably Harry Collins (1985), have provided detailed accounts of how the meaning and interpretation of an experimental replication are highly negotiable and often controversial. Blake engaged in precisely this kind of interpretive exercise when she accorded a sufficient degree of similarity to all of the epidemiological studies in question, which allowed them to be characterised as a failure to replicate Hardell's findings. Notwithstanding this view, it would have been open to Blake to dismiss Hardell's work even if the Inskip and Muscat studies had supported his findings. The studies could have been distinguished, drawing upon Blake's categories, on the grounds that they tested laterality 'by other means' and with a 'smaller number of people'.

Blake's critiques of the testing and replication also demonstrate the way post-*Daubert* visions of science, coupled with a tough gatekeeping ethos, can be used to restrict the entry of (novel) scientific claims. One of the general features of the mobile-telephone health debate (and no doubt many other controversies around certain risks to health) has been the difficulty in achieving standardisation of study methodologies and establishing what types of scientific studies should be accorded weight in ascertaining causation (Berger 1997; Edmond and Mercer 2000; Mercer 2002; Miller 2003). While simplistic images of the sciences are de rigueur in legal formulations and contexts (exemplified in *Daubert*), the 'real world' science is considerably more complex. Current research into the health effects of mobile telephones exemplifies this complexity in a manner that might inform our understanding of Blake's *Newman* decision.

The World Health Organisation (WHO) is currently running an international epidemiological study examining the medical records of cancer patients while endeavouring to establish their past mobile-telephone use. Whatever its findings, this study will be vulnerable to future legal or methodological deconstruction by claims that its results embody an unscientific recall bias. The study's *retrospective* approach has already encountered criticism from epidemiologists who favour *prospective* methods. Prospective methods tend to start by monitoring phone use and then track future health outcomes. Prospective studies, interestingly, have limited relevance to current mobile-phone tort litigation (or regulation), as they often take decades to complete. Furthermore, even if prospective studies did indicate a *positive* correlation between adverse health effects and mobile phone use, they may be vulnerable to challenge unless (future) plaintiffs can identify physical causal mechanisms that explain why mobile phones appear to be harmful. The debate over possible causal mechanisms for mobile telephone health problems also suffers from entrenched theoretical disagreements and a lack of acceptance around protocols for experimental work (Stewart 2000; Swicord 2003). Somewhat ironically, these entrenched theoretical disagreements are some of the factors originally motivating policymakers, such as the WHO, to oversee the retrospective epidemiological studies (Graham-Rowe 2003).

The extended gatekeeping undertaken by Blake in *Newman* illustrates how *Daubert*-inspired quests to establish scientific truth at the pretrial stage of litigation may assist in discouraging ongoing legal scrutiny of intransigent scientific controversies involving uncertain risks. Another example is Erica Beecher-Monas's critique* of the Eighth Circuit's handling of science issues in *Wright Willamette Industries, Inc.*†

## Judges as Managers

Another judicial abuse of *Daubert* arises from the fact that it allows courts to clear their dockets. *Daubert* motions present a tempting opportunity

* Erica Beecher-Monas, *The Heuristics of Intellectual Due Process: A Primer for Triers of Science*, 75 N.Y.U. L. Rev. 1563, 1636–48 (2000) (enclosed).

† *Wright v. Willamette Indus., Inc.*, 91 F.3d 1105 (8th Cir. 1996).

for judges to dispose of cases without risk of appellate review. As Professor McGarity states*:

> After more than a decade of experience with *Daubert*, it is now clear that the lower courts have applied it vigorously to exclude expert testimony.† Since the plaintiff ordinarily has the burden of proof in tort litigation, this aggressive invocation of the judge's new role as guardian of the purity of scientific evidence has had a disproportionate impact on plaintiffs. With remarkable speed, judges have gone far beyond throwing the clinical ecologists out of the courtroom.‡ Impressed by artful defense counsels' smoke screens, they are now excluding testimony of well-regarded experts.§ A plaintiff's attorney must prepare not only to establish an expert's qualifications, but also to convince a sceptical trial judge that the testimony supports a scientifically reliable conclusion based upon scientifically reliable data and that the conclusion fits the legal requirements for establishing cause-in-fact.¶ If the plaintiff's attorney fails, everyone goes home, and no one knows whether the plaintiff was a victim of cruel fate or of the defendant's possibly unconscionable conduct. If the attorney succeeds, the judge and jury must sit through days of confusing and conflicting expert testimony, at the end of which the jury may still decide that the plaintiff did not carry the burden of proof.

* Thomas O. McGarity, *On the Prospect of 'Daubertizing' Judicial Review of Risk Assessment*, 66 Law & Contemp. Probs. 155 (2003).

† Jeffry D. Cutler, *Implications of Strict Scrutiny of Scientific Evidence: Does* Daubert *Deal a Death Blow to Toxic Tort Plaintiffs?* 10 J. Envtl. L. & Litig. 189, 214 (1995). ('[I]t doesn't take a rocket scientist to figure out that a four or five part test including "general acceptance" as one factor will be more difficult to meet than a test based on "general acceptance" alone'.)

‡ In the mid-1980s, the science of clinical ecology appeared to be the answer to the causation conundrum for plaintiffs' attorneys. Professor Elliott described the phenomenon as follows: 'For a price, some clinical ecologists will testify that exposure to even very small amounts of a wide range of chemicals suppresses the immune system, thereby weakening the body's ability to ward off disease. This weakening, in turn, allegedly makes the plaintiff vulnerable to virtually all diseases known to humankind, including "nervousness," "malaise," and other conditions that present only subjective symptoms'. E. Donald Elliott, *Toward Incentive-Based Procedure: Three Approaches for Regulating Scientific Evidence*, 69 B.U. L. Rev. 487, 490 (1989); see also Peter Huber, *A Comment on Toward Incentive-Based Procedure: Three Approaches for Regulating Scientific Evidence by E. Donald Elliott*, 69 B.U. L. Rev. 513, 515 (1989). Even before *Daubert* refined their screening function, the courts never seriously entertained claims based heavily upon the testimony of clinical ecologists.

§ For an excellent example, see Professor Beecher-Monas's thorough and devastating critique of the post-*Daubert* Eighth Circuit Court of Appeals' opinion in *Wright v. Willamette Indus., Inc.*, 91 F.3d 1105 (8th Cir. 1996). Erica Beecher-Monas, *The Heuristics of Intellectual Due Process: A Primer for Triers of Science*, 75 N.Y.U. L. Rev. 1563, 1637 (2000).

¶ See Harvey Brown, *Eight Gates for Expert Witnesses*, 36 Hous. L. Rev. 743 (1999) (detailing eight 'gates' through which a proponent of expert testimony must navigate in order to demonstrate that the testimony is admissible); see also Beecher-Monas, *supra*.

### Understanding the Cases

*Daubert* requires case-specific decisions that are often hard to extrapolate to different cases. *Daubert* decisions often turn on the quality of a particular presentation, which is generally hard to glean from the text of judicial opinion. As a result, it is important to focus on the *Daubert* elements and to produce proof tailored to the court's sophistication on each of these elements.

## Legal Background

### Overview

Enforcement of environmental rights is both relatively old and relatively new. Some of these rules, such as the public trust doctrine, date back to the time of Justinian.* Although experts do not need to become legal historians, it is worthwhile to understand the different regimes of law applicable to a particular problem as well as their interplay.

### Private Environmental Law

#### *Toxic Torts*

'Toxic torts' refers to any private environmental claims regardless of the specific common law theory used, such as strict liability, nuisance, and trespass. They are a relatively underpursued way to address environmental problems. Toxic torts are an effective way to sue for environmental damages to property and person. This is because complex environmental issues can be simplified under a toxic tort claim, making for more effective litigation. This risk of exposure to toxic tort damages has been credited with deterring certain polluting activity.

#### *Strict Liability*

Strict liability remains the default choice in many toxic tort cases. Strict liability describes a set of legal concepts that hold one party responsible for damages caused to another without any showing that the liable party is 'at fault' in causing the damages. The most relevant applications may be non-natural and hazardous uses of land, as well as certain classes of products. This arises when a landowner or occupier brings on the land something that is not a matter of natural usage and that escapes, causing injury to the land, person, or property of his neighbour. The party is strictly liable, irrespective

* For example, Allan Kanner, *The Public Trust Doctrine,* Parens Patriae, *and the Attorney General as the Guardian of the State's Natural Resources,* Duke Envtl. Law & Policy 57 (2005).

of all acts of due care.* There is a good example in U.K. law with *Empress Cars (Abertillery) v NRA*.

### *Nuisance and Trespass*

Private tort law, especially common law rights nuisance and trespass, has long been used to promote environmental concerns. Indeed, what is now called environmental law has its roots in ancient air and water cases dealing with questions of nuisance and riparian rights. Both society at large, via the sovereign's public trust responsibilities, and individuals could exercise property-type rights against polluters. Those early cases recognised types of legal right to a clean environment based on a theory of property rights. While these rights were later balanced against the needs of an industrialising society to pollute, the earliest cases granted close to an absolute right to be free from pollution intrusion. Much of that damage (e.g., interference with use and enjoyment) reflects core environmental values, especially those that cannot be reduced to a market price, such as loss-of-use damages.

The common law of strict liability has changed in response to changing conditions in American society. The storage and disposal of toxic chemical waste pose the same threat to health and welfare today as the detonation of dynamite and impoundment of waters posed in years past. The Court finds that strict liability in New Mexico is not confined to blasting, and that plaintiff has alleged sufficient facts supporting a characterisation of defendant's hazardous waste activities as abnormally dangerous.

For example, in *William Aldred's Case*,† the plaintiff filed suit against his neighbour for the erection of a pig sty near his house, claiming that the sty fouled his air and blocked his windows. The court in that case applied the principle of *sic utere tuo alienum non laedes*: 'so use your own property so as not to injure your neighbours'. The court rejected the defendant's argument that 'the building of the house for hogs was necessary for the sustenance of man', refusing to accept the social utility defence for allowing a nuisance. The court did note that the injury complained of, under nuisance principles, could not be a trivial one.‡ Thus, in the context of property holders, the early common law guaranteed absolute protection from concrete, nontrivial environmental damage.

* Many states have subjected to strict liability hazardous waste that causes injury to surrounding property. *Cities Serv. V. State*, 312 So.2d 799 (Fla. App. 2nd Dist. 1975). *See Ashland Oil, Inc. v. Miller Oil Purchasing Co.*, 678 F.2d 1293, 1308 (5th Cir. 1982); *State Dep't of Envtl. Protection v. Ventron Corp.*, 468 A.2d 150 (N.J. 1983); *Bridgeton v. B.P. Oil, Inc.*, 369 A.2d 49 (N.J. Super. 1976); *McLane v. Northwest Natural Gas Co.*, 467 P.2d 635 (Or. 1970); *Langan v. Valicopters, Inc.*, 567 P.2d 218 (Wn. 1977). *See also Schwartzman, Inc. v. Atchison, T. & S. F. Ry.*, 842 F.Supp. 475, 479 (D.N.M. 1993).

† 9 Coke 57b, 77 Eng. Rep. 816 (K.B. 1611).

‡ *Id.*

As noted, the Industrial Revolution brought enormous change to the early environmental law of nuisance.* With the emergence of industry, the courts began to look beyond the substantiality of the plaintiff's injuries to the nature and reasonableness of the defendant's enterprise.† In *St. Helen's Smelting Co. v. Tipping,*‡ a copper smelter was emitting 'noxious vapours' onto the plaintiff's estate, damaging trees and shrubs and bothering the occupants. The court granted an injunction but in dictum paved the way for a new utilitarian analysis.

> [B]oth English and American Courts relaxed their protection of the more passive uses of property, and, in effect, subsidized developing industry by forcing the cost of its by-products on society in general. In this atmosphere, the rationale of *St. Helen's Smelting* was misconstrued as establishing a balancing of utilities doctrine that overrode the previous *sic utere tuo* guarantees of the common law to the passive property holder. In addition, the relaxation of the *sic utere tuo* guarantee removed an important incentive for industrial control of pollution by improved technology.§

This balancing of the utilities doctrine remains important in the United States.¶

A similar evolution can be seen in the law pertaining to the use of water. The early case law guaranteed an absolute right to a riparian landowner. The court in *City of Richmond v. Test,*** for example, applied the 'natural flow' rule. Under this rule, downstream riparians enjoyed an absolute right to the flow of the stream undiminished in quality or quantity. Later cases modified the rule to a 'reasonable use' rule, wherein all riparian owners had a right to the reasonable use of the water.†† The courts apply a multifactorial analysis to what constitutes 'reasonable use'. This analysis includes a consideration of the social use and the extent of the harm,‡‡ not unlike the utilitarian nuisance analysis. In water law, as in property law, early courts recognised the right to be free from unreasonable pollution.

However, environmental degradation to private property during and after the Industrial Revolution continued under a belief that the benefits

* Larry D. Silver, *The Common Law of Environmental Risk and Some Recent Applications,* 10 Harv. Envtl. L. Rev. 61, 74 (1986).

† See *Hole v. Barlow,* 140 Eng. Rep. 1113, 1118 (C.P. 1858) (nuisance would not lie if offending was legal, reasonable, and carried on in an appropriate place).

‡ 11 Eng. Rep. 1483 (H.L. 1865).

§ Daniel R. Coquillette, *Mosses from an Old Manse: Another Look at Some Historic Property Cases about the Environment,* 64 Cornell L. Rev. 761, 820 (1979).

¶ See Paul M. Kurtz, *Nineteenth Century Anti-Entrepreneurs Nuisance Actions—Avoiding the Chancellor,* 17 Wm. & Mary L. Rev. 621, 670 (1976).

** See *City of Richmond v. Test,* 48 N.E. 610 (1897).

†† See *City of Hampton v. Watson,* 89 S.E. 81(1916).

‡‡ William Goldfarb, *Water Law,* 15 (2nd ed. 1988).

of the products produced outweighed the costs to human health and the environment. Despite the private property issues at stake, the courts were willing to allow a reasonable amount of pollution to interfere with the landowner's rights.

### *Emotional Distress*

Negligent infliction of emotional distress is a new concept in toxic tort cases, and has yet to gain acceptance in most jurisdictions.* In those jurisdictions that do allow compensation for negligent infliction, usually some physical injury or threat of serious physical injury must accompany the emotional distress.† Where no physical injury has manifested, it is not clear whether subclinical and subcellular levels of changes constitute physical injury to which emotional distress damages may attach.‡ In one toxic tort case, damage to the immune system caused by a change to the bone marrow was deemed to constitute physical harm sufficient to support a cause of action for personal injury.§ Some courts do allow for recovery in the absence of injury if the plaintiff demonstrates sufficient evidence of a nonphysical injury, such as fear of acquiring cancer.

In *Collier v. Simpson Paper Co.*,¶ the Ninth Circuit provided a guideline for such a claim. To establish tortious inducement of fear of contracting cancer given the absence of physical injury or illness, appellants may only recover for fear of contracting cancer if:

1. due to appellee's negligence, appellants were exposed to a toxic, cancer-causing substance; and

* See *McClenathan v. Rhone-Poulenc Inc.*, 926 F.Supp. 1272, 1276 (S.D.W.Va. 1996) (suggesting that plaintiffs could prevail on a theory of intentional infliction of emotional distress but not on negligence).

† *Potter v. Firestone Tire & Rubber Co.*, 863 P.2d 795 (Cal. 1993); *Robinson v. United States*, 175 F.Supp.2d 1215, 1228 (E.D. Cal. 2001).

‡ Conrad G. Tuohey & Ferdinand V. Gonzalez, *Emotional Distress Issues Raised by the Release of Toxic and Other Hazardous Materials*, 41 Santa Clara L. Rev. 661, 696–98 (2001). See *Potter v. Firestone Tire & Rubber Co.*, 863 P.2d 795, 809, n. 10 (Cal. 1993) ('No California cases address whether impairment of the immune system response and cellular damage constitute "physical injury" sufficient to allow recovery for parasitic emotional distress').

§ *Duarte v. Zachariah*, 22 Cal. App. 4th 1652 (Cal. App. 3rd Dist. 1994) (finding that overdose of negligently prescribed medication impaired ability of bone marrow to produce blood platelets; platelets are the component of blood that causes the blood to clot in response to a wound or cut), cited in Conrad G. Tuohey & Ferdinand V. Gonzalez, *Emotional Distress Issues Raised by the Release of Toxic and Other Hazardous Materials*, 41 Santa Clara L. Rev. 661, 696–98 (2001).

¶ *Collier v. Simpson Paper Co.*, No 97-15101, 1997 U.S. App. LEXIS (9th Cir. Dec. 24, 1997).

2. appellants fear cancer based on the knowledge, corroborated by reliable medical or scientific opinion, that it is more likely than not that the appellants will develop cancer in the future due to the toxic exposure.*

### *Medical Monitoring*

Medical monitoring is a form of damages that realistically, consistently, and adequately addresses the needs of toxic tort victims. The method involves collecting and disbursing funds that enable individual plaintiffs to receive ongoing diagnostic evaluations. These evaluations are designed to detect the presence of an exposure-related disease at an early stage in its development, in order to reduce substantive treatment costs and better preserve the victim's health. Some courts enthusiastically support medical monitoring, while others have rejected it. Increased risk alone is usually not enough to warrant medical monitoring.† Those courts that do not require plaintiffs to establish that future injury is reasonably certain to occur may call for a showing that clinical examination is medically appropriate.‡

Under CERCLA (the Comprehensive Environmental Response, Compensation, and Liability Act), there is no private remedy of medical monitoring.§ However, this does not always defeat a separate tort cause of action. The court in *In re Paoli R.R. Yard PCB Litig.*¶ recognised a common-law cause of action for medical monitoring costs under Pennsylvania law.

## Public Environmental Law

### *Common Law Public Trust Doctrine*

From these roots and from the early private environmental law emerged the public trust doctrine. At the heart of the doctrine is the concept that some natural resources are too important to the public good to be privately held. Rather, these resources must be preserved for public use and enjoyment. Further, the public trust doctrine anticipates that the state, under the legal fiction

* *Collier v. Simpson Paper Co.*, No. 97-15101, 1997 U.S. App. Lexis 36334, at *11 (9th Cir. Dec. 24, 1997). See also *Abuan v. General Elec. Co.*, 3 F.3d 329 (9th Cir. 1993) (affirming summary judgement against class in toxic tort case where members did not present medical and scientific evidence that accidental exposure to toxic chemicals increased risk of future injury or disease or that they were presently injured).

† See *Plummer v. Abbott Labs.*, 568 F.Supp. 920 (D.R.I. 1983); *Betts v. Manville Injury Personal Trust*, 588 N.E.2d 1193, 1218 (Ill. 1992); *Morrissy v. Eli Lilly & Co.*, 394 N.E.2d 1369, 1376 (Ill. 1979).

‡ See, for example, *Miranda v. Shell Oil Co.*, 7 Cal. Rptr.2d 623 (Cal. Ct. App. 1992); *In re Paoli R.R. Yard PCB Litig.*, 916 F.2d 829, 851 (3rd Cir. 1990), *cert. denied, sub nom. General Electric Co. v. Knight*, 499 U.S. 961 (1991); *Werlein v. United States*, 746 F.Supp. 887 (D. Minn. 1990), vacated in part, 793 F.Supp. 898 (1992).

§ *Daigle v. Shell Oil Co.*, 972 F.2d 1527, 1533–35 (10th Cir. 1992).

¶ *In re Paoli R.R. Yard PCB Litig.*, 916 F.2d 829, 851 (3rd Cir. 1990).

of the owner and later under an exercise of a variant of its police power, will hold and protect these resources for these purposes. In practice, however, the use of the public trust doctrine has been somewhat limited.

Unfortunately, while public trust suits have seemingly unlimited potential and have occasionally proven useful in the field of environmental law, they have never reached the type of prominence suggested by commentators, who go on to suggest that inadequate damage awards in several *parens patriae* and public trust precedents may be to blame.* Others take a contrary view and believe this doctrine will fully compensate for environmental damage.†

Historically, public environmental law has statutory and common-law components. The public trust doctrine to natural resources held in common has ancient roots in both civil- and common-law jurisdictions.‡

The public trust doctrine allowed the government to bring suit in its trustee capacity to protect some natural resources. Public nuisance provided a measure of protection for resources such as air and water that are held by no one. Up until the late 1960s, however, government regulation of the environment for the public health, either in common-law causes of action or in statutory causes of action, did not adequately protect the public's interest.

### *Common Law Public Nuisance Doctrine*

Most legal systems have a concept of 'public nuisance'. The American version is summarised aptly by the Restatement of Torts:

> (1) A public nuisance is an unreasonable interference with a right common to the general public.
> (2) Circumstances that may sustain a holding that an interference with a public right is unreasonable include the following:
> (a) whether the conduct involves a significant interference with the public health, the public safety, the public peace, the public comfort or the public convenience, or
> (b) whether the conduct is proscribed by a statute, ordinance or administrative regulation, or

* Frank B. Cross, *Natural Resource Damage Valuation,* 42 Vand. L. Rev. 269, 278 n. 41 (1989) (citing *Note, Defining the Appropriate Scope of Superfund Natural Resource Damage Claims: How Great an Expansion of Liability?* 5 Va. J. Nat. Resources L. 197, 201–02 (198) (authored by Thomas Newlon).

† See, for example, Allan Kanner, *The Public Trust Doctrine,* Parens Patriae, *and the Attorney General as the Guardian of the State's Natural Resources,* 16 Duke Envtl. L. & Policy Forum 57 (2005).

‡ Going back to early English common law, the sovereign and landowners had a strong legal basis to protect the public trust, including forests, rivers, and beaches. Similar normative material exists in code jurisdictions and dates back at least to Justinian. See, for example, *Illinois Central Railroad Co. v. Illinois,* 146 U.S. 387 (1892).

(c) whether the conduct is of a continuing nature or has produced a permanent or long-lasting effect, and, as the actor knows or has reason to know, has a significant effect upon the public right.*

Public nuisance has become a major component in the field of environmental law.

### *Citizen Suits*

While there are a variety of methods under which to pursue environmental claims, citizen suits are among the most effective. Twelve federal environmental acts enable private action in the form of citizen lawsuits.† There are two basic kinds of citizen suits in environmental law: (1) suits by private citizens against the executive branch of the government, usually the EPA, alleging that the government has not carried out a mandatory duty in implementing an environmental law, and (2) suits by private citizens for injuctive relief intended to enforce agency regulations against third parties.

### *Natural Resources*

In summary, the field of environmental law has developed along two tracks: private and public. Property owners, asserting the rights that they believed their ownership gave them, were successful in abating a certain amount of pollution and damage to their own natural resources or air and water, which cannot be owned. Industry and other individuals were able to pursue their activities up to the point that the individual property owner was unreasonably harmed. Public actions for protection of the public trust, public nuisance, and later complete environmental regulation attempted to balance industrial and modern technological interests against the public's right (individually and collectively) to natural resources held in common.

The weak link in this system is in protection of natural resources (as earlier defined) in which the public may have no direct health concern or that are not under the control of any one person and may go unregulated. What was and is needed is a cause of action for natural resource damages. 'The law of damages consists of the rules, standards, and methods used by the courts

* Restatement (Second) of Torts § 821B (1979).

† Citizen suits appear in at least 12 separate environmental laws: CAA § 304, 42 U.S.C. § 7604 (1982), the Federal Water Pollution Control Act (FWPCA), 33 U.S.C. § 1365 (1982), the Marine Protection, Research, and Sanctuaries Act (MPRSA), 33 U.SC. § 1415(g) (1982), the DPA, 33 U.S.C. §4911 (1982), the Endangered Species Act (ESA), § 11(g), 16 U.SC. § 1540(g) (1982), the DPA, 33 U.S.C. § 1515 (1982), RCRA § 7002, 42 U.S.C. §6972 (1982 & Supp. V 1987), SWDA § 1449, 42 U.S.C. § 300j-8 (1982), the Surface Mining Control and Reclamation Act (SMCRA), § 520, 30 U.S.C. § 1270 (1982), CERCLA, 42 U.S.C. § 9659 (Supp. V 1987), the Emergency Planning and Community Right-to-Know Act (EPCRA), §326, 42 U.S.C. § 11046 (Supp. V 1987), and the Outer Continental Shelf Lands Act (OCSLA), 43 U.S.C. § 1349(a) (1982).

for measuring in money the compensation given for losses and injuries'.* In his treatise on the law of damages, J.G. Sutherland states,

> The consequences of an act which is an invasion of another's right may be arrested; in some cases partial restoration is practicable. But unless compensation can be made as a substitute for that to which a party is entitled, and of which he has been more or less deprived, there will be an irreparable injury, and a corresponding failure of justice.†

In accordance with this principle, 'every infraction of a legal right causes injury'.‡ For a violation of a legal right, therefore, availability of damages is presumed.§ However, Sutherland also recognises that

> (t)he law does not give a private *remedy for* anything but a private wrong. A public wrong, though the perpetrator of it may be subject to prosecution by the public, may also have the nature and consequences of a private wrong, and be actionable as such in behalf of a person who sustains an injury differing in kind from that which the public at large suffers.¶

Inexplicably bound to the issue of the availability of natural resource damages, therefore, is the question of whether harm to the environment is a legally cognizable cause of action for an individual, state, or federal government, or the environment itself. The two most important natural resource laws are CERCLA and the Oil Pollution Act, 1990 (OPA).

### *Future Trends in Natural Resource Damages*

The cause of action for natural resource damages has come a long way since William Aldred sued to require his neighbour to take down a pig sty. Private law rights of nuisance and tort continue to provide a degree of protection for the landowner's and the individual's environment. The public trust doctrine and public nuisance have been expanded from preventing the alienation of navigable waterways and sewage dumping to protecting more natural resources and human health from a wider scope of harms. Finally, natural resource damage statutes, passed by the state and federal governments, have begun to legitimise and expand the cause of action to a point where courts are more likely to recognise and apply it.

* Charles T. McCormick, Damages § 1 (1975).

† J.G. Sutherland, *A Treatise of the Law of Damages* § 1 (2nd ed. 1893).

‡ *Id.* at § 2.

§ *Id.*

¶ *Id.* at § 4.

However, modern environmental laws have not been extensively used to protect natural resources.* Structurally, CERCLA and its state analogues allow natural resource claims to sit until after site remediation claims are fully addressed. Accordingly, until recently, natural resource damage claims were rarely included in early Superfund cases, which tend to focus on immediate site remediation concerns. However, the inclusion of natural resource damage actions in Superfund cases is becoming more routine in some jurisdictions and is expected to become far more so. In the aftermath of the 1989 *Exxon Valdez* oil spill and with passage of OPA, companies whose operations may lead to natural resource liability now face a new political willingness and stronger laws and regulations to prosecute these claims.

Several federal statutes define natural resources and natural resource damages and authorise federal, state, or local officials to assess and collect damages related to natural resource injury. The most important federal environmental laws pertaining to natural resource damages are CERCLA, as amended,† and the Clean Water Act (CWA),‡ as amended, through OPA.§ Other federal laws that contain natural resource damage provisions include the Marine Sanctuaries Act¶ and the Trans-Alaska Pipeline Authorization Act (TAPAA). All these statutes authorise natural resource trustees to recover compensatory damages for injury to, destruction of, or loss of natural resources resulting from a release of a hazardous substance or a discharge of oil into navigable waters.

The federal government has legislated causes of action for natural resource damages. The beginning of the federal cause of action for natural resource damages came in TAPAA.** The TAPAA opened the door for later, more comprehensive environmental damage provisions by providing that the oil pipeline company would be 'strictly liable to all damaged parties, pub-

* See, for example, *Artesian Water Co. v. New Castle County,* 851 F.2d 643, 650 (3d Cir. 1988). There has likewise been a revitalisation of *parens patriae,* for example, *Idaho v. Southern Refrigerated Transp., Inc.,* No 88-1279 1991 U.S. Dist. Lexis 1869, at *12–*14 (D. Idaho, Jan. 24, 1991) (allowing state *parens* suit as alternative to CERCLA NRD for damage to wildlife and sport fish).

† See Comprehensive Environmental Response, Compensation, and Liability Act of 1980, 42 U.S.C. § 9607(a) (1994 Supp. III 1997) (providing that responsible parties shall be liable for 'damages for injury to, destruction of, or loss of natural resources, including the reasonable costs of assessing such injury, destruction, or loss resulting from such a release').

‡ Clean Water Act, §311(f)(4)(5), 33U.S.C. §1321 (f)(4),(5).

§ The Oil Pollution Act of 1990, 33 U.S.C. § 2702(a) (1994 Supp. IV 1998) (providing that 'each responsible party for a vessel or a facility from which oil is discharged, or which poses a substantial threat of a discharge of oil...is liable for...damages [including damages to natural resources] that result from such an incident').

¶ National Marine Sanctuaries Act of 1988, 16 U.S.C. §1443(a)(1).

** Pub. L. No. 93-153 (Title 11), 87 Stat. 584 (1973) (codified as amended 43 U.S.C. §§ 1651-55 (1982)).

lic or private, without regard to fault for such damages, and without regard to ownership of any affected lands, structure, fish, wildlife, or biotic or other natural resources relied upon by Alaska natives, Native organizations, or others for subsistence or economic purposes'.* Obviously, the applicability of this statute is limited regionally (to the area where an oil pipeline would be constructed; from wells in northern Alaska to ports in the southern part of that state) and in scope (providing only for cleanup costs and economic damages to humans). However, TAPAA was the first step in codifying the common law concept of natural resource damages.

The shortcomings of the TAPAA were improved upon with the passage of the Deepwater Port Act of 1974 (DPA).† The DPA provided that the secretary of transportation 'may act on behalf of the public as trustee of the natural resources of the marine environment to recover for damages'.‡ The DPA defined 'damages' to mean damages 'suffered by any person, or involving the natural resources of the marine environment, or the coastal environment'.§ The concepts of the public trust over natural resources and the government's rights and responsibilities pursuant to that trust are evident in this statute.

The CWA¶ continued the evolution of the natural damages concept in federal law, applying it to a much broader base. 'Congress got its opportunity to apply the new model in a major-league setting....Unlike the limited applicability of the DPA, the FWPCA [CWA] liability provisions apply generally to oil spills throughout United States waters'.** The CWA states, 'Sums recovered [in a natural resource damages action] shall be used to restore, rehabilitate, or acquire the equivalent of such natural resources'.†† This language, unlike the TAPAA, more clearly anticipates a cause of action for natural resource damages separate from the effect of those damages on humans. The CWA also expanded who could sue, extending the scope of potential plaintiffs from the federal government alone to the states.‡‡

The concept of a federal cause of action for natural resource damages was given its greatest expansion under Superfund (CERCLA). Under that statute, governmental entities are entitled to recover money damages for the release of a hazardous substance that causes an 'injury to, destruction of, or loss of natural resources'.§§ The statute defines 'natural resources' as

* 43 U.S.C. § 1653(c)(1) (1982).

† Pub. L. No. 93-637, 88 Stat. Ann. 2126 (1975) (codified at 33 U.S.C. §§ 1501-1524 (1982)).

‡ 33 U.S.C. § 1517(i)(3) (1982).

§ 33 U.S.C. § 1517 (m)(2) (1982).

¶ The Federal Water Pollution Control Act, Pub. L. No. 95-217, 91 Stat. 1566 (1977) (codified as amended at 33 U.S.C. §§ 1251-1376 (1982)).

** Barry Breen, *Citizen Suits for Natural Resource Damages: Closing a Gap in Federal Environmental Law*, 24 Wake Forest L. Rev. 851, 857 (1989).

†† 33 U.S.C. § 1321(f)(5) (1982).

‡‡ 33 U.S.C. § 1321(f)(5) (1982).

§§ 42 U.S.C. § 9607(a)(4)(C) (Supp. IV 1986).

> land, fish, wildlife, biota, air, water, ground water, drinking water supplies, and other such resources belonging to, managed by, held in trust by, appertaining to, or otherwise controlled by the United States...any State or local government, any Indian tribe, or if such resources are subject to a trust restriction on alienation, any member of an Indian tribe.[*]

A significant factor in assessing the CERCLA natural resource damage provision is the explicit exclusion of petroleum from the definition of 'hazardous substance'.[†] However, importantly, Congress also expanded the concept of who could sue to now include local governments.[‡] Also striking is the breadth of the definition of natural resources, including resources protected under the public trust doctrine as well as those subject to regulation by way of public nuisance law.

Finally, the amendments to the Marine Protection, Research, and Sanctuaries Act,[§] though a much smaller program than Superfund, expand natural resource liability for damage to designated areas by *any* means.[¶] These amendments are the latest step in the development of the federal cause of action for natural resource damages. They represent, although on a limited scale, the public trust doctrine taken to its logical extreme.

However, many problems with the cause of action for natural resource damages persist. Despite the growth of the concept, the lack of cases that apply the cause of action supports the proposition that the cause of action for natural resource damages has so far failed to coalesce into a legitimate and effective tool in protecting all of the nation's natural resources.

At present the concept of natural resource damages is a hodgepodge of state and federal statute and common law. One problem with the cause of action is its lack of uniformity and comprehensibility. Moreover, each of the states may have a different statute for natural resource damages, further complicating potential liability. As a result, any given harm to natural resources may or may not be fully redressable. The type of harm may not be one covered by CERCLA. The natural resource damaged may not be protected under a public trust theory. If both of these theories fail, the state may or may not have a natural resource damage statute, which may or may not be applicable to the situation. In addition, theories of liability aside, the government, as the sole party able to bring suit, may choose not to. If the government does

[*] 42 U.S.C. § 9601(16) (1982 7 Supp. IV 1986).

[†] See 42 U.S.C. § 9601(14) (Supp. IV 1986).

[‡] 42 U.S.C. § 9607(f) (1982 & Supp. V 1987). See *City of New York v. Exxon Corp.*, 697 F.Supp. 677 (S.D. N.Y. 1988); *Mayor of Boonton v. Drew Chem. Corp.*, 621 F.Supp. 663 (D. N.J. 1985) (both holding that §9607(f) allows local governments to sue for natural resource damages).

[§] Pub. L. No. 100-627 (1988).

[¶] *Id.*

bring suit and wins, the amount recovered may not adequately compensate for the loss.

A re-examination of the case of the *S. S. Zoe Colocotroni* with some variations more clearly illustrates the problems with the cause of action for natural resource damages. A captain dumps 5000 tons of crude oil into the water that subsequently destroys a bay. The first problem is who will sue. The damage was to public land held by the state as well as to the transitory wildlife that inhabits the bay. Certainly, the government has standing to sue as an individual landowner. Suing on behalf of transitory wildlife, however, or on behalf of the natural resources themselves is problematic. The doctrine of public trust gives theoretical support to the cause of action, but may not be a legally cognizable cause of action. Moreover, political factors may influence the government to not sue at all, leaving the natural resources without a remedy. In the case of the *S. S. Zoe Colocotroni* spill, the government of Puerto Rico decided to sue and was granted standing based on its police power to regulate the area and the existence of a statute that allowed suit. At least 12 states would not be as fortunate if the same event were to occur. In many cases no one could sue.

Linked to the question of standing is the question of a theory on which to sue. In the example case, CERCLA liability is not available because the spill is petroleum and petroleum is explicitly exempted from the definition of hazardous substances in CERCLA. The CWA is available for oil spills but may provide incomplete relief. Also, in cases where the damage is not to water, the CWA does not apply. With adequate federal protection lacking, we turn to state and common-law remedies.

If state and federal statutes prove inadequate, the state could attempt to use the theory of public trust. The public trust doctrine serves as an excellent theoretical model for the cause of action but proves impractical in application. If a court does recognise the public trustee concept, it may still limit its application based on the natural resource at stake. Thus, again, transitory natural resources and resources not held by anyone will not be redressed. In the alternative, public nuisance requires a continuing violation and does not allow for the recovery of damages. It would be inadequate in most settings.

Finally, there is the problem of *compensation* should the government win. Puerto Rico chose to forgo its remedies as an individual landowner because the measure of damages would not have compensated for the loss. The adequate measure of damages for loss of natural resources is a complex question that cannot be adequately addressed in this chapter. It should be noted, however, that different federal and state statutes and common-law remedies may not provide full relief.

Due to all of these problems—standing, an adequate theory on which to sue, inadequate scope of protection, and incomplete remedies for natural resource destruction—the natural resource damage action is dramatically underused. One commentator notes, 'Despite the power of the natural resource cause of action, the experience to date has been one of missed opportunities'.*

## Environmental Justice

Environmental justice redresses the purposeful locating of waste sites and noxious industry in areas that are predominantly inhabited by low-income minorities.† Torts may eventually prevail on an environmental justice theory, but torts generally concern compensating harms rather than rights to preventing the harm. Judicial standards protecting minority and low-income neighbourhoods from toxic torts are still nascent.‡

## Conclusions

The concept of natural resource damages has underpinnings in both private and public law. State and federal natural resource damage statutes and the common law related to the public trust and public nuisance doctrines are a collective recognition of each individual's right to be free from unreasonable environmental degradation of his own land. It is only with the rise of industry and technology that degradation of natural resources, held in common or by no one, becomes what amounts to an individual invasion. In his treatise on damages, McCormick states:

> For a past and completed invasion of the owner's land due to the wrongful conduct of another, as in cases of direct trespass or of fire, flood, or destructive fumes, the owner recovers not only for the past injury to the land and its products, but also for any future loss which is reasonably certain to accrue from the same past invasion.§

Thus, in the case of *Indiana Pipe Line Co. v. Christianson*,¶ the plaintiff was entitled to recover all damages past and future from the leakage of oil that had escaped from a pipe and injured the plaintiff's land.** Should not an individual, as part owner of natural resources held in common; the federal,

* Breen, 24 Wake Forest L. Rev. 851, 867 (1989).

† See Exec. Order No. 12898, 40 C.F.R. 1.70 (February 11, 1994) (Federal Action to Address Environmental Justice in Minority Populations and Low-Income Populations).

‡ Craig Arnold, *Planning Milagros: Environmental Justice and Land Use Regulation*, 76 Denv. U. L. Rev. 1, 67 (1998).

§ McCormick, Damages at § 127(1).

¶ 123 N.E. 789 (1919).

** See, for example, *Blankenship v. Kansas Explorations*, 30 S.W.2d 471 (1930).

state, and local governments, as trustees of resources held in common or protectors of those held by no one; or the environment for its own sake be entitled to the same protection given an individual landowner? Congress, states, and individuals are increasingly answering yes. What then can be done to correct the problems with the cause of action?

This chapter has illustrated that the theoretical underpinnings of the cause of action permit a great deal of expansion. The concept of natural resource damages has not even begun to realise its full potential as a tool in environmental protection. Commentators have proposed various solutions ranging from citizen suits under current federal statutes* to an expanded use of state common law remedies.† It is hoped that, as government and the public recognise more environmental rights, the legitimacy of the cause of action will increase, and more cases will be brought giving legal protection to molluscs and mangroves.

## References


Berger, M. A. (1997) Eliminating general causation: Notes toward a new theory of justice and toxic torts. *Columbia Law Review* 97: 2117–52.

Collins, H. (1985) *Changing order: Replication and induction in scientific practice.* Chicago: Univ. of Chicago Press.

Edmond, G., and D. Mercer. (2000) Litigation life: Law science knowledge construction in (Bendectin) mass toxic tort litigation. *Social Studies of Science* 30: 265–316.

Graham-Rowe, D. (2003) Special Report: Mobile Phone Safety. *New Scientist* 179: 12–13.

Mercer, D. (2002) Scientific Method Discourses in the Construction of EMF Science. *Social Studies of Science* 32: 205–33.

Miller, C. R. (2003) Novelty and heresy in the debate on nonthermal effects of electromagnetic fields. In *Rhetoric and incommensurability.* Waterloo, Ontario: Department of English, University of Waterloo.

Stewart, W. (2000) Independent Expert Group on Mobile Phones (IEGMP) Mobile Phones and Health. Chilton: National Radiological Protection Board, UK.

Swicord, M. (2003) Interview: It's good to talk. *New Scientist* 179: 46–8.


* See Breen, 24 Wake Forest L. Rev. 851 (1989).

† See Cynthia Carlson, *Making CERCLA Natural Resource Damage Regulations Work: The Use of the Public Trust Doctrine and Other State Remedies,* 18 ELR 10299 (1988).

# Index


## A

*ABT Summer* spill, 123
*Abuan v. General Elec. Co.*, 368
Acenaphthene, 50, 51
Acenaphthylene, 50
Acyclic isoprenoids, 90
Adamantanes
  diagnostic ratios, 91–92
Additives, 49
AED. *See* Atomic emission detection (AED)
*Aegean Sea* spill, 123
Aerial photographs, 3
Air pollution
  episode, 327–330
  models, 312
Air quality, 310
*n*-Alkane(s), 117
  ions for monitoring, 119
  structure, 118
  weathering check using, 95–97
Alkyl lead additives, 49
Aluminum, 16
  in normalisation techniques, 261
AMBI
  for detecting spatial and temporal changes in estuarine system, 227–233
  for detecting spatial and temporal gradients in relation to submarine outfall, 233–238
  for detecting spatial gradients in oil field exploitation, 238–243
  ecological models for, 226–227
Americium, 16, 21
*Amoco Cadiz* spill, 123
*An Tai* spill, 124
Analysis, 8–9
Anthracene, 50
*Arrow* oil spill, 93
*Artesian Water Co. v. New Castle County*, 372
*Ashland Oil, Inc. v. Miller Oil Purchasing Co.*, 365
Asphaltenes, 56
*Atlantic Empress* spill, 123
Atomic emission detection (AED), 61

## B

Background *vs.* baseline, 4–6
Belief, 1
Benz[a]anthracene, 50, 51
Benzo[a]pyrene, 51
Benzo[b]fluoranthene, 51
Benzo[e]pyrene, 51
Benzo[k]fluoranthene, 51
Beryllium, 16
*Betts v. Manville Injury Personal Trust*, 367, 368
Bicadinanes, 120
Biodegradation, 121–122
  mechanism, 154
  qualitative evidence, 146–147
  of sequence of intermediaries, 152–154
Biomarker(s), 79–94
  bicyclic, 86
  crude oil, 117, 126
    ions used for monitoring, 119
  diagnostic ratios, 89–94
  distribution and quantification, 79–86
  historical perspectives, 115–117
  sterols as, 131
  target, in crude oils and refined products, 84–85
  weathering using, 98–99
Biphenyl, 50
*Blankenship v. Kansas Explorations*, 376
*Botryococcus braunii*, 83

Box model, 317–318
*Braer* spill, 123
*Bridgeton v. B. P. Oil, Inc.*, 365
Bunker C type fuel, 56

## C

Caesium, 16, 21
  isotopes, 20–21
Canadian Council of Ministries od the Environment (CCME), 47
Carbazole, 51
Carbon, 16
Carbon preference index, 62
Case studies, 25–31
*Castilla de Bellver* spill, 123
CCME. *See* Canadian Council of Ministries od the Environment (CCME)
Chemical mass balance, 335–336, 339–340
  positive matrix factorisation *vs.*, 338, 344
Chernobyl accident (1986), 20
Chlorine, 16
Chrysenes, 48, 50
*Cities Serv. v. State,* 365
Citizen suits, 370
*City of Hampton v. Watson,* 366
*City of Richmond v. Test,* 366
Clean Water Act (1972), 372
Cobalt, in normalisation techniques, 263
*Collier v. Simpson Paper Co.*, 367, 368
Compound-specific isotope ratio analysis (CSIA), 132
  applications, 156–157
  for identification of source signatures, 139–142
  limitations, 154–155
  quantitative interpretation, 147–152
Comprehensive Environmental Response, Compensation, and Liability Act (1980), 368, 372
  Superfund, 373
Contaminant(s), 10–11
  knowledge of, 6
  pathway, 10
  sink, 10
  source, 10
  sources, stable isotopes in tracking, 137–144
Crude oil, 56
  biomarkers, 117
  chemical composition, 62–67
  fate of, 49
  Fourier transform ion cyclotron resonance MS, 47
  organic nitrogen hydrocarbons in, 51
  percentage in spills, 46
  worldwide consumption, 44
CSIA. *See* Compound-specific isotope ratio analysis (CSIA)

## D

*Daigle v. Shell Oil Co.*, 368
Data presentation, 10
*Daubert v. Merrell Dow Pharmaceuticals, Inc*, 354–357
Daughter-excess dating, 19, 20
Deepwater Port Act (1974), 373
Denaturing gradient gel electrophoresis (DGGE), 202–209, 213
  derivatives, 202
  forensic science and, 208
Developing case, 9–10
DGGE. *See* Denaturing gradient gel electrophoresis (DGGE)
Diamantanes, diagnostic ratios, 92
Diamondoids, ions used for monitoring, 119
Diasteranes, diagnostic ratios, 90–91
Dibenz[a,h]anthracene, 51
Dibenzo[ghi]perylene, 51
Dibenzothiopenes, 48, 50
Diesel fuel, 56, 68, 69, 127
Dimethyldiethyl lead, 49
Dimethylsulphide, 172
Dispersion modelling, applications, 327–334
Distillates, 67–68
*Duarte v. Zachariah,* 367
Dynamic head space, 179

## E

Ecological quality, use of indicator species in assessing, 224–226
Emission sources, 311–312
Emotional distress, 367–368
*Empress Cars (Abertillery) v. NRA,* 365
Environment Canada Oil Spill Research Program, 56
Environmental justice, 376
Environmental laws, 370

Environmental Liability Directive, 4
EPA. *See* U. S. Environmental Protection Agency (EPA)
*Erika* spill, 46, 126, 143
Ethanol, 49
Ethyl tertiary butyl ether (ETBE), 49
Eulerian framework, 317
Events, 3–4
Expert process, 354–364
 goal of liberalising, 354–360
 judges, 360–363
 underlying problem, 360
 understanding cases, 364
Expert witness report, 11–12
*Exxon Valdez* spill, 46, 75, 83, 124, 372

## F

Factor analysis, 336
*Fillebrown v. Steelcase,* 359
Flame ionization detector, gas chromatography with, 47
Fluorancene, 50
Fluorenes, 48, 50
Foot-and-mouth disease epidemic, 330–334
Fourier transform ion cyclotron resonance MS, 47
*Frye v. United States,* 354

## G

Gas chromatography, two-dimensional, 60–61
Gas chromatography-flame ionization detector (GC-FID), 47
Gas chromatography-isotope ratio mass spectrometer (GC-IRMS), 114, 132
Gas chromatography-mass spectrometer (GC-MS), 47
 purge-and-trap, 57, 133
 selected ion mode analysis (*See* Selection ion mode (SIM))
 semivolatile, 57
Gasoline, 66–67
 biomarkers and, 130
Gaussian model(s), 318–327
 air dispersion, 326–327
 basic dispersion equation, 318–319
 Pasquill-Gifford dispersion parameters, 320–325
 plume rise, 319–320
 uncertainty, 325
 validation, 325–326
GC-FID. *See* Gas chromatography-flame ionization detector (GC-FID)
GC-IRMS. *See* Gas chromatography-isotope ratio mass spectrometer (GC-IRMS)
GC-MS. *See* Gas chromatography-mass spectrometer (GC-MS)
*General Electric Co. v. Joiner,* 357
Geological maps, 3
Geomarker, 131
Geostatistics, 297–304
 anisotropy, 302–304
 checklist for, 304
 kriging, 301–302
 variogram and semivariogram analysis, 298–301
Gravimetric methods, 56

## H

Halocarbon(s), 172
 degradation and fate, 176–178
 lifetimes, 173
 marine natural sources, 176
 production and historical uses, 175–176
 sources, 173, 176
*Haven* spill, 123
*Hawaiian Patriot* spill, 123
Head-space analysis, 179, 180
Heavy fuel oils, 70
Henry's law constants, 173–174
*Hole v. Barlow,* 366
Hopane(s), 117, 119
 structure, 118
Hydrocarbon(s)
 biogenic *vs.* petrogenic, 64
 group analysis for crude oils, 65
 individual saturated, 48
 nonmethane, 172
 petrogenic *vs.* biogenic, 64
 petrogenic *vs.* pyrogenic, 78–79
 pyrogenic *vs.* petrogenic, 78–79
Hydrocarbon fingerprinting, 43–112
 case study, 99–106
 methodologies, 47–61
  Nordtest, 58–60

tiered analytical approach, 51, 56–58
two-dimensional GC, 60–61
weathering check using, 94–99
Hydrogen, 16

## I

*Idaho v. Southern Refrigerated Transportation,* 372
*Illinois Central Railroad Co. v. Illinois,* 369
Indenol[1,23-cd]pyrene, 51
*Independenta* spill, 123
*Indiana Pipeline Co. v. Christianson,* 376
Iodine, 16
isotopes, 20
*Irenes* spill, 123
Iron, in normalisation techniques, 262
Isoprenoid(s), 117
ions used for monitoring, 119
structure, 118
Isotopes, 20–21
fractionation, 135–137
geochemistry, 134–137
intrinsic effect, 135
stable, 132
in contaminant attenuation studies, 145
in tracking contaminant sources, 137–144

## J

*Jakob Maersk* spill, 123
Jet fuel, 69
Judges, 360–363

## K

*Katina P* spill, 123
*Khark 5* spill, 123
*Kumho Tire v. Carmichael,* 357

## L

Lagrangian framework, 317
*Langan v. Valicopters, Inc.,* 365
Law Public Nuisance Doctrine, 369–370
Law Public Trust Doctrine, 368–369
Lead, 5–6, 16
isotopes, 19–20
Legal framework, 4
Liberalising proof, 354–360
Light distillates, 67–68
Lithium, in normalisation techniques, 261–262
Lubricating oil, 70–71

## M

Maps, 3
Marine Protection, Research, and Sanctuaries Act (1972), 374
Marine systems
detecting environmental impact gradients in, 226–243
ecological tools for assessing impacts in, 221–224
multimetric indices, 222–223
multivariate and modelling approaches, 223–224
univariate indices, 221–222
Mass spectrometer, 47. *See also* Gas chromatography-mass spectrometer (GC-MS)
*McClenathan v. Rhone-Poulenc Inc.,* 367
*McLane v. Northwest Natural Gas Co.,* 365
Medical monitoring, 368
Merchant Shipping and Maritime Security Act (1997), 4
Methanol, 49
Methyl dibenzothiphenes, 76–77
Methyl phenanthrenes, 76
Methyl tertiary butyl ether (MTBE), 49
Methyltriethyl lead, 49
Microbiology, 195
molecular, 195
techniques, 197
Midrange distillates, 68
*Miranda v. Shell Oil Co.,* 368
Model(s)
air dispersion, 326–327
box, 317–318
Gaussian, 318–327
nuclear accident, 327
uncertainty, 325
Modelling, 312–327
dispersion, 327–334
framework, 316–318

guidelines, 313–314
type of models, 314–316
*Morrissy v. Eli Lilly & Co.*, 368
MTBE. *See* Methyl tertiary butyl ether (MTBE)
Multimetric indices, 222–223
Multivariate methods, 281–297
cluster analysis, 281
discriminant analysis, 281
factor analysis, 281
multidimensional scaling, 281
PCA, 281, 282–292 (*See also* Principal components analysis (PCA))
PLS, 281, 292–297 (*See also* Partial least squares (PLS))
statistics from, 9–10

## N

Naphthalene, 48, 50
National Marine Sanctuaries Act (1988), 372
Natural resources, 370–371
defined, 373–374
future trends in damages, 371–376
Nipisi spill, 66
Nordtest methodologies, 58–60
basis of final assessment, 59–60
inconclusive, 60
nonmatch, 60
positive match, 59
probable match, 59
Normalisation techniques, 254–271
application of, 265–271
comparison of, 264
elemental, 256–263
choice of elements, 260–263
granulometric method, 255–256
physical segregation, 254–255
postextraction method, 263, 265
*Nova* spill, 123
Nuclear accident model, 327
Nuisance, 365–366

## O

*Odyssey* spill, 123
Oil hydrocarbon fingering methodologies, 47–61
Oil Pollution Act (1990), 372
Oil spills, 46, 64
major worldwide, 123
sesquiterpanes in, 88
Oil weathering, 94–95
Oleanane, 117
Open-system progressive evaporation, 136
Organic carbon, in normalisation techniques, 263
Ozone depleting substances, 172

## P

*Parens patriae*, 372
Partial least squares (PLS), 281, 292–297
check list for, 296–297
overlap between signatures, 295–296
variance, 293, 294
Pasquill-Gifford dispersion parameters, 320–325
PCA. *See* Principal components analysis (PCA)
PCR. *See* Polymerase chain reaction (PCR)
Perylene, 51
in beach and subtidal sediments, 66
Petroleum
biomarkers (*See* Biomarker(s); Sterane; Terpane)
consumption, 44
input into the sea, 46
refined products, 67–71
worldwide demand and supply, 45
Phenanthrene, 48, 50, 51
Plume rise, 319–320
Plutonium, 16
Polonium, 16
Polycyclic aromatic hydrocarbons (PAH), 47
cluster analysis, 75–76, 77
fingerprints, 71–79
isomer, 75–76, 77
recommended diagnostic ratios, 72, 73–75
in situ genesis, 66
weathering using, 97–98
Polymerase chain reaction (PCR), 201–202
DNA fingerprinting techniques based on, 202
limitations of, 211–212, 214
Positive matrix factorisation, 337–338, 342–344
chemical mass balance *vs.*, 338, 344

Potassium, 16
*Potter v. Firestone Tire,* 367
Preparation, 2–3
*Prestige* spill, 46, 123
Principal components analysis (PCA), 188, 281, 336–337, 340–342
 checklist for, 292
 loading factors, 283
 mean centring to unit variance, 288
 normality and transformations, 287
 score, 284
 using proportions and closure, 288–290
 zeros in data analysis, 291–292
Proving the case, 11
Public environmental law, 368–376
Purge-and-trap GC-MS, 57, 133, 179
Pyrene, 50
Pyridine, 51

## Q

Quality assurance, 9, 48, 57
Quinoline, 51

## R

Radionuclide(s)
 application for tracing and dating, 22–25
 artificial, 15, 20–21
 classification, 17
 current and future roles in environmental forensics, 31–33
 decay rates, 16
 fission and activation products, 20–21
 in groundwater studies, 28–30
 half life, 16
 inputs to environment, 17
 natural, 17–20
 in normalisation techniques, 260–261
 in sediment dating, 30–31
 sources, 16–17
 as tracer of reprocessed uranium, 28
 transuranic, 21
 ubiquitousness, 15
 as water mass tracers, 27–28
Radium, 16
 in groundwater studies, 29
 isotopes, 17–19
Radon, 16
 in groundwater studies, 29–30
 isotopes, 19–20
Rayleigh fractionation model, 136
Receptor modelling, 334–338
 in action, 339–344
Ribosomal RNA, 200–201
Ribosomes, 199–200
Rubidium, 16
Ruthenium, 16

## S

*S. S. Zoe Colocotroni* spill, 375
Sample(s), 6–8
 analysis, 8–9
 bias, 7
 media, 7
 number of, 7
 quantity of medium, 7–8
 security, 8
 storing, 184
Sampling, 6–8
 cost, 8
 design, 184
Scandium, in normalisation techniques, 262
SCD. *See* Sulphur chemiluminescence detection (SCD)
*Schwartzman, Inc. v. Atchison, T.,* 365
*Sea Star* spill, 123
Secular disequilibrium, 19
Sediment(s), 64, 184–186, 251–252
 advantage, 252
 composition, 252
 estuarine, 267
 fluvial, 269
 marine, 265–267
 organic matter in, 64
 physical properties, correlation between contaminant concentration and, 253 (*See also* Normalisation techniques)
 salt marsh, 61
 signatures, 185–186, 187
 variability, 253
Selected ion mode (SIM), 48, 56
Semivolatile GC-MS, 57
Sesquiterpane(s), 86–88, 117
 diagnostic ratios, 91
 GC-MS, 86, 87
 ions used for monitoring, 119
 in oil spills, 88
 structure, 118
Shannon index, 207

Shewhart plots, 9
Single-stranded conformation polymorphisms (SSCPs), 202, 209–210, 213
Sites, 3
Solid-phase microextraction (SPME), 180–186
Solvent extraction, 179
Source, 10–11
  apportionment, 334–344
Spill(s), 3–4, 46. *See also* Oil spill(s)
  crude oil percentage in, 46
  dating contaminant, 157–160
SPME. *See* Solid-phase microextraction (SPME)
SSCPs. *See* Single-stranded conformation polymorphisms (SSCPs)
*St. Helen's Smelting Co. v. Tipping,* 366
Stable isotopes, 132–157
*State Dep't of Envtl. Protection v. Ventron Corp.,* 365
Statistical significance, 281
Statistics, 9–10
  comparative, 278–281
  errors in, 279–281
    systemic and random, 279
    type 1 and type 2, 279–281
  methods in, 278–297
    comparative, 278–281
    multivariate, 281–297 (*See also* Multivariate methods)
  variance in, 282
Sterane(s), 52–55, 117
  diagnostic ratios, 90–91
  distribution of compounds, 82
  ions used for monitoring, 119
  methyl, 119
  monoaromatic, 119
  structure, 118
  triaromatic, 119
Sterols, 131
Strict liability, 364–365
Strontium, 16
Sulphur chemiluminescence detection (SCD), 61

## T

T-RFLP. *See* Terminal restriction fragment length polymorphisms (T-RFLP)
Target analytes, 48
Technetium, 16
  environmental behavior, 21
  isotopes, 20
Temperature gradient gel electrophoresis (TGGE), 202, 213
Terminal restriction fragment length polymorphisms (T-RFLP), 202, 208, 210–211, 213
Terpane(s), 52–55, 117
  diagnostic ratios, 90
  distribution of compounds, 81
  ions used for monitoring, 119
  pentacyclic, 119
  structure, 118
  tetracyclic, 117, 119
  tricyclic, 117, 119
Terphenyl, 51
Tertiary amyl methyl ether (TAME), 49
Tetramethyl lead, 49
TGGE. *See* Temperature gradient gel electrophoresis (TGGE)
Thin layer chromatography, 56
Thorium, 16
Time-of-flight mass spectrometer (TOF-MS), 60, 61
TOF-MS. *See* Time-of-flight mass spectrometer (TOF-MS)
*Toole v. Baxter Healthcare Corp,* 359
Topographical maps, 3
*Torrey Canyon* spill, 122, 123
Total petroleum hydroacrbons (TPH), 49
Toxic torts, 364
Trans-Alaska Pipeline Authorization Act (1973), 372
Trespass, 365–366
Triaromatic steranes, 92
Trimethylethyl lead, 49
Triterpanes, 119
Truth, 1

## U

U. S. Environmental Protection Agency (EPA), 9, 47
Univariate indices, 221–222
Unpaired-mean group analysis (UPMGA), 205
Unresolved complex mixtures (UCM), 49, 61
UPMGA. *See* Unpaired-mean group analysis (UPMGA)

Uranium, 16
reprocessed, 28
*Urquiola* spill, 123

## V

Volatile aromatic hydrocarbons (VAH), 49. *See also* Halocarbon(s)
analysis, 178–188
data interpretation, 186, 188
fibre selection guide, 181
seawater, 181–184
sediment, 184–186
in coastal seawater, 184
PCA and, 188

## W

Waste oil, 71
Water Framework Directive, 4
Weathered diesel fuel, 69
Weathering check
using hydrocarbon fingerprints, 94–99
using *n*-alkane and GC traces, 95–97
using PAHs, 97–98
*Weisgram v. Marley Company,* 358
*Werlein v. United States,* 368